# LOVING DR. JOHNSON

# LOVING DR. JOHNSON

HELEN DEUTSCH

THE UNIVERSITY OF CHICAGO PRESS

CHICAGO & LONDON

HELEN DEUTSCH is professor of English at the University of California, Los Angeles. She is the author of *Resemblance and Disgrace: Alexander Pope and the Deformation of Culture* and coeditor of *"Defects": Engendering the Modern Body.*

THE UNIVERSITY OF CHICAGO PRESS, Chicago 60637
THE UNIVERSITY OF CHICAGO PRESS, LTD., London

*Printed in the United States of America*

14 13 12 11 10 09 08 07 06 05   1 2 3 4 5

ISBN: 0-226-14382-1

Library of Congress Cataloging-in-Publication Data

Deutsch, Helen.
Loving Dr. Johnson / Helen Deutsch.
p. cm.
Includes bibliographical references and index.
ISBN 0-226-14382-1 (alk. paper)
1. Johnson, Samuel, 1709–1784. 2. Johnson, Samuel, 1709–1784—Influence.
3. Authors, English—18th century—Biography. I. Title: Loving Doctor Johnson.
II. Title.
PR3533.D46 2005
828'.609—dc22
2005004645

∞ The paper used in this publication meets the minimum requirements of the American National Standard for Information Sciences–Permanence of Paper for Printed Library Materials, ANSI Z39.48–1992.

FOR MICHAEL AND THEODORE

AND IN MEMORY OF

JOHN L. TRAUGOTT (1921–2004)

# CONTENTS

ILLUSTRATIONS

# ACKNOWLEDGMENTS

Anecdote and author love, as Boswell's *Life* acknowledges, are communal endeavors; thus this book came into being through many lucky sparks of conversation and association. I am deeply grateful to Jerome Christensen for the orange peel; to Wayne Gochenour for James Merrill on the orange peel and to Steven Yenser on Merrill, the orange peel, and "Oranges"; to Elin Diamond for bringing Beckett's work on Johnson to my attention and to Michael North for his thoughts on Beckett; to Lynn Enterline for her recollection of Dido at the heart of the matron reference; to Page duBois for simulacra and icons, and for associating to "*nugarum contemptor*"; to Kate Marshall for Heidegger's jar and to Lorna Clymer for pointing me in that direction; to Sean Silver for the epigraph to *Pale Fire;* to Michael Colacurcio for *Our Old Home;* to Lowell Bowditch for crucial early tutoring on ancient romance; to Al Braunmuller for his thoughts on the Ephesian matron and her history; to Jonathan Crewe for *Foucault's Virginity;* to Felicity Nussbaum for Inkle and Yarico; to Daniel Brownstein and Anita Guerrini for guidance in the history of anatomy; to Joshua Scodel for Goldsmith's *Retaliation;* to Lee Edelman for Hart Crane; to Robert Folkenflik for "nervous"; to Kathleen McHugh for Lydia Davis's *Samuel Johnson Is Indignant;* to Jay Clayton for *England, England;* to Dale Peterson for help with Nabokov; to Paul Sheats for his memory of Walter Jackson Bate; to Leo Damrosch for thoughts on annihilation; to William Pritchard for Q. D. Leavis; to Bill Warner for his reflections on an earlier Johnsonian era; and to Rick Barney for asking the right question at the right time. The students in my graduate seminars on anatomy, the Johnsonian sentence, and especially the anecdote contributed to this book in ways too numerous to mention.

Thanks to Toni Bowers, Lorna Clymer, Patricia Fumerton, Claudia Johnson, Lawrence Lipking, David Mitchell, Vincent Pecora, Raymond Stephanson, Janet Sorenson, Mary Terrall, Blakey Vermeule, and Sharif Youssef for providing important forums for presenting parts of the manuscript. Bob Erickson, Anita Guerrini, Bill Warner, and the students at the University of California, Santa Barbara's Early Modern Center gave valuable responses to an earlier version of chapter 4. Luke Bresky, Stephen Dilks, Anita Guerrini, Nicole Horejsi, Kate Marshall, Sean Silver, Cheryl Wanko, and Richard Wendorf shared unpublished work that facilitated my thinking. Felicity Nussbaum's conversation and critique gave this project the most auspicious of beginnings; her collegiality and that of Joseph Bristow, Jonathan Grossman,

Anne Mellor, Max Novak, Barbara Packer, Jenny Sharpe, Malina Stefanovska, Eric Sundquist, and Mary Terrall sustained me throughout. Wendy Belcher, Diane Divoky, Jane Gallop, Corrinne Harol, Heather James, Ann Kibbie, Jonathan Kramnick, Alice Wexler, and Matthew Wickman read parts or all of the manuscript and provided invaluable responses. Margaret Doody, Paul Hunter, and Joseph Roach were sources of ongoing support. This book would never have been written had it not been for such fertile and fortuitous interchanges.

Some friends gave so much to both this book and its author that it's impossible to detail their contributions. In countless embodied and disembodied conversations, Sharon Achinstein, Eva Cherniavsky, Joe DiMuro, Jennifer Fleissner, Deanna Kreisel, Rachel Lee, Chris Looby, David Wong Louie, Claire McEachern, Kathleen McHugh, Valerie Smith, and Wendy Wall heard more than they needed to about this book's progress—I thank them for their patience and support. The generosity and inspiration of Page duBois made a new scholarly openness possible for me, while Vivian Sobchack's excitement and engagement with this project enabled me to find its true shape. I am grateful to them both for their insights, their fearlessness, and their examples. Lowell Gallagher's translation of Beckett's French was just a small part of his contribution—his imagination, originality, and sheer brilliance still keep me going. Lorna Clymer, Jayne Lewis, and Julia Stern were each in their different ways invaluable readers, interlocutors, and, above all, indispensable friends. Claudia Johnson has been this book's ideal reader from the start and has helped to find it an ideal audience. Estelle Shane has overseen and enabled the growth of both this book and its author.

I am extremely grateful to Thomas Wortham and the Department of English at the University of California, Los Angeles, and to Peter Reill and the UCLA Center for Seventeenth- and Eighteenth-Century Studies for providing essential and unflagging research support (often at the eleventh hour) and an ideal environment in which to work. This support took the most miraculous form in a series of research assistants—Norman Jones, Allison Kroll, Meredith Neuman, Johanna Schwartz, Elliott Visconsi, and, most notably, Debra Bronstein, Kimberly Garmoe, Anne Myers, and Sean Silver—without whose labor this book would not exist. My heartfelt thanks to the English department staff—especially Nora Elias, Rick Fagin, Jeanette Gilkison, and Doris Wang—and to the staff of the Center for unfailing and cheerful assistance. Yearly grants from the UCLA Council on Research, along with a travel grant from the American Philosophical Association and an NEH fellowship for a year in residence at the Huntington Library, made this book possible. Thanks are due to the staffs at the William Andrews Clark Memorial Library (especially the wonderful Jennifer Schaffner); the Huntington Library; the Beinecke Rare Book and Manuscript Library; the Wellcome Library; the Library of the Royal College of Physicians, London; and the Rare Book and Manuscript Library of Columbia University. Special thanks to Simon Chaplin, curator of the Hunterian Museum of the Royal College of Surgeons of England; Annette French of the Samuel Johnson Birthplace Museum, Lichfield; Natasha McEnroe of Dr. Johnson's House, London; and Professor A. P. Payne of the Hunterian Museum, Glasgow, for help with Johnsoniana. I am grateful to Julian A. Garforth of the Archive of the Beckett International Foundation at the University of Reading for his generous assistance with "Human Wishes" and for his permission (along with that of Edward Beckett of the Beckett Estate) to reproduce Beckett's sketch and to quote from his unpublished notebooks, as well as to Gordon Turnbull, the general editor of the Yale Boswell Editions, for permission to publish page 1012 of the manuscript of Boswell's *Life of Johnson.*

My special thanks go to the University of Chicago Press, especially to the anonymous readers of the manuscript for their tremendously helpful and detailed responses, to Erin DeWitt for her intelligent and attentive copyediting, to Randolph Petilos for unfailingly efficient assistance, and to the amazing Alan Thomas for believing in this book from the beginning and for years of formative conversation culminating in the title that pulled everything into place.

Permission has been granted to reprint part of chapter 1, which was originally published as "Doctor Johnson's Autopsy," *The Eighteenth Century: Theory and Interpretation* 40, no. 2 (Summer 1999): 113–27, © Texas Tech University Press, 1999; an earlier version of chapter 2, which was published as "The Author as Monster: The Case of Doctor Johnson," in *"Defects": Engendering the Modern Body,* ed. Helen Deutsch and Felicity Nussbaum (Ann Arbor: University of Michigan Press, 2000), 177–209; and part of chapter 3, which appears as "'Thou Art a Scholar, Speak to It, Horatio': Uncritical Reading and Johnsonian Romance," in *Polemic: Critical or Uncritical,* ed. Jane Gallop (New York: Routledge, 2004), 65–102.

Heartfelt thanks to my family, especially to Betty Capaldi, Ruthanne Deutsch, and Barbara Meranze. This book belongs to the two dearest to me, without whom nothing would seem possible, and with whom everything (especially love) is imaginable: Michael and Theodore Meranze. Words can't express how grateful I am to them and for them, every day.

INTRODUCTION

# THE BEGINNING, IN WHICH NOTHING IS FOUND

Mr. Thrale's attentions and my own now became so acceptable to him, that he often lamented to us the horrible condition of his mind, which he said was nearly distracted; and though he charged *us* to make him odd solemn promises of secrecy on so strange a subject, yet when we waited on him one morning, and heard him, in the most pathetic terms, beg the prayers of Dr. Delap, who had left him as we came in, I felt excessively affected with grief, and well remember my husband involuntarily lifted up one hand to shut his mouth, from provocation at hearing a man so wildly proclaim what he could at last persuade no one to believe; and what, if true, would have been so very unfit to reveal.

HESTER LYNCH THRALE PIOZZI, *Anecdotes of the Late Samuel Johnson, LL.D., during the Last Twenty Years of His Life*

The names of many greater writers are inscribed on the walls of Westminster Abbey; but scarcely anyone lies there whose heart was more acutely responsive during life to the deepest and tenderest emotions. In visiting that strange gathering of departed heroes and statesmen and philanthropists and poets, there are many whose words and deeds have a far greater influence on our imagination; but there are very few whom, when all has been said, we can love so heartily as Samuel Johnson.

LESLIE STEPHEN, *Samuel Johnson*[1]

Now to my word.
It is "Adieu, adieu, remember me."
I have sworn't.

*Hamlet,* 1.5.110–12

The Samuel Johnson who has lived in the Anglo-American popular imagination for the last two hundred years is as familiar, as clubbable, as a childhood uncle, as charming and commodifiable as the image of his cat Hodge that festoons the entries of a recent popular abridgement of his *Dictionary.*[2] He is known for his quotable conversation and for his emblematic gestures—kicking the stone to refute Berkeley thus, comparing women preachers to dancing dogs, toasting the next insurrection of slaves in the West Indies, rejecting an anonymous gift of new shoes as a proud but destitute young Oxford scholar, admonishing Hester Thrale to sympathize with the poor who love the smells of Porridge Island, praying annually for his dead wife, standing in the rain at Uttoxeter market to do penance for long-past neglect of a paternal command. This Johnson is portable—there is at least one *Johnson Handbook* and a pocket book of his insults[3]—and he is reassuring. This is the Johnson who comes to Leslie Stephen's mind as he views the walls of Westminster Abbey, the Johnson whose acutely responsive heart inspires others to love him heartily. This book, whose subject is that love, will give many examples of Johnson the familiar. But Stephen, let's not forget, is walking through an illustrious graveyard, and each of the categories with which he characterizes that eclectic and "strange gathering" of the dead applies to Johnson. Thrale's anecdote reminds us that the popular idealization of a man who comes to encapsulate all the famous English dead is shadowed by a darker version, as immortality is shadowed by mortality.

Hester Lynch Thrale—the biographer who lived with Johnson for intermittent short periods during almost twenty of his happiest years in the realm of domestic privacy that he himself considered the stuff of the genre—provides the tonal counterpoint to Stephen's celebration of the hearty communal love of a celebrity who is public, indeed national, property. "The business of the biographer"—Johnson had written in his *Rambler* 60 "on the dignity and usefulness of biography" for October 13, 1750, sixteen years before the moment Thrale records—"is often to pass slightly over those performances and incidents, which produce vulgar greatness, to lead the thoughts into domestick privacies, and display the minute details of daily life, where exterior appendages are cast aside, and men excel each other only by prudence and by virtue."[4] In the case of the biography of an author—as Johnson himself, a major innovator of the form in his prefatory *Lives of the Poets,* well knew—there is an added obligation to balance the moral effects of the work against the inevitable mortal shortcomings of an individual life.[5] No biographer of Johnson experienced this potential conflict between "vulgar greatness" and "minute details" more intensely than Thrale, in whom Johnson confided "a Secret far dearer to him than his Life," along with a padlock and fetters to use in the event that the secret—uncontrollable madness—should become public. Such private knowledge, she wrote in her journal, "contradict[s] the Maxim of Rochefoucault, that no Man is a Hero to his Valet de Chambre.—Johnson is more a Hero to me than to any one—& I have been more to him for Intimacy, than ever was any Man's Valet de Chambre."[6] Having witnessed the horrific scene she describes in my epigraph, Hester is left alone by her husband in order "to prevail on him to quit his close habitation in the court, and come with us to Streatham, where I undertook the care of [Johnson's] health, and had the honour and happiness of contributing to its restoration."[7]

Characterized by later biographers as mother-substitute, potential wife, and dutiful daughter to Johnson, Thrale, who married (in a Roman Catholic ceremony) the Italian singing master Gabriel Piozzi after her husband's death and shortly before Johnson's own death, became for many an ungrateful betrayer of both her famous friend and her country.[8] Johnson himself provided the foundation for this construction of Thrale, responding to the news of her marriage with these famous words:

> If I interpret your letter right, you are ignominiously married, if it is yet undone, let us once talk together. If You have abandoned your children and your religion, God forgive your wickedness; if you have forfeited your Fame, and your country, may your folly do you no further mischief.[9]

His final letter to her is fraught with the emotional vulnerability that characterizes Stephen's portrait:

> When Queen Mary took the resolution of sheltering herself in England, the Archbishop of St. Andrew's attempting to dissuade her, attended on her journey and when they came to the irremeable Stream that separated the two kingdoms, walked by her side into the water, in the middle of which he seized her bridle, and with earnestness proportioned to her danger, and his own affection, pressed her to return. The Queen went forward.—If the parallel reaches thus far; may it go no further. The tears stand in my eyes.

The errant geography of this comparison is striking—the Catholic Mary Stuart takes refuge in England to her ultimate destruction, while the Protestant Hester Thrale, for the duration of Johnson's life, remains in Italy, "seduce[d]" in Johnson's view, by "phantoms of imagination" and desire.[10] Johnson uses this historical anecdote to persuade Thrale to return home, but the anecdote renders home as threateningly unattainable. We will return to the anecdote and the idea of wandering throughout what follows. For now I want to focus on Johnson's moment of sentiment, his contemplation of the anecdotal conflation of past and present, Mary Stuart and Thrale, in an acknowledgment of loss: "The tears stand in my eyes." Thrale from that time on was dead to Johnson; as he told Frances Burney, "I drive her quite from my mind.

If I meet with one of her letters, I burn it instantly."[11] Her abandonment earns him a permanent place in future sentimental imaginations of the eighteenth century.[12]

My delineation of this double Johnson—the clubbable authority at the center of public masculine feeling and the tortured soul at the limit of private feminine sympathy—is not new.[13] What is new is my consideration of this dividedness as essential to a brand of author love unique to its historical moment yet productive of both a literary and extraliterary tradition. These two passages, in other words, delineate a structure of desire, a desire that has possessed many readers of Johnson to know the author himself. While Stephen's Johnson is transparent, his heart laid open (with the anatomical undertones common to the eighteenth-century discourse of sentiment of which Stephen is an heir) to inspire a seemingly universal response, Thrale's Johnson is almost completely opaque. Stephen's Johnson inspires the heartiness of manly sentiment. He speaks the quotable language that the Johnsonian scholar George Birkbeck Hill describes, in the preface to his compilation of the *Wit and Wisdom of Samuel Johnson* (1888), as one of utter lucidity: "He neither lives in a mist nor does he ever try for a single moment to throw one round him. He thinks clearly, and he states with perfect clearness what he thinks. He always knows what he means to say, and his readers and hearers always understand what he does say."[14] Thrale's Johnson speaks his own dark secret so harrowingly that her husband involuntarily raises his hand to silence him.[15]

Stephen's Johnson inspires an impersonal love that cements community; Thrale's Johnson, potentially mad, almost monstrous, threatens such community with the indecipherability of his own desires. While Johnson's contemporary and fellow Lichfield denizen, the poet Anna Seward, disparaged his "last and long-enduring passion" for Thrale as "composed equally perhaps of cupboard-love, Platonic love, and vanity tickled and gratified from morn to night by incessant homage," that passion continues to pose to critics and novelists alike the awkward question of that great heart's possible erotic need.[16] In (and beyond) the Victorian imagination, the public Johnson remains proximate to the realm of erotic desire but never part of it; the 1884 "subject picture" of W. P. Frith, in which Johnson hands the great actress Sarah Siddons into her carriage, illustrates this perfectly (fig. 1).[17] Johnson is the picture of a gentleman in this image because he performs pure manners; marked by what Thrale in her journal called the "stigma" of class, he can only imitate well-bred behavior. Awkward, ungainly, and plainly dressed, he similarly purifies masculine gallantry of its underlying sexual interestedness. He is, in short, sentimental art without nature.[18] His emotion is always disinterested, unattached

Figure 1. William Powell Frith, *Dr. Johnson and Mrs. Siddons* (1884). Reproduced by courtesy of the trustees of Dr. Johnson's House.

to one object, and thus capable of moving many others, as it did Thomas Carlyle—who called him an English hero and "*ultimus Romanorum*"—to great rhetorical heights: "Within that shaggy exterior of his there beat a heart warm as a mother's, soft as a little child's. . . . Tears trickling down the granite rock: a soft well of Pity springs within!"[19] This manly Johnson with a heart of gold has governed the profession of eighteenth-century English letters for two centuries. Thrale's lives on in literature.

When Henry Thrale raises his hand to silence Johnson from "provocation" at his "wild" self-condemnation, his gesture is one of disavowal. He leaves the room, ordering his wife to persuade their friend, in his frenzy, to leave his bachelor confinement. The speculation about the potential erotic tie between Johnson and Thrale—the padlock and fetters he entrusted to her care have been interpreted as possible signs of masochistic tendencies rather than fear of madness—is also a kind of disavowal, a fending off of a much greater threat: namely, that the heroic man of letters, the writer and talker of perfect clarity, might speak an unintelligible private language. Rather than reassure, his words, unspeakable words that Mr. Thrale cannot stand to hear, might alienate, cause shock, fear, and pain. The fact that they are lost to us now is what gives Thrale's anecdote its unique power.

## "Be another": Author Love and Anecdote

> I love anecdotes. I fancy mankind may come, in time, to write all aphoristically, except in narrative; grow weary of preparation, and connection, and illustration, and all those arts by which a big book is made. If a man is to wait till he weaves anecdotes into a system, we may be long in getting them, and get but few, in comparison of what we might get.
>
> SAMUEL JOHNSON, as quoted by James Boswell, *Journal of a Tour to the Hebrides,* August 16, 1773

> Where then shall Hope and Fear their Objects find?
>
> SAMUEL JOHNSON, *The Vanity of Human Wishes*

This book, originally to be titled *Dr. Johnson's Autopsy,* began with my curiosity about why the exemplary eighteenth-century Englishman of letters was dissected, probably against his living will. While it has led me down many unexpected paths, that curiosity remains unsatisfied; it has resulted, instead, in my recognition—and profession—of love, both the love of literature and the love of Johnson, of which the autopsy was a complex and conflicted expression. In its service, I want in the rest of this introduction to consider not the anecdotal form in which Johnson endures but rather the material remains of Samuel Johnson's autopsy and my own and others' ongoing search for them. That autoptic desire to see the thing itself, a seeing by oneself and of oneself (as the word's etymology indicates) is not, as it turns out, so different from the familiar introductory impulse to focus upon an anecdote, since Johnson's selected remains, claimed by the surgeons and carefully preserved as "preparations," have yet to be found. One of the many functions of the anecdote—like the anatomical preparation, the result of a personal selection and preservation process—is to stand in for the lost body.

I am inclined too, I must admit, to begin with a form of apology, or at least explanation, for my use of the "I" in this introduction and periodically throughout what follows. My turn to the personal voice is a turn away from the certainty of linear argument and toward this book's more literary anecdotal logic of association, allusion, and affection. This book owes its essayistic and emotional impulses in no small degree to my position as a woman attempting to participate in and to understand a largely all-male form of author love so passionately institutionalized as the proper form of literary authority that its roots in the subjective (as a form of love and thus of personal inclination or choice) and in the corporeal (as the preservation of a singularly eccentric authorial body) have been all but forgotten.

This "I" is hardly unaware of a variety of histories into which it fits, most recently the (oft-criticized) call in the mid-1980s, issuing most prominently from the Duke school, for feminist criticism in the personal voice. In my bringing of my own and Johnson's persons into view, I owe something to the work of critics such as Marianna Torgovnick, who in *Gone Primitive* includes a personal writing exercise in which she imagines the anthropologist Malinowski's body. Relocating Malinowski's "authority to speak, his basis for generalization," in the particularized anecdotal, personal realm of the diary he kept during his early fieldwork in Australia—a realm in which his body figures in vexed opposition and connection to those of his primitive objects of study—Torgovnick insists "that we reverse the ethnographer's traditional gaze and look inward toward the ethnographic authority, not just out at the Primitive Other—that we use our eyes and imaginations to make Malinowski an anthropological exhibit as pointed and meaningful as any other."[20] In a parallel economy of disavowal, behind the professional authority that until recently legislated both scholarly deportment and the agenda of an eighteenth-century British canon now in the process of being radically redefined, is a body so radically particular as to have been labeled monstrous in its own time.

If Torgovnick figures herself as an ethnographer in her scrutiny of Malinowski's body, in my recovery of Johnson's body I figure myself as an anecdotal collector. In doing so, I ally myself with another feminist famous for her attention to the personal, Jane Gallop, who in her recent *Anecdotal Theory* calls for and exemplifies a new mode of theory—at the intersection of feminism, deconstruction, and psychoanalysis—that by taking the "merely anecdotal" seriously would open up theory to its own historical moment, its own embodiedness. "The usual presupposition of theory," writes Gallop, "is that we need to reach a general understanding, which then predisposes us toward the norm, toward a case or model that is prevalent, mainstream. To dismiss something as 'merely anecdotal' is to dismiss it as a relatively rare and marginal case. Anecdotal theory would base its theorizing on exorbitant models." Linking exorbitance with "the excessive, romantic, perverse, unreasonable, and queer," Gallop declares her own debt to Derrida's championing of an exorbitant method in *Of Grammatology* as an exemplary attempt "to get outside the metaphysical closure that sequesters theory from the real."[21] The case of Johnson—at once mainstream and marginal, imagined as universal yet constructed almost entirely of anecdotal evidence, inspiring varieties of desire that might easily be branded excessive, romantic, perverse, unreasonable, and queer—demonstrates the ways in which, in the case of British literature (a canon often enlisted in defense against abstract [French] theory), what is most central, most generally known, is simultaneously most exorbitant.

I want to relocate my movement toward the personal—a movement I align with the wandering of romance—in a history of literary response that begins with Boswell at the head of a host of collectors of Johnsonian anecdote. This host, as any reader of the *Life of Johnson* and the periodical press of the time is aware, was hardly united; and throughout what follows, anecdotes by Thomas Tyers, John Hawkins, and (most symbolically significant for my purposes) Hester Thrale among others will provide alternatives to Boswell's attempt to monopolize the Johnsonian monument. Still, my use of the personal voice is at once an attempt to distinguish myself from abstract certainties about Johnson himself (certainties that mark the male Johnsonian tradition) while joining the tradition of author worship that Boswell most definitively began. That tradition begins with an interest in the other so complete as to be self-effacing, and a reliance on one's subjective experience of that other so complete as to be grandiose. In his careful preservation of the most minutely embodied particulars of his hero, in other words, Boswell manages to draw excessive attention to himself.[22] John Wolcot, perhaps the most brilliant of the many contemporary critics of Boswell's penchant for anecdotes, hilariously articulates the danger of egotism that comes with attention to the detailed corporeality of one's object of devotion:

> I see thee stuffing, with a hand uncouth,
> An old dry'd whiting in thy Johnson's mouth,
> And lo! I see, with all his might and main,
> Thy Johnson spit the whiting out again.
> Rare anecdotes! 'tis anecdotes like these
> That bring thee glory, and the million please!
> On these, shall future Times delighted stare,
> Thou charming haberdasher of small ware!
> STEWART and ROBERTSON, from *thee,* shall learn
> The simple charms of HIST'RY to discern:
> To *thee,* fair HIST'RY's palm, shall LIVY yield,
> And TACITUS, to BOZZY, leave the field![23]

Throughout this satire of Boswell's *Journal of a Tour to the Hebrides* (which served as a preview to the *Life*), Wolcot, a parodic visionary looking through Boswell's anecdotal lens into the recent past, continually evokes that past as future spectacle—I see, and so will many after me. But what he sees is not the stuff of epic prophecy, or even the material of proper history. He sees the coarse details of the body—the great Englishman's tour of Scotland encapsulated in Boswell's force-feeding and Johnson's rejection of a piece of dried fish.[24] In Wolcot's ironic logic, Boswell's popularity reduces the great historians of the Scottish Enlightenment to mere shopkeepers, retailers of "small ware." The biographer's grandiosity in the recording and mass distributing of Johnsonian trivia cheapens even Tacitus and Livy, while eclipsing them only in egotism. Wolcot articulates the threat that published anecdote poses not only to the reputation of those it exposes to intimate scrutiny, but also to literary and historical value. If details like these are important enough to record, then nothing is sacred, on the one hand, and nothing is meaningless, on the other.

Boswell thus makes history a joke:

> JOE MILLER's self, whose page such fun provokes,
> Shall quit his shroud, to grin at Bozzy's jokes!
> How are we all with rapture touch'd, to see
> *Where, when,* and at *what hour,* you swallow'd *tea*!
> How, *once,* to grace his Asiatic treat,
> Came haddocks, which the RAMBLER could not eat.
> Pleas'd, on thy book thy SOV'REIGN's eye-balls roll,
> Who loves a gossip's story from his soul! (10)

Boswellian anecdote's affinity with the joke (the contemporary comparison with Joe Miller, a popular compiler of jests, is a common one) is similarly unsettling—both reveal a slippage in "reality," an absurdity in trivial things as they are, and in the usual ways that we order those things in narrative, that we laugh at in order to return to some semblance of the normal.[25] Johnson's friend Charles Burney despaired, for example, over the dearth of biographers capable of stylistically rising to the occasion afforded by Johnson's death: "Tommy Tyers is such a quaint, feeble, fumble-fisted writer, as is only fit for Mother Goose's tales—& Boswell with all his Anecdotes, will only make a story book, a kind of Joe Miller the 2d without address or dignity of introduction or application in relating his bon mots."[26] The anonymous author of *A Poetical Epistle from the Ghost of Dr. Johnson* (1786) to four of his biographers has Johnson himself condemn the anecdotal method he recommends in my epigraph:

> The pleasing task I therefore shall decline;
> To sate the publick taste be solely thine:
> Who, while weak Authours, studious of connection,
> Lose many an hour in profitless reflection,
> Before they venture to take quill in hand,
> Thy best ideas always can command;
> And, knowing books are made with pen and ink,
> Keep ever writing on, but never think. (15–16)
>
> NOTE: This abundantly confirms the account Mr. Boswell has given of Dr. Johnson's opinion *how books should be made;* though it seems *rather odd* that he never made them *so* himself. It certainly is the most *expeditious mode;* and as Dr. Johnson *advises it to be pursued,* who can blame his friend Bozzy!
>
> . . . Rejoice my *dull brethren*! We authors have the *sanction* of Dr. Johnson for *writing any how;* it is *recorded* by the great Mr. Boswell, and he has gloriously *set us the example*! (15)[27]

While Wolcot intends to diminish Boswell's biographical project by dismissing it as a joke (and thus to dismiss the king, who enjoys such anecdotes, as a gossip), and while contemporary critics deplored the anecdotal method as replicating a joke book in its lack of studied connection, I treat jokes and anecdotes in what ensues—Boswell's, Wolcot's, Tyers's, Petronius's, my own—as profitable and often profound diversions, productive opportunities for reorienting thought.

Thomas Tyers, to give my most important example—Johnson's first biographer, Burney's first target, compulsive anecdotalist and alluder, and the only contemporary to speculate in print as to Johnson's likely horror at the thought of the autopsy—turns to the story of the Ephesian matron in Petronius for his justification, asking in a transposed and translated moment from the Latin, "What do the dead care about the living?"[28] As it turns out, the story—immensely popular and as old as Homer—is one of the most serious and multivalent jokes in the history of literature. Playing on the gendered relation between life and death, virtuous exemplarity and individuality, monumental loss and erring life, Petronius's anecdote and its afterlife structure this book's attitude toward Johnson's monument. As Freud puts it, "A new joke acts almost like an event of universal interest; it is passed from one person to another like news of the latest victory."[29] Petronius's joke, passing through a variety of hands in the history of Western literature, was mind-altering news to me.

In paying serious attention to the whimsy of the anecdote, this book replicates some of the same risks that Boswell was faulted for; it is thus not a monograph in any conventional sense, nor does it focus solely on Johnson or on Boswell—rather it considers the relationship between the two authors and the anecdotal form that they jointly author, a form that precedes and goes beyond their collaboration in the *Life* but nevertheless still remembers them. The relationship between Boswell and Johnson is one of the most symbiotic in British literature—for better or worse, as many have commented over the centuries, neither would have his literary afterlife without the other. The collaboration that enabled Johnson to live on as the quintessential character of the author independent of his texts thus paradoxically undermines the mystique of individual authorship.

Still, as the writer most (though far from solely) responsible for the construction of the Johnsonian monument that has endured to this day in Johnson societies around the world and (until recently) in literature classrooms around the country, Boswell provides this book with its model and impetus. I admire Boswell for daring to love Johnson with such egotistical selflessness, for making his reverence—rather than his knowledge—exemplary. Emerson's exhortation serves me well as a critic:

> Compromise thy egotism: Who cares for that, so thou gain aught wider and nobler? Never mind the taunt of Boswellism: the devotion may easily be greater than the wretched pride which is guarding its own skirts. Be another: not thyself, but a Platonist; not a soul, but a Christian; not a naturalist, but a Cartesian; not a poet, but a Shaksperian.[30]

How might Boswell put this? Be not a moralist, not a melancholic, not a philosopher, but a Johnsonian. "*Strongly impregnated with the Johnsonian aether*,"[31] Boswell abandons pride for devotion, defensive self-assertion for self-effacing worship; speaking of the many who supplied him with anecdotes in his advertisement to the first edition, Boswell likens their efforts to "the grateful tribes of ancient nations, of which every individual was eager to throw a stone upon the grave of a departed Hero, and thus to share in the pious office of erecting an honourable monument to his memory," a monument of "innumerable detached particulars" that is still his sole construction.[32] This grandiose modesty had its drawbacks, as the portrait of Boswell as artless toady that emerged with help from Thomas Macaulay and Thomas Carlyle in the Victorian era reminds us. Yet Boswell's devotion seems to me, though it is perhaps unfashionable to say so, worth honoring. I differ with Emerson, and with Boswell and his Johnsonian heirs, only—and hugely—in this: the idealized other in the name of whom one abandons the meager certainty of self can never be found or known. (Even Emerson goes on to say that such self-effacement cannot last forever; rather it will produce, indeed nurture, new and independent intellectual life.) As this book's range of treatments of repetitions, allusions, imitations, and parodies of Johnsonian anecdotes will show, Boswell's devotion presupposes and preserves an original Johnson. I focus instead on the variety of versions of Johnson—and of author love—that gave rise to Boswell's monument and that his monument has produced. Boswell deals in Platonic icons; I deal in literary simulacra.[33] But like Boswell, I have been moved to write by the exceptional love, in some of its phases nothing less than a secular religion, that the figure of Samuel Johnson has inspired, and thus to reflect upon what the love of authors has to do with the love of literature. In the service of such meditation, I want to turn briefly to several anecdotes, all of which hinge on jokes.

## Devoted Infidelity

The first is the tale of the Ephesian matron, to which I will return in more detail in chapter 4. In this story, a wife renowned for her virtue mourns her dead husband by refusing to leave his tomb, refusing to eat or drink, in essence giving up her life along with his own. A nearby soldier, guarding a crucified corpse, hears her weeping, enters the tomb, offers her sustenance, and ultimately seduces her. In his absence, the corpse is stolen. The soldier, knowing that the penalty for his negligence is to replace the corpse on the cross with his own body, threatens to kill himself. The matron offers her husband's corpse as a substitute in order to save her new lover's life.

In his *Satyricon,* Petronius turns this story of exemplary love for a corpse turned to life-affirming lust into a joke with an irreverent punch line: "Next

Morning every one wonder'd, how a dead Man should be able to find his Way to the Cross."[34] The laughter of his audience is as much at the misogyny the story inspires in one enraged male listener (who is also a cuckold)—who argues that the matron should have donated her own body and restored her husband to the tomb—as it is at the possibility of resurrection of the dead (the latter recent Christian news to the *Satyricon*'s sophisticated pagan world in the first century A.D.). And it is to the love of the dead in the study of literature that I now turn.

I was flattered when as a beginning young assistant professor, I was asked by my senior colleague, a prominent and deservedly praised Johnsonian, if he might read my recently completed dissertation. We made a lunch date to discuss it—a baggy monster that considered the literary career as a mode of imitation in Horace, Pope, and Johnson, and that contained reflections on the Johnsonian phenomenon that are continued in this book. When I arrived at the mailroom to meet my colleague, he looked crestfallen. Clearly I had disappointed him. We sat down to a meal the awkwardness of which was encapsulated in his reproach, "But I thought you *liked* these authors!"

I was flummoxed. Most pressing at that moment was my sense that his remark, so indicative of the sad tenor of our discussion, meant, "But I thought you liked me!" That concern took me, like many other young women in the profession of eighteenth-century letters, years to overcome. But more important is the way in which his words struck me, even at that moment, as true. In some way this strange praise, at once offered and withdrawn, was appropriate. I *did* like, even love, the authors I wrote about. If I was not conveying that love successfully or, worse, if I was conveying contempt in its place, then what was I doing? Advertising my own intelligence at their expense? Dismissing what I most valued? (The central caveat by which I had determined to live as teacher and critic was never to presume to be smarter than the authors whose work I taught and wrote about.) Was criticism really about liking, even loving authors? If so, was the only way to love Johnson to be a Johnsonian rather than a critic of Johnsonians? Would I really be welcome in that world, the image of the profession as it was before women joined it? How could I abandon my position as interested outsider—how could I stop being curious? Did loving an author entail declaring that sort of impossible allegiance?

If the previous story is a joke on gender gone sad, my next two anecdotes are jokes that position gendered embodiment against its masculine idealization. The first has circulated through pockets of the profession like news of the latest victory (although whose victory is unclear). A friend told me a story about a young woman interviewing for a position in eighteenth-century

English literature. The hiring committee was entirely male. They asked her the predictable question: How would you teach a survey of the period? In the midst of her answer, one interrupted to ask: "Where is your Johnson?" She responded, with some confusion, "I don't have one." Here Johnsonian authority plays itself out in a joke that reverses the standard formula: the joke's usual object is also its originator, at once excluded from the proper linguistic register and able to turn that register to improper use in an admission of her own lack. In this gendered scene of absence both canonical and physical, professional power is embodied, reinforced, and threatened, rehearsing, on a comic scale, the drama of disavowal of their hero's death that Johnsonians perform as they gaze upon the author's body in an attempt to keep him forever whole, alive, and before their eyes. (This almost certainly apocryphal anecdote resonates interestingly with the true story an anonymous reader of an early version of this manuscript reported of nightmares she had as a first-year graduate student "that I was undergoing a personality metamorphosis and turning into Boswell.")

The next joke (and the book contains others) was on me. When I presented an earlier version of chapter 3 on the eucharistic dimensions of the autopsy and its aftermath for a 2002 English Institute panel on the subject of "Author Love," I was asked during the question period about my own investment in the panel's topic. After all, I was analyzing the Johnsonian phenomenon as an anthropologist would, rather than participating in it myself. What did I think the future of author love was going to be in the academy? The question seemed to me to be *the* question, and one that I couldn't fully answer. I replied tentatively that it was now considered in many circles retrograde to teach Age of Johnson classes, that the phenomenon I was examining—as the young woman's lack of a Johnson also reveals—was almost obsolete. And it was obsolete, though I did not say this then, in part because of the almost entirely male nature of the love for Johnson that had excluded scholars such as the young woman from participating in and inevitably transforming the nature of that love.[35] I added that I felt it was now acceptable, even hip, to love literature, which had not been the case ten years ago. Struggling for words, I added, "What I'm trying to do is open him up". . . and the room exploded into laughter.

Open him up, indeed. There in the same halls that had witnessed Walter Jackson Bate's famous Johnson lectures—lectures that had apparently involved some impassioned imitation of the great man's speech—my own curiosity, itself autoptic in nature, was exposed at that moment, along with my alliance with the Ephesian matron's choice of life over death, pleasurable exchange over monumental fixity.[36]

## Anecdotal Immortality

To point to the bodies that inform the canon of English literature is to insist on that canon's basis in mortality and mutability; to be able to do so is not unlike saying that the emperor has no clothes. To speak what has largely gone without saying is dependent upon being positioned neither fully within nor fully without the emperor's court. Like Tyers invoking the Ephesian matron story in justification of the autopsy, like the matron herself, I have chosen to undermine Johnson's monument by foregrounding the materiality of the corpse upon which it was built. Like the matron, I've undone the melancholic denial of the patriarch's death—a denial that imagines a husband's afterlife through his widow's fidelity, a denial that in Johnson's case takes the form of masculine melancholic identification—for the sake of literary life. The Johnsonian plot against which I position myself here (in which a version of the matron appears as Gertrude and also as Hester Thrale) is that of *Hamlet,* a plot of filial mourning and reparation to which I allude throughout these pages, and that Boswell himself significantly summons up when Johnson's first appearance in the *Life,* through a bookstore-door glass darkly, is announced, ironically enough by an actor turned bookseller, "Look, my Lord, it comes!" This plot of paternal loss has recently taken an intensely nostalgic turn, mourning more than Johnson's death. The Johnsonian tradition to which I am indebted and of which I am a historian is at our particular point in time at once bereft and threatened—much as Boswell was when he created a Jacobite Johnson in his 1791 *Life* who would embody a lovable brand of English authority—as it evokes an uncannily familiar ghost once synonymous with the profession of eighteenth-century letters. A sampling of recent panel titles indicates this current state of Johnsonian affairs: "Johnson at the Millennium," "Is There Room *in* Samuel Johnson for the New Eighteenth Century?" and the one we might ponder longest, "Whatever Happened to the Age of Johnson?" (One panelist's response, however ironic, said it all: "Ageless.")

Ageless, yet flesh. Johnson's case is the crowning example of the ways in which bodily difference has in fact distinguished the English literary canon—itself a product of the eighteenth century and to some extent of Johnson's authority as the author of the prefatory but nonetheless prescriptive *Lives of the Poets*—as a catalog of authorial monsters and paradoxically representative oddities. (One of Johnson's many sobriquets was in fact "Oddity.")[37] By the late eighteenth century, the exceptional man of feeling and letters, the type of which Johnson was a paragon, had rewritten the standard equation of sound mind in sound body along the new diseased lines of the culture of sensibility.[38] In their haunted disavowal of Johnson's mortality and in their re-interment of Johnson in their individual breasts, Johnson's devotees both confront and avoid the double nature, material and immortal, of Johnson's body. I propose to take the profession in yet another direction by repeating an old pattern with a difference; my goal is to trace the path of an embodied anecdotal afterlife that takes the English canon once epitomized by Johnson beyond male melancholia toward other directions.[39]

Julian Barnes's satiric novel *England, England,* in which the Isle of Wight is turned into an amusement park offering the delights of "Old England," provides one such alternative. One of the park's most singular attractions, the actor hired to play Johnson inspires numerous complaints due to his uncivil disruption of the "Dining Experience" of which he is meant to be the star (other actors play members of the Club in order to mediate Johnson's potential effect on his audience, and even a bibliophilic toady has been hired to provoke amusing putdowns). In a scene that ranges from comedy to tragedy, as he is called on the carpet by management, it becomes clear that the actor—who has changed his legal name to "Samuel Johnson"—plays his part too well. Rather than playing jolly old "Dr. Johnson," as per his contract, he enacts—indeed, becomes—the darker and quirkier Johnson who haunts the *Life,* the depressed and despairing Johnson of the *Prayers and Meditations,* the Johnson whom Mrs. Thrale glimpsed in extremis, and whom Samuel Beckett embraced as "always with me."[40] The female executive who interviews him (and whose shoe he is compelled to remove, in an emblematic movement captured in an anecdote from Hawkins, conflating intentional desire and involuntary tic, to which we will return in chapter 2) is left strangely moved, "as if, after long search, she had found a kindred spirit," while feeling that "he had behaved as if she were less real than he was." When she considers hiring a "new Johnson" or "rethink[ing] the whole Dining Experience with a different host," she despairs, "Hadn't Old England produced any wits who were . . . *sound*?"[41]

As Barnes's episode elucidates, Johnson scholars will recognize the "Dr. Johnson" of my title as a misnomer and national fantasy, historically untrue (for the first part of his life at least) to what Johnson would have called himself. This tension between the "Dr. Johnson" of popular imagination and the "true" Samuel Johnson that scholars attempt to recover, indeed defend, from such misattributions plays a crucial role in my exploration of anecdotal form. As some of the shrewdest readers of the *Life* have noticed, the interplay (with which this book began) in and beyond Boswell's text between comic and tragic versions of Johnson—between "Dr. Johnson" and "Samuel Johnson"—is what gives an otherwise relatively plotless story its narrative impetus.[42] My title is meant to highlight the many forms and objects of love—the complex interaction of desire, fantasy, narcissistic misrecognition, and unsettling confrontation with the alien—that the idea of Johnson has inspired.

In what follows, my historicist impulses to locate and define specific varieties of Johnsonian devotion—Thrale or Tyers or Hawkins vs. Boswell, amateur vs. academic, American vs. British, Romantic vs. Victorian, to name only a few possible varieties—are at war with my equally strong sense of an uncanny recurrence of Johnson as familiar. While I'm aware of irreconcilable differences (between, say, Boswell's Johnson and Coleridge's, Macaulay's and Bertrand H. Bronson's, J. C. D. Clarke's and Donald Greene's), this study has led me toward startling continuities. Bertrand H. Bronson and Donald Greene—great pioneers in the field of Johnson studies defending the literature of their period against the prejudices of the New Criticism, on the one hand, and the popular fantasies begun with Boswell and Macaulay, on the other—may deplore the avatar of "Dr. Johnson," but they are forced to admit its power. The polite denizens of Lichfield society who continue to crown Johnson's statue in the town's market square with a laurel wreath each year are inspired by different motives than the groups of academics who attend panels on Johnson at national and international conferences, but these groups do intermingle—Donald Greene, for example, served as president of the Lichfield society to whom he quoted Samuel Beckett (who himself made a pilgrimage to Johnson's Lichfield birthplace) on *his* undying love for their mutual hero. The ritual at Lichfield, we might also add, involves five minutes of silent reflection, as did the annual London meetings of the belletrists, statesmen, and academics who called themselves the Johnson Club at the turn of the nineteenth century. At the annual meetings of the Johnson Society of Southern California, by contrast, professors mingle with collectors in a common appreciation of good meat and music performed on period instruments—but they all meet to honor Samuel Johnson. In keeping with their locale, the Southern California Johnsonians do not wear black tie, but the East Coast Johnsonians do.

Inspiring both spiritual communion and moral outrage, literary Johnsonians have preserved their hero in anecdotal detail, just as medical Johnsonians have preserved his corpse in parts. Both are inspired by author love, a kind of secular religion based on the necessary insufficiency and self-transcending power of the printed text. Johnsonian communities vary across time and place—the festivity of the self-styled "Brethren" of the fin-de-siècle Johnson Club; the nostalgic Oxbridge camaraderie of early twentieth-century British scholars such as Walter Raleigh and R. W. Chapman; the gentlemanly curiosity of the Royal College of Surgeons dining out at their annual London meeting on details from the manuscript of Johnson's autopsy; the professional historicism of American scholars of eighteenth-century England who found in Johnson and Boswell both a scholarly treasure trove and corporeal supplement to the text-based disembodiment of the New Criticism; the politeness of the Lichfield citizenry listening to the novelist Dame Beryl Bainbridge deliver an after-dinner lecture depicting Johnson's nervous breakdown (drawing on the same anecdote of Thrale's with which we began); the sociability of the guests—amateurs and academics alike—of the Los Angeles lawyer and literary collector who hosts the annual cocktail party of the Johnson Society of Southern California in his Beverly Hills home.

American devotion to Johnson, whether amateur or academic, differs from British commemoration (parodied so trenchantly in Barnes's novel): the American president did not give the troops in the trenches copies of Johnson's collected works,[43] as Britain's queen did in World War I, nor did American critics at midcentury boast—as John Bailey did in his oft-reprinted volume on Johnson for the Home University Library—of the fact that even New York cabbies quoted Samuel Johnson, whether or not they had read Boswell.[44] American Johnsonophilia (a personified subset of Anglophilia), it would seem, has more of an acquisitive and elitist than a nationalist and populist cast, and more of a material base in the collection and institutionalization of the manuscripts that gave birth to the Johnsonian phenomenon in the first place. We might think here of A. S. Byatt's scathing and somewhat homophobic portrait in *Possession* (1990) of the meticulously self-maintained American academic, living in the Southwest with his mother, whose obsession with collecting the ephemera—both manuscript and material object—of Victorian authors reduces both to sterile commodities devoid of intellectual or spiritual worth.

But there is a dimension to the collector's impulse that goes beyond possession toward something less easy to condemn. I prefer the words of James Osgood in his introduction to the limited edition of the catalog for *The R.B. Adam Library Relating to Dr. Samuel Johnson and His Era, Printed for the Author at Buffalo, New York:*

> Dr. Johnson remarks, in one of the letters of this collection, that a rare book is not safe "in a scholar's talons." Perhaps he was thinking how he once picked up a fine old folio *Petrarch* in Garrick's library, raised it absently over his head, and let it fall crashing to the floor. Whether or no, he touches here upon an imperfect sympathy that has sometimes divided the scholar from the collector. Yet here is a great scholar who has now become a favorite of collectors. Rarer and more expensive

> game is to be had, of course—quarto plays and seventeenth century worthies that demand "a total defiance of expense." But collectors choose Dr. Johnson not for the rarity, but for the love of him—of his wit, his geneality, his hunger for humanity, his triumphant reconciliation of mundane with spiritual things. It is the man they seek and the atmosphere in which he lives. Such is the force and charm of Dr. Johnson that one sometimes seems to see certain of his personal qualities mysteriously conveyed through his relics, and reappearing in those who preserve them.[45]

This passage begins by nostalgically evoking a Johnson who used books rather than revered them. A variety of scenarios evoking Johnson's voracious appetite and implementation of the printed word come to mind: Johnson holding a book under the table while he eats dinner, devouring first one, then the other, like a dog holding a bone in reserve; Johnson holding a book so close to his squinting eyes that he is breaking its spine; Johnson beating the bookseller James Osborne over the head with a folio edition. The power of these images—anecdotes that themselves unite material with spiritual things in their bringing of their subject to momentary life—far outweighs the collector's usual reverent impulse to protect books and manuscripts from those who see them as vehicles for gratification of scholarly appetites rather than things in themselves. What motivates collectors of Johnson's works is the love of Johnson *himself,* in an impulse that merges the scholar with the books he abuses. The collector of Johnsoniana is driven not by a desire for rarity but rather a need to seek "the man and the atmosphere in which he lives." This is a spiritual desire, inspired by Johnson's union of "mundane with spiritual things" in his own work and person (it is important that Osgood does not specify which), and by the mystical "force and charm" that allows his followers to summon him through his "relics," and even thus to embody him themselves.[46]

There is a story to be told here, one that would involve the Byzantine transatlantic saga of the recovery and purchase of the Boswell papers and the eventual creation of the monumentally impressive Johnson/Boswell editing factory at Yale University; the purchase of many of the Boswell papers, along with the Adam Library by the American bibliophile and scholar (whose focus was on Hester Thrale), Mary Hyde, Viscountess Eccles, and the building of her own collection of Johnsoniana, second only to Yale's, recently donated to Harvard's Houghton Library. This story would not only give me the opportunity to read Hyde's play on the "colorful" collector Ralph Isham, who first bought the Boswell papers discovered at Malahide Castle, "Levee at Fifty-Third Street," about which I am extremely curious.[47] It would also link the love of Johnson to the formation of the profession of eighteenth-century British letters in America and would consider, among other things, the relation of Johnson's embodied presence to the invisible mid-twentieth-century intelligence communities of the American New Criticism so provocatively analyzed by William Epstein. But this is not the history I'm going to tell.[48]

## Forgetting to Remember

Instead, I choose to focus on a story that repeats itself with a difference, that of the ambiguity, multiplicity, and surprising continuity of Johnsonian desire, and to admit to the personal, exorbitant nature of this focus. This returns me to my fourth epigraph and to Johnson's love of anecdote, what he defined in his *Dictionary* as "something yet unpublished; secret history," and to which he added in the fourth edition, "a minute passage of private life."[49] Johnson's "fancy," his pleasurable fantasy, is of a time when the burdens of narrative—the concerns of preparation, connection, illustration, of context, of, in short, history—are discarded forever. Instead of the "big books" with which the great author (largely because of the *Dictionary* and paradoxically because of the *Life*) was so often associated, we would have anecdotes—the smallest possible version of narrative; instead of the labor to ground anecdote in coherent historical connection, we would have the pregnant ambiguity of aphorism. What would we gain from such an abdication of scholarly responsibility and gratification of fancy and curiosity? We would get more anecdotes. Rather than waste time "weav[ing] anecdotes into a system," we would glean them before they disappeared. Johnson conveys here a newly acquisitive sense of urgency, of the fleeting evanescence of the present, germane to his own work as a biographer. Boswell took this injunction, and Johnson's authorial example, powerfully enough to heart to turn it into a new form of biography, told primarily in "scenes" of conversation enlivened by and interspersed with the most ambitious collection of anecdotes perhaps ever published.[50] There have been times, I confess, as I have worked to complete this book, when I have found myself possessed by a similar urge to leave nothing out, an urge that had its material complement in my compulsion (as we shall see) to find lost Johnsonian objects. I found myself, in other words, replicating Boswell's overwhelming desire to preserve Johnson by collecting his conversation in its impossible totality:

> We cannot, indeed, too much or too often admire his wonderful powers of mind, when we consider that the principal store of wit and wisdom which this Work contains, was not a particular selection from his general conversation, but was merely

> his occasional talk at such times as I had the good fortune to be in his company; and, without doubt, if his discourse at other periods had been collected with the same attention, the whole tenor of what he uttered would have been found equally excellent.[51]

Had everyone been as diligent in his collection of conversational anecdotes as the biographer, the truth that he uncovered from "occasional talk" would have been proved—or so Boswell imagines—universal.

Despite the tautology of Johnson's fantasy that mankind will come to "write all aphoristically, except in narrative," I take his effusion on behalf of anecdotes as my starting point for narrating Johnsonian desire. This desire—both Johnson's own for an anecdotal record of a life that must be "seen before it can be known"[52] and that of Johnsonians for their object—is plotless, recurring, always returning to the past; it is exorbitant, and it is queer. I am indebted here to Claudia Johnson's ongoing work on the history and multiplicity of meanings of Jane Austen as "cultural fetish." Many of the early twentieth-century British Janeites that Claudia Johnson analyzes—"committed to club rather than domestic society," to a queer form of literary reproduction rather than the heterosexual imperative, and to a love of tangential detail rather than the teleology of plot—were also Johnsonians.[53] Some of the same queer impulses toward male community outside of domesticity, linked to a common nostalgia for a past Britain, fuel the love of Johnson. He seems to be Austen's male counterpart, living a single literary life (the queerness of his household almost a parody of domesticity) free from both the novel genre and the tyranny of the novel's plot. Even Johnson's own novel, if we can consider *Rasselas* a novel, refuses a love story, as do twentieth-century fictions about Johnson's thwarted love for Thrale—Samuel Beckett's Samuel Johnson, to give the most extreme example, is impotent.

The example of Austen provides a particularly productive contrast to the brand of author love that Johnson has inspired: both have helped to shape the profession of English letters in Anglo-America—but while Austen love is inherently exclusive, Johnson love intends to be inclusive.[54] While, as Claudia Johnson has shown, rising middle-class British academics shunned the frivolous Austen love of aristocratic critics who preferred the effusions of detail to manly moral analysis, the love of Johnson has supported both those same elite Janeites and professional literary critics alike. While the love of Austen, whatever form it takes, is inherently proprietary, marking distinctions between proper and improper reading, the love of Johnson veils such distinctions, masquerading in good faith as potentially universal.[55] And while the love of Austen can be queer (though Claudia Johnson is right to point out that such queerness has been erased from historical memory), the queerness of Johnson love, as I have begun to show, allows it to endure as objectless, open to all who respond to the great man's great heart.[56]

This reciprocity, dependent upon the sublimation of an author's distinctive artifice in order to render him or her commonplace and communal property, is echoed in Johnson's own description of his response to the last stanzas of Thomas Gray's 1751 "Elegy Written in a Country Churchyard," a poem that he claimed "abounds with images which find a mirror in every mind, and with sentiments to which every bosom returns an echo."[57] Gray's extreme allusiveness, his work's status as a ready-made commonplace book for the common reader, also provides the classic example of the emergence of the modern academic version of what John Guillory, following Pierre Bourdieu, has termed "cultural capital." For Guillory, characterizing the literary milieu of the early eighteenth century as one of "a *confusion* of aristocratic and bourgeois cultural norms, and an *ambivalence* in the relation of vernacular to classical literacy" (a confusion that can take the form of appropriations of aristocratic gentility by bourgeois men of letters like Gray and Johnson), Gray's poem is published at "virtually the last moment at which literary culture can sustain a discourse of polite letters in the vernacular without establishing this discourse in the schools."[58] But Johnson's brand of cultural capital, both as critic of the "Elegy" and personification of the canon, is at once related to but distinct from Gray's. Just as Johnsonians like Stephen commemorate their hero by abstracting him from the realm of corporeal and psychic particularity, so Johnson, both critic and poet, who found Gray objectionable for his excessive pedantry and disdain of the literary marketplace, in his praise of the final two stanzas of the "Elegy" (the key text for Guillory and the most stylistically accessible of Gray's works), forgets precisely the learning that gives Gray his distinction, and thus marks his own critical authority in the name of the "common reader." Gray's apparent universality, and illusion of originality, is a stylistic effect, Guillory claims (in an argument that complements William Epstein's emphasis on the poem's affinity with New Critical abstraction and disembodied objectivity), of

> the *systematic linguistic normalization of quotation* [emphasis Guillory's]. . . . By this means both classical literary works and older works of English literature are absorbed into the poem in a linguistically homogeneous form, the proto-"Standard English" of mid-eighteenth-century England. . . . The cento of quotable quotations which *is* the poem thus generates a reception-scenario characterized by the reader's pleased recognition that "this is my truth," while at the same

> time concealing the fact that this pleasure is founded upon the subliminal recognition that "this is my language."[59]

Similarly, readers of Johnson and Boswell recognize their common truth in Johnson the man, rather than Johnson the writer.[60]

In a striking counterexample to the rise of Johnsonian communion with the ideal of the author, Eric O. Clarke has traced the critical industry that arose from textual editing of Shelley's literary corpus to the original worship of his corpse. As Clarke traces the history of Shelley love, the author's queer body gives way to the flawed text in need of editorial correction; the worship of Shelley's corpse is sublimated into the fetishization of his texts in scholarly editing.

> The fetishist believes that what an object represents "lives" in that object, yet the very fact that it must be re-presented by a substitute object implicitly acknowledges the original's absence. This contradictory attitude approximates the fantasy involved in recreating the presence of an author through imagining a fully present intention organizing an authoritative text.[61]

Despite its queer particularity, Shelley's case still seems more familiar to us in its replacement of loving devotion to the author's body with scholarly attention to the text. Johnsonian worship, by contrast, operates by a disavowal not (as Clarke has claimed in Shelley's case) of the text's flaws, but of the text itself. For Johnsonians, neither the author's mortal body nor his book is sufficient. Texts must remain imperfect and incomplete, open to "life" and the living author. Johnson "himself," rather than Shelley's governing intention in the text, becomes the fetish. This difference is not unrelated to the differing sexual politics of Shelley love and Johnson love. In Clarke's account, Shelley love becomes too queer as definitions of masculinity change over the course of the nineteenth century. Thus, in Shelley's case, the fetishization of the text allows the poet's readers, like Freud's fetishizer of a jock strap, to have their homoerotic cake and deny it too. For the Johnsonian "Brethren," the manly love of an unthreatening and desexualized father is much less dangerous; indeed, it is enabling. The magnificent scholarly achievement of the Yale edition of Johnson's *Works* notwithstanding, Johnsonians must devote themselves, whether in commendation or refutation, to the endlessly proliferating and eternally present-tense genre of the anecdote, extratextual, and semi-corporeal supplement on the margins and footnotes of scholarly editions. The pressing question for the study of Johnson becomes not a decision on a textual variant, but rather a confrontation with the form that summons up the here and now of the embodied speaking voice, the form that, as epitomized by Boswell's *Life,* demands communion and response.[62]

## Exemplary Eccentricity

The genre that makes a science of the fusion of living author with dead letter, praised by twentieth-century New Historicists and nineteenth-century antiquarians as "uniquely refer[ring] to the real," is that of the anecdote.[63] Like the anecdote, my story is based in a particular historical moment that frames the autopsy and Johnson's particular exemplification of the mind/body problem—I recur to that moment but also watch it repeat and by repeating, transform itself. Beginning with the first collection of Johnsonian conversation in 1776, *Johnsoniana; or, A Collection of Bon Mots, &c. by Dr. Johnson and Others*—which Johnson termed a "mighty impudent thing,"[64] noting that its sales outnumbered those of his own *Journey to the Western Islands of Scotland*[65]—the anecdote and its closely related form, the aphorism, have served as powerful fetishes that have allowed Johnson to endure, in speaking person, beyond the grave or the page. The same physical detail that marked Johnson as flawed, mortal, and of his historical moment thus keeps his image alive, distinguishing him as one of the most memorable literary characters of this or any other age. Johnson's surprise at the public's preference (disputed by Boswell) for the anonymous *Johnsoniana* over a travel narrative in which he thought he had "told the world a great deal that they did not know before"[66] points to the kind of desire that is my subject: not the well-known eighteenth-century British predilection for novelty but rather that culture's equally strong and equally novel craving for the familiar, for the presence, at once uncanny and reassuring, of a personal authority much closer to home than a monarch could ever be, and much more lovable than dangerous French abstractions of liberty. For Boswell, as well as for fellow clubman Edmund Burke in his *Reflections on the Revolution in France* (1790), institutional authority had to be "embodied . . . in persons; so as to create in us love, veneration, admiration, or attachment."[67] As Boswell puts it in his advertisement to the second edition of the *Life* (1793), Johnson's concrete example, "his strong, clear, and animated enforcement of religion, morality, loyalty, and subordination, while it delights and improves the wise and the good, will, I trust, prove an effectual antidote to that detestable sophistry which has been lately imported from France, under the false name of *Philosophy*."[68] Anecdote served as sentimental antidote conveying personable British authority as prophylactic against abstract French theories of liberty.

The anecdote also structures the cultural poetics that informed this author as a familiar, representative, yet also monstrously unique body at the end of the eighteenth century. How is his ghost our own inheritance? Whether preserving his corpse in bodily fragments or universalizing his peculiarities in an anecdotal present, medical and literary Johnsonians cling to their hero's aberrations in something of the same way he kicks the stone in Boswell's famous anecdote to refute Berkeley thus. The story that ensues therefore also forms a chapter in the history of the cultural function of bodily difference and its relation to new forms of modern celebrity.

Samuel Johnson in his embodied, diseased, and paradoxically representative particularity can thus be seen to occupy a charged and liminal position in a variety of cultural narratives. Even some of the most recent revisionist work on this figure, though unconcerned with issues of embodiment, continues to battle over the history of the literary profession with Johnson as its representative icon. In its suspension of Johnson's identity as author between classical humanist and modern hack, such work rewrites his particularity on the level of style. A 1784 letter from Edinburgh to the *Gentleman's Magazine* shortly after the author's death begins this particular strand of the Johnsonian narrative, contrasting (in terms reminiscent of Guillory's description of the ambivalence of Gray's cultural moment) the solid learning of the past with the new professionalized age of commercialized superficial style:

> It was the misfortune of Johnson and of his contemporaries, to be born as it were out of the due time, and to survive the age of Erudition, which he himself enriched and adorned; and he saw, and may of these still see, laborious attention to the unfolding of the principles of science and of literature yielding to the flimsy ornaments of style, where point and antithesis, embroidered with metaphor, lord it over argument, and where hypothesis wages war a second time with true philosophy, and we shall soon see, I fear, a compleat victory obtained by News-papers, Magazines . . . Translations, Abridgments, Beauties, Reviews, and Fugitive Pieces, with the light Summer Infantry to compleat the rout over the heavy-armed Legion of the Learned.[69]

J. C. D. Clarke's *Samuel Johnson: Literature, Religion and English Cultural Politics from the Restoration to Romanticism* is germane from this perspective in its claim to reverse "the orthodoxy which has dominated" analysis of Johnson's life and thought "for over thirty years."[70] Clarke writes specifically against what he identifies as a Whiggish critical tradition that imagines Johnson as a commonsensical, largely apolitical embodiment of Englishness and the vernacular canon. Such an image, he argues, is a narcissistic projection of the newly solidified profession of English letters of the nineteenth and twentieth centuries. Clarke's Jacobite Johnson (the subject of much recent debate in Johnson studies, as well as of a 1923 novel, *Midwinter,* by a former Canadian government official)[71] is a passionate and belated inheritor of a dying Anglo-Latin tradition of classical humanism that the Whig version of literary history has largely forgotten. Clarke's argument is doubly useful for my purposes: it attempts to reveal the political and historical stakes that inform Johnson's mythic persistence, while it nevertheless affirms Johnson's usefulness as icon for his age, however the politics and history of that age are rewritten. If style in the 1784 letter was the enemy of erudition, in Clarke's related account, style also becomes the vehicle for political truth. The question remains, however: Why should it matter so much whether or not Johnson was a Jacobite? Why is Johnson's "oddity" representative?

From my perspective, Clarke's concern with unmasking Johnson's own aesthetic allegiances and those of his worshippers as political constitutes a late-twentieth-century revision of an eighteenth-century debate. This contemporary controversy over Johnson's representative Englishness, initially fueled in large part by his authorship of the *Dictionary,* resulted—as we shall see in chapter 2—in the simultaneous stigmatization of Johnson's Latinate prose (beginning with a host of contemporaries from different political and social perspectives, including James Thomson Callender, Elizabeth Montagu, and Horace Walpole, and culminating in the Romantic condemnation of Coleridge and others) and Englishification of his character (beginning with Boswell and most famously marked thereafter in Macaulay and Carlyle).[72]

From a different angle but with similar intent, Lawrence Lipking attempts to redress the murder of the author by postmodern literary theory in his recent *Samuel Johnson: The Life of an Author,* by bringing to life a Johnson who lives (almost) solely in his texts. Lipking's Johnson, a writer resolutely opposed to the Shakespearean cult of the author/genius that characterized his age, is a sort of imago immanent in his works, at once ancient and modern, "partly a hack yet also partly an eternal ideal," who aspired to authorship not by nature but by conscious choice.[73] In both Clarke's and Lipking's cases, however much the Johnsonian myth is historicized and refuted, Johnson remains the mirror in which would-be arbiters of the profession of eighteenth-century English letters see themselves. It is my hope to make the very frame of singularity that has characterized Johnson studies up until now, and that has allowed Johnson studies to appeal to amateur and professional alike, my subject, and thus to see beyond it.

The more I study Johnson and the Johnsonian following he has inspired (a group, it will soon become obvious, of which I am a member), the more it seems to me that Johnson has never fully died, even though (or perhaps because) his corpse and the record of its autopsy have been the subject of greater medical attention than that of perhaps any other famous author. In a parallel paradox, the writer many consider the father of modern literary criticism and the English canon—instrumental in founding modern concepts of originality, authorship, exemplarity, and the common reader—has also been lauded as the last bastion of a neoclassical scholastic community besieged by the anonymous and democratizing world of print capitalism. More dramatically, as the anecdote from Thrale with which we began reminds us, the man whom the Home University Library termed the paradigm of English sanity was plagued famously by fears of madness, death, and, worse, an eternity of damnation—by what we might call an existential terror of moral and literal endlessness. As G. M. Trevelyan put it—in his analysis of a nineteenth-century ethos of a British national culture that carried on, in Kevin Hart's phrase, "the tradition that history was related to literature"—Johnson was "the most abnormally English creature God ever made."[74] Neither dead nor alive, neither original nor derivative, a monument fraught with flawed particularity, Johnsonian authority uncannily repeats itself in the service of both individuality and community, giving rise to a secular religion that has lasted, however nostalgically, however embattled, however diminished, for over two hundred years.

## Singular Science

In considering the nature of the anecdote, the genre that possessed the eighteenth-century literary world like a mania, and that was most responsible for giving Johnson literary life, I had occasion to wonder about its relationship to the photograph.[75] The anecdote has a visual dimension, while both genres freeze narrative by a complex temporality that at once mimics and forestalls death. Roland Barthes's anecdotal study of the photograph, *Camera Lucida,* provided me with an unexpected source of inspiration.

Barthes's study, for all of its influence, refuses to offer a theory of the photograph. Vowing at the start to treat only those images that move him, that wound him, personally, Barthes narrates this decision anecdotally:

> Then I decided that this disorder and this dilemma, revealed by my desire to write on Photography, corresponded to a discomfort I had always suffered from: the uneasiness of being a subject torn between two languages, one expressive, the other critical; and at the heart of this critical language, between several discourses, those of sociology, of semiology, and of psychoanalysis—but that, by ultimate dissatisfaction with all of them, I was bearing witness to the only sure thing that

> was in me (however naïve it might be): a desperate resistance to any reductive system.[76]

Barthes's resistance to system is familiar to admirers of Johnson—clear your mind of cant!—while his sense of "being a subject torn between two languages" is also my own. Like Barthes, I write out of an uncritical, even reverential, desire to account for the power of my own response to certain Johnsonian texts and anecdotes, a desire for particularity that would defy linear certainty for something associative, allusive, something more like art. In that ambition, I fall far short of the brilliance of, to give one example of this book's trajectory, Nabokov's crazed author-lover Charles Kinbote in that homage to Boswell's *Life, Pale Fire,* who takes Barthes's embrace of his own subjectivity to its creative limit.

Barthes resolves to make his own "protestation of singularity into a virtue" first by insisting on the validity, indeed the potential representative "truth" of his own contingent choices:

> So I resolved to start my inquiry with no more than a few photographs, the ones I was sure existed *for me.* Nothing to do with a corpus: only some bodies. In this (after all) conventional debate between science and subjectivity, I had arrived at this curious notion: why mightn't there be, somehow, a new science for each object? A *mathesis singularis* (and no longer *universalis*)? So I decided to take myself as mediator for all Photography. Starting from a few personal impulses, I would try to formulate the fundamental feature, the universal without which there would be no Photography.[77]

This study similarly has nothing to do with a corpus, only a corpse. In what follows, I find myself in this book perpetually returning to the same paradoxical intersection of personally resonant anecdotal particulars and the ever-elusive larger possibility of truth. Whether it be Barthes's *mathesis singularis* or what Theodor Adorno characterized as Walter Benjamin's "magical positivism," I, too, am increasingly drawn to the elusive possibility of truth embodied by "things in themselves," by the things, that is, preserved in anecdotal form, that have the power to wound me and others before and after me. This puts me in sympathy with the words of the American critic Joseph Wood Krutch, who said in a radio interview in 1942, "Boswell's Johnson is more continuously and perfectly Johnson than Johnson ever was himself, but it is still the essence of Johnson, not something else."[78] The thing in itself, truer in art than in life.

Which brings me back to Johnson's lung. For eighteenth-century literary and medical men of letters alike, Johnson's body was an object of spectacular importance, portrayed in ink, paint, marble, and wax by Ozias Humphrey, Joseph Nollekens, John Opie, Samuel Percy, Joshua Reynolds, and others;[79] puzzled over in Johnsoniana by a multitude of biographers, including Hester Thrale, John Hawkins, Arthur Murphy, Frances Reynolds, and most definitively James Boswell (who was canonized by many readers as the very author of Johnson himself); autopsied and preserved in parts and in print, most strikingly exemplified by the image of a lung attributed to Johnson in Matthew Baillie's 1799 *Series of Engravings* accompanying his 1793 textbook *Morbid Anatomy*. When I first learned that Johnson had been autopsied, I was particularly intrigued to find that his lung, along with a kidney and a slice of his scrotum, had all been preserved. The lung, said to have been cataloged in the collection of surgeon William Cruikshank, was thought to have served as the original for the illustration to "Emphysema" in Matthew Baillie's anatomy text. A considerable medical literature, I discovered, beginning with the first number of the *London Medical Journal* in 1849, is devoted to speculating not only as to the proper diagnosis for Johnson's final illness based on the autopsy report, but also as to whether or not the illustration in Baillie's text, reproduced in twentieth-century medical textbooks, was indeed a picture of Johnson's actual lung (fig. 2).[80] I came to read this medical speculation over fragments of Johnson's corpse, as we shall see in chapter 1, as the literal fleshly equivalent to the anecdotal preservation of the living Johnson in encapsulated fragments of time.

Yet somehow fascination with the autopsy itself, not just as a metaphor, persisted and spread over the course of my research. The bodies of authors are perhaps the ultimate memento mori for literary doctors writing in venues like the *Lancet*'s "Xhumation" section, and I learned of a variety of other authors whose remains had been puzzled over, including Ben Jonson, Jonathan Swift, Percy Shelley, and Samuel Richardson. John Milton's corpse, in a curiously commercial version of revolutionary violence in 1790, was unearthed, torn to pieces, sold as relics, and verified as genuine (gender being a particular concern) by medical professionals.[81] Johnson seems to me to be perhaps the most powerful such example, partly because of the ways in which—with the help of Boswell leading a host of other anecdote collectors, and at a historical moment of emergent professionalism and nationalism—he came to epitomize more than any other author a certain embodiment of Englishness, partly because the medical evidence in his case is so complete (the manuscript record of the autopsy still exists and is currently on display at Dr. Johnson's House museum in London), and partly because the autopsy itself has had such a complicated and varied afterlife in print in both medical and literary forums. What I was

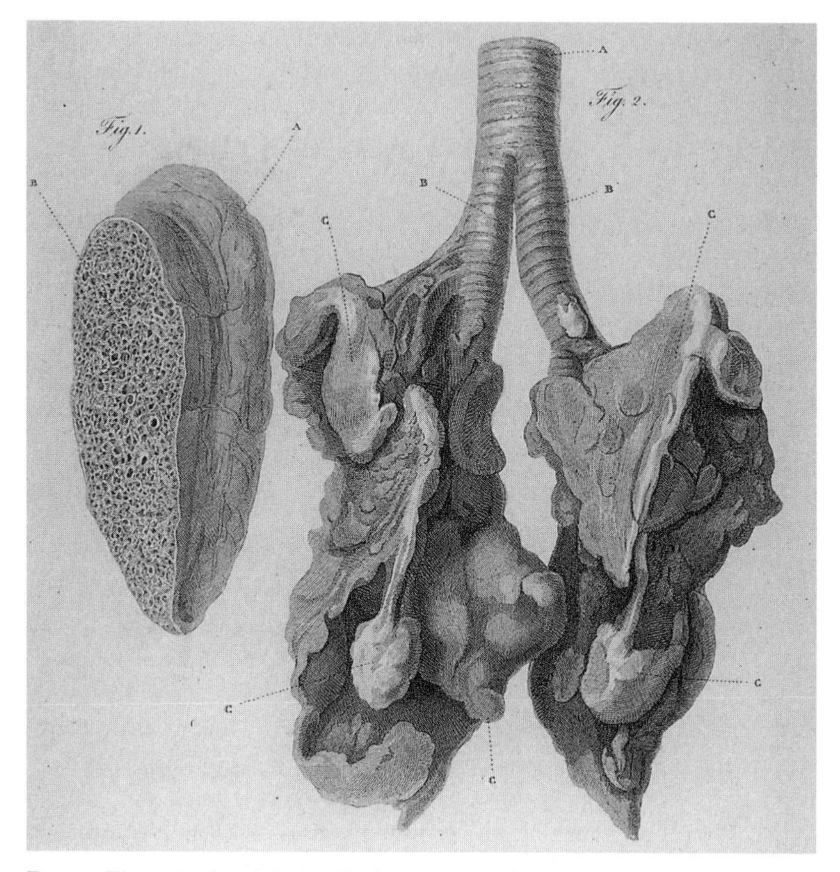


Figure 2. Illustration from Matthew Baillie, *A Series of Engravings, Accompanied with Explanations, which are Intended to Illustrate* The Morbid Anatomy of Some of the Most Important Parts of the Human Body (London, 1803), figure 1, plate 6, fascicule 2. For the past two hundred years, scholars have debated whether the original of the image on the left was Samuel Johnson's lung. Reproduced by permission of the Wellcome Library, London.

not expecting was the interest with which my search for the lung was received by curators of a variety of archives, including the Hunterian anatomy museums in Glasgow and London, as well as Dr. Johnson's House. Eager curators of each of these collections have searched for the lung throughout Britain and as far abroad as St. Petersburg (where Cruikshank's collection took up its final residence). It has never been found. I wonder, if it exists, does the label on the jar read "emphysema in a 75-year-old man"? Or does it proudly bear the author's name? The photograph of John Hunter's preservation of a lung demonstrating "emphysema" is the closest we will ever come (fig. 3).

Michael Paterniti's road narrative and memoir, *Driving Mr. Albert: A Trip*

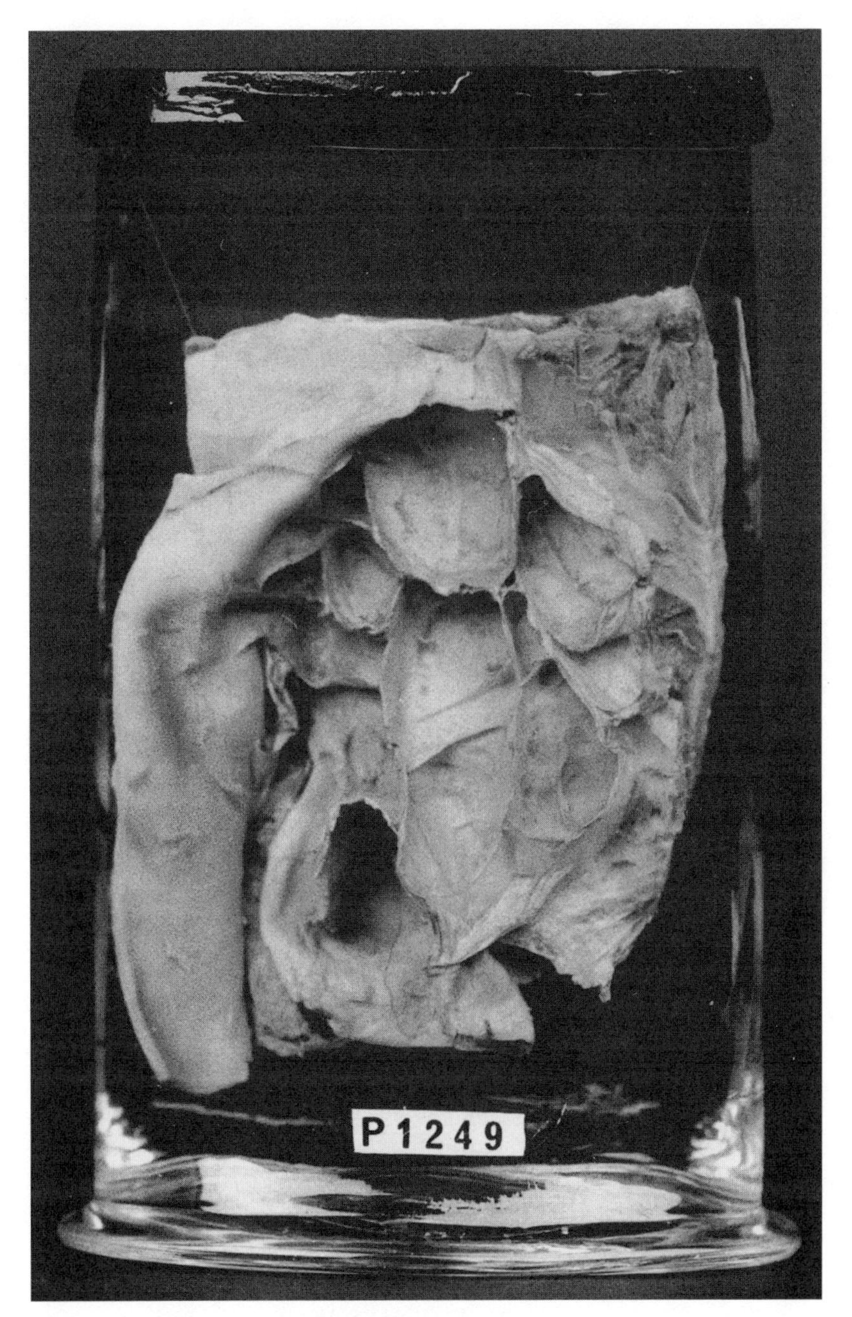


Figure 3. Lung illustrating emphysema from the eighteenth-century surgeon John Hunter's original collection. Object number RCSHC/P 1249. By permission of the Royal College of Surgeons of England.

*Across America with Einstein's Brain,* confirms that my search for the lung is not entirely unique; the remains of the famous dead continue to fascinate a variety of readers and writers. While my mission was informed by the desire to locate a lost anatomical object, that is where Paterniti's story begins. The goal of his cross-country sojourn is to restore that object to its rightful owner, if such a person can be said to exist. The first appearance of Einstein's brain (which was kept illegally by the doctor who performed the autopsy, Thomas Harvey, and is returned at journey's end to Einstein's granddaughter in Berkeley) is the stuff of simultaneous bathos and sublimity:

> Harvey appeared from the darkness with a big cardboard box in his hands. He set it down, and one at a time pulled out two large glass cookie jars full of what looked to be chunks of chicken in a golden broth: Einstein's brain chopped into pieces ranging from the size of a turkey neck to a dime. A swirling universe unto itself: galaxies, suns, and planets. It seemed to glow. And the old man, he stood over it, transfixed, nodding. In his face was a sudden nakedness, an expression of awe, of the soul manifesting itself on the surface of the body.

The sight of the brain, glowing like a sacred relic, embodying in its chicken-soupy (Jewish?) fragmentedness an Einsteinian universe, the thing in itself, transforms the mortal viewer into a visible demonstration of "soul."[82]

My frustrated search for the lung extended to a variety of other lost or inaccessible Johnsonian artifacts—the scrapbook of Johnsonian ephemera painstakingly assembled by the antiquarian Samuel Lysons for his friend Hester Thrale, somewhere in the bowels of the Columbia University rare book room but yet to be located except on microfilm; the letter from the Hyde Collection describing the taking of a death mask from the corpse, which the archivist—himself, as it turns out, a lost connection—was unable to find; the original notebooks in which Samuel Beckett transcribed the autopsy record from a medical journal as he researched his first and unfinished play, "Human Wishes," which I was only able to read in nearly illegible Xeroxed form. The universe was perhaps trying to tell me something about the impossibility of originals, something I was reminded of in the form of the joke that entitles a recent collection of short stories by Lydia Davis, *Samuel Johnson Is Indignant.* A friend had given me the book and I eagerly turned to the title story, only to find the following:

> SAMUEL JOHNSON
> IS INDIGNANT:
> that Scotland has so few trees.[83]

That was it for Lydia Davis on Samuel Johnson. This is anecdote reduced to its barest narrative form, a joke about absence. The punch line brings Johnson's indignance, a word worthy of him, summoning him to animated conversational life, down to size. He is indignant over an absence, Scotland's barren landscape rendered in monosyllables and lowercase letters. Much ado about next to nothing.

Yet the absence of trees in Scotland leads to one of the most famous presences in the British cultural imagination—Thomas Trotter's print of the Englishman Samuel Johnson in traveling dress, bearing the equivalent of a royal scepter, an oak stick (fig. 4). In the *Philadelphia Public Ledger* for February 20, 1916, H. B. Brougham, unsure "whether I should rejoice to sit through an evening at a meeting of the American Samuel Johnson Club"—the proposed founding of which he has just heard—contemplates the antiquarian spectacle of Johnsonian collectors in America with some help from this image and his own imagination:

> With a mind rendered reckless from physical distress and poverty, and made ferocious at affronts of stupidity and low-mindedness; by passion a Tory and by reason a tolerator of Kaiserdoms and communes; the drudge of a Cave and the scourger of a Chesterfield; tutor of the genius Garrick and mentor of the genius Boswell, the figure of the muttering, gesticulating doctor had seemed vivid enough to me in the pages of the great biography. But in a rare old print of Johnson, reproduced on this page, the whole man stands revealed in a flash of wit. He is in the Hebrides with Boswell. Besides using his oak stick as a staff, it was notched, showing feet and inches, in order that Johnson might, if he chose, use it to take dimensions of ruins or any interesting object during his famous journey. He told Boswell he intended to give the stick to a museum in London, so that future generations might gaze on it as a memento of their journey, when he suddenly became aware that his staff was missing. Boswell, speaking up for the honesty of the Scottish people, told him[,] "Be sure, sir, any one who gets this stick will return it to you." Johnson glanced at the stark landscape, bare even of shrubbery, and said grimly, "No, no, my friend; it is not to be expected that any man in Mull who has got it will part with it. Consider, sir, the value of such a piece of timber here!"[84]

In order to appreciate the need for a Johnson club (here conflated, in the most common Johnsonian mingling of author with character with text, with Boswell's vivid evocation of an eccentrically embodied "muttering, gesticulating" original),[85] this reader needs something more: an image that supplies an epiphany. A famous anecdote from Boswell's *Tour to the Hebrides* is thus captured by the "rare old print," in which "the whole man stands revealed in a flash

Figure 4. "Dr. Johnson in his Travelling Dress," drawn and engraved by Thomas Trotter (1786). Reproduced by permission of the Huntington Library, San Marino, California.

of wit." With one glance at this image, reproduced for readers of the article, the author sees the virtues of both antiquarianism and Johnsonianism. All of Johnson's intellectual and textual contradictions for this 1916 American sensibility—his conservative tolerance of political extremes, his enslavement to the press and defiance of patrons—coalesce in the sight of the author as English curiosity in the savage highlands of Scotland, aware of his own rarity, his status as museum piece epitomized by the lost—or, due to the aforementioned rarity of trees that his own desirability raises to aesthetic heights, stolen—oak staff.

If the anecdote of Johnson's oak stick serves as a kind of icon for the narrative of the *Tour*,[86] then the image serves as an icon for the anecdote. At yet another level of remove, the stolen stick restored to view in Trotter's portrait condenses both image and anecdote into the ultimate lost-and-found object of fantasy, a fetish for Johnson himself, collectible for centuries to come. The staff functions as a literal piece of England (oak being the mark of British empire building, as it is in, say, Pope's *Windsor Forest*), sign as well, in its measuring capacity, of the author's enlightened affinity for the rational rule and evaluation of a subordinate colony. Yet it also functions as metonymy for Johnson's status as oddity to the Scots in his own time and to generations to come at home and abroad. This artifact from the past, restored to twentieth-century eyes in a contemporary image reprinted in a modern newspaper, thus functions as at once Enlightenment antidote to and twentieth-century object of wonder. It signifies and supplements history. As Boswell tells and Brougham remembers the story, Johnson himself becomes complicit in the fantasy of the author as staunch embodiment and defender of English objectivity in a foreign land and British colony. Such commonsensical authority, the image demonstrates, becomes all the more enduring once lost and aestheticized. Its loss lends a particular piquancy to this American's affection for an English tyrant rendered all the more lovable by his harmlessness. The desire engendered by such loss, as self-reflexive as the curiosity that renders the inquiring Johnson its greatest object, is common to both author and reader alike. And it is at the heart of Davis's joke, a joke that makes an interesting accompaniment to the lost Johnson of the young woman's job interview.[87]

My initial disappointment about the lack of a story about Johnson in Lydia Davis's book led to a larger insight about the ways in which all anecdotes about Johnson—and his indignance among a host of other endearing qualities—are linked through loss. In another brief story, "Happiest Moment," Davis evokes the inverse of the anecdote's evocation of loss, its revivification of presence:

> If you ask her what is a favorite story she has written, she will hesitate for a long time and then say it may be this story that she read in a book once: an English language teacher in China asked his Chinese student to say what was the happiest moment in his life. The student hesitated for a long time. At last he smiled with embarrassment and said that his wife had once gone to Beijing and eaten duck there, and she often told him about it, and he would have to say the happiest moment of his life was her trip, and the eating of the duck.[88]

Here the anecdote is less about the fetishistic restoration of a lost phallus (as in the case of the oak stick) than it is about questioning such integrity by collapsing the boundaries between subjects and their experiences. The story itself begins in the anecdotal tense of future perfect—"if you ask her . . . she will hesitate"—like all anecdotes, it is one that has been told and retold numerous times. In the case of this anecdote about an anecdote, this particular story will repeat itself again and again—if the question is asked (as is the case with a joke or riddle), the answer will be the same. The author-figure in this story, when asked to recall a favorite story she has written, offers instead one she "read in a book." Her favorite narrative is an exercise in translation—a dialogue between an English language teacher and a Chinese student—that repeats the initial request for a favorite story, this time rephrased as a desire for that most minimal of stories, an anecdote; the student is asked "to say what was the happiest moment in his life." His answer repeats that of the author both in its initial hesitation and in its deferral to another's story. The happiest moment of his life is his wife's oft-repeated anecdote of going to Beijing and eating duck. Like photographs, anecdotes if repeated often enough can replace, even become more vivid than, "actual memory." The Chinese student's "happiest moment" embraces the absence that underlies the very concept of "actual memory," while offering instead an immediacy that links writers and readers, storytellers and listeners, in bonds of love, desire, and imaginary experience, even imaginary consumption.

Davis's story encapsulates the power of Johnsonian anecdotes to bring their object to life even as they acknowledge his absence. Whether I consider in what follows Boswell's anecdote of the orange peel, or the vivid details of Johnson's twitching, gesticulating body, or the final moments of his embattled death, I am similarly engaging in a kind of communion with an unobtainable reality, a literary communion perhaps even more moving because these stories could never be fully real, could never be my own.

## Eloquent Flesh

> . . . Poor Dr. Lawrence had long been his friend and confident [*sic*]. The conversation I saw them hold together in Essex-Street one day in the year 1781 or 1782, was a melancholy one, and made a singular impression on my mind. He was himself exceedingly ill, and I accompanied him thither for advice. The physician was however, in some respects, more to be pitied than the patient: Johnson was panting under an asthma and dropsy; but Lawrence had been brought home that very morning struck with the palsy, from which he had, two hours before we came, strove to awaken himself by blisters: they were both deaf, and scarce able to speak besides; one from difficulty of breathing, the other from paralytic debility. To give and receive medical counsel therefore, they fairly sate down on each side a table in the Doctor's gloomy apartment, adorned with skeletons, preserved monsters, &c. and agreed to write Latin billets to each other: such a scene did I never see! 'You (said Johnson) are *timidè* and *gelidè*'; finding that his friend had prescribed palliative not drastic remedies. It is not *me,* replies poor Lawrence in an interrupted voice; 'tis nature that is *gelidè* and *timidè.* In fact he lived but few months after I believe, and retained his faculties still a shorter time.
>
> HESTER LYNCH THRALE PIOZZI, *Anecdotes of the Late Samuel Johnson*

Hester Thrale's "singular impression" arises as much from the scene she witnesses—two men near death, nearly mute, conversing through writing beneath the memento mori tableau of anatomical science, "skeletons, preserved monsters, &c.," a tableau they are perilously close to joining. Equally striking is the mutual authority of doctor and patient, who prescribe for each other in "Latin billets." This book begins at the eighteenth-century moment when, paradoxically, the professions of literature and medicine had emerged, yet the competing authorities of aesthetic subjectivity and scientific objectivity had not yet fully diverged. This confluence is particularly striking in Johnson's case when science's object is the mind: Johnson's portrait of the mad astronomer in *Rasselas,* for example, oppressed by his belief that he controls the weather, was treated as medical evidence and even as medical case history by "mad-doctors" Thomas Arnold in *Observations on . . . Insanity* (1782–86) and John Haslam in *Observations on Madness and Melancholy* (1809); while his conclusion to the astronomer's story, titled by the editors "On the Uncertain Continuance of Reason," is included as an entry in Richard Hunter and Ida Macalpine's *Three Hundred Years of Psychiatry.* A detailed biographical headnote to this entry identifies Johnson as a "man of letters" and describes him with a personal tone exceptional in the volume as "a melancholic and hypochondriac all his life who . . . had deep insight into troubles of mind and penned observations—

many of them self-observations—which would do credit to a modern psychiatrist."[89]

But Johnson's authority extends to the body as well. Though "panting under an asthma and dropsy," the literary doctor in his eagerness to prolong life cannot help himself from exclaiming in protest of Lawrence's "palliative . . . remedies," insulting (and diagnosing?) him in Latin, the language of medical authority, as timid and cold. Lawrence, barely able to complete his sentence, replies by abdicating his authority to "nature."

This scene of desperate patient and hesitant doctor, mutually oppressed by mortal bodies, mutually immersed in classical culture as men of letters but separated by Johnson's defiance of "nature," will be reenacted repeatedly in Johnson's final days. The man whom biographer John Hawkins gave the last words of a Roman gladiator, "Jam moriturus," vowed on his deathbed, "I will be conquered; I will not capitulate."[90] In Boswell's account, Johnson corrected his physician Dr. Brocklesby's Latin quotation of Juvenal and later addressed him with a passage from Shakespeare's *Macbeth* (act 5, scene 3) beginning, "Can'st thou not minister to a mind diseas'd . . . ?" and was responded to in kind: "Therein the patient must minister to himself." In William Windham's narrative, the patient Johnson taught his physicians how to heal: in his doctors' absence, Johnson scarified himself in three places, "two in the leg, &c.," castigating another attending physician, Cowper's Dr. Heberden, whom he had referred to not long before as "*ultimus Romanorum,* the last of the learned Physicians," with being "*timidorum timidissimus*" when he expressed dismay at the wound.[91] In John Hawkins's account, the strength of the suffering author's words seem to wrest the lancet from the surgeon's hand in an attempt to prolong life: "When Mr. Cruikshank scarified his leg, he cried out—'Deeper, deeper;—I will abide the consequence: you are afraid of your reputation, but that is nothing to me.'—To those about him, he said,—'You all pretend to love me, but you do not love me so well as I myself do.'"[92] These are the moments to which this book will repeatedly return.

In the last letter Johnson sent to Boswell, entreating a response to alleviate the loneliness of his dying, he writes, "The water is now encreasing upon me."[93] This phrase offers a literal description in eighteenth-century medical terms of his disease—which was called at the time a dropsy, and which we would now term the result of heart failure, kidney disease, or pulmonary edema—as the writer feels his body fill with fluid. In that final desperate attempt at self-scarification in order to rid himself of this fluid and thus to prolong life, in a gesture that will resonate throughout this book, Johnson will wound himself fatally, precipitating his death. I want to consider the "water" that Johnson speaks of, the water that "encreases" internally, yet is evoked dramatically as an external force that threatens to drown his epistolary voice, to silence him forever.

As with his turning of "*timidè* and *gelidè*" back upon Dr. Lawrence, this poignant merging of the literal and metaphorical returns us to the lung—removed at the autopsy and preserved as a pathological specimen of emphysema. Why else might the lung have been removed? Why not the heart, which the surgeons noted in the autopsy record was enlarged, and which one contemporary biographer commented was "as if analogous to the extent and *liberality of his mind*"?[94] Why not the brain? (It is curious, and perhaps evocative of a certain prohibitive awe, that Johnson's skull was not opened.) What, in other words, might the lung have meant to the doctors?

If we turn to the literature and medicine of antiquity, which through the legacy of Galen still informed the thinking of eighteenth-century medical men, we will find that the lungs—synonymous with the Greek *phrenes,* and among the crucial organs of the Latin *praecordia* or *pectus*—were considered along with the heart to be the corporeal habitation of the soul.[95] Uniting mind with body, as intricate in their conducting of spirit as the convoluted brain, the lungs were thought to be the seat of consciousness, thought, intention, emotion, and language. In their healthy state, the lungs are dry; as one ancient authority put it, "The dry soul is wisest and best." When moist "with sleep or wine," the *phrenes* become "inefficient."[96] Even more suggestive, when we think of Johnson fighting off the waters of death, is the ancient belief "that in grief or yearning the relevant parts of the body 'melt' and as they 'diminish' there issues liquid. This thought must have been inspired in part by the tears and wasting of grief. The heart and lungs, as the parts chiefly concerned in emotion, were supposed to 'melt.'"[97] Both the Greeks and the Romans also "believed that the mind in the lungs was in direct relation to the native liquid there, the blood, and that water or wine, alien liquid, when drunk, went to the lungs, and the power in it possessed or displaced the mind there." To be overtaken with water, then, could be a form of mourning or a form of madness. It could also be understood as a disease of love.[98]

I do not mean by this speculation, which the classicist Richard Onians takes to forensic levels of brilliance, to read the minds of Johnson's doctors. Instead, I want to suggest that, like the chicken-soup universe of Einstein's brain emitting its aura of mystery, Johnson's lung is at once a thing in itself, and a metaphor for being that unites ideas of body and mind. The image of Johnson's lung (see fig. 2), marked by the body that it replicates, is a talisman or relic, unattainable yet repeatedly reproduced, that initiates us into the mysteries of another's grief, another's fear (famously in Johnson's case bordering on madness) at the prospect of death, an other who is an author who has written to reassure us in grief and fear, in whom we try without fully succeeding to see ourselves, whom we, in short, love.

## Death's Divagations

This book meditates on fragments of the unattainable anecdotal real that have endured throughout the past two hundred years of Johnson's celebrity. It is haunted by two complementary plots of mourning—one the "joke" of the Ephesian matron's turn from devotion until death to the embrace of life, the other the tragedy of Hamlet, mindful of his father's ghost, disgusted at "sullied" flesh, and contemplating the weary, stale, flat, and unprofitable uses of the world. I use these plots to meditate on two trajectories that stem from Johnson's death: one the traditional Johnsonian phenomenon, beginning with Boswell and living on (though perhaps nearing its end) in and beyond the academy, whose mission is to preserve and honor the author himself; the other a literary afterlife, marked by predominantly masculine melancholy but devoid of a fixed original, that links Johnson, while transforming him, to the disparate company of James Merrill, Vladimir Nabokov, Nathaniel Hawthorne, Beryl Bainbridge, Saul Bellow, and Samuel Beckett.

Chapter 1, "Johnsonian Romance," delineates and investigates the autopsy as the founding event of both the Johnsonian monument (which Boswell imagined as a kind of tomb for the great man's collected relics) and of a Johnsonian community united by desire for individual conversation, by means of the anecdote's eternal present, with the living author. Chapter 2, "Style's Body: The Case of Dr. Johnson," shows that the body upon which the Johnsonian monument is based was in fact in perpetual motion, marred by a series of convulsive starts, tics, and gesticulations that put the agency of the representative author into question. Juxtaposing anecdotal evidence with visual documentation (in the form of portraiture), this chapter also positions the body of Johnson's style (itself characterized as both monstrously repetitious and monumental) against anecdotal traces of Johnson's physical moving, ticking body that endure on the pages of literary biographies and medical journals. Chapter 3, "'Look, my Lord, it comes': Uncritical Reading and Johnsonian Communion," investigates the desire that fuels Johnsonian commemoration (in critical texts as well as in communal rituals and museums) as profoundly uncritical, indeed religious, in its imagination of communion with the author's spirit. The spiritual nature of such communion is evident at the moment of its inception in the parodies it inspires, a late Enlightenment repetition with a difference of Petronius's pagan critique of Christianity. Both critiques hinge on the material remainder—corpses, excrement—of worshipful fantasies of transcendence.

Chapter 4, "The Ephesian Matron and Johnson's Corpse," by tracing the elaborate literary path of Thomas Tyers's allusion to the story of the Ephesian matron in attempted justification of the autopsy, uncovers an ancient romance that undoes the very idea of a monument to a man who is irreplaceable; what is left, instead of a lost and mourned original, is the singular repetition of romance's erring. Nevertheless, even in Petronius's story, traces of identity persist—his joke's punch line depends on the audience's ability to recognize the dead man's face while laughing at the thought of his resurrection. This chapter returns to the book's beginning with a reconsideration of the anecdote's unique relationship to the "real," along with the differently gendered stakes in anecdote of Boswell and Thrale. The concluding chapter, "Coda: Anecdotal Errancy—Three Authors," traces an alternate version of Johnsonian afterlife in texts by Nathaniel Hawthorne, Vladimir Nabokov, and Samuel Beckett. We end where we began, at the moment of the autopsy, of desire provoked by death, of the body preserved in its final moment, of that body's lost breath.

CHAPTER ONE

# JOHNSONIAN ROMANCE

HAMLET. To what base uses we may return, Horatio! Why, may not imagination trace the noble dust of Alexander, till he find it stopping a bung-hole?
HORATIO. 'Twere to consider too curiously, to consider so.
HAMLET. No, faith, not a jot, but to follow him thither with modesty enough, and likelihood to lead it; as thus: Alexander died, Alexander was buried, Alexander returneth into dust, the dust is earth, of earth we make loam, and why of that loam, whereto he was converted, might they not stop a beer-barrel?

*Hamlet*, 5.1.196–205

## Autopsy as Romance

To consider Samuel Johnson's autopsy at all is to consider perhaps "too curiously." To consider it as a romance is to wander more dangerously, tracing the paths of desire, disavowal, and fantasy that link the legendary greatness of an Alexander to the loam stopping a beer-barrel, to the scatological reminder of a "bung-hole." The flesh in Hamlet's pun is at once plugging up a void and itself a void, charged with reminders of its status as excrement, the soul's leftover. Caught between the same extremes of fascinated repugnance at mortality and idealization of a lost hero, Johnsonians reverse Hamlet's curious, decidedly immodest history of Alexander in order to keep their hero whole, even as they cut him into parts.[1]

How is Samuel Johnson's autopsy a romance? The anecdote at the heart of that question is contained in the manuscript record of Johnson's autopsy and its subsequent publication history. Before staging that history in search of the anecdote, I want to begin with the Johnsonian phenomenon that is the autopsy's legacy, its enduring ghostly double. We can see it, at its most sentimental, in Thomas Macaulay's fantasy vision:

In the foreground is that strange figure which is as familiar to us as the figures of those among whom we have been brought up, the gigantic body, the huge massy face, seamed with the scars of disease, the brown coat, the black worsted stockings, the grey wig with the scorched foretop, the dirty hands, the nails bitten and pared

> to the quick. We see the eyes and mouth moving with convulsive twitches; we see the heavy form rolling; we hear it puffing; and then comes the "Why, sir!" and the "What then, sir?" and the "No, sir;" and the "You don't see your way through the question, sir!"
>
> What a singular destiny has been that of this remarkable man! To be regarded in his own age as a classic, and in ours as a companion! To receive from his contemporaries that full homage which men of genius in general received only from posterity! To be more intimately known to posterity than other men are known to their contemporaries! That kind of fame which is commonly the most transient is, in his case, the most durable. The reputation of those writings, which he probably expected to be immortal, is every day fading; while those peculiarities of manner and that careless table-talk the memory of which, he probably thought, would die with him, are likely to be remembered as long as the English language is spoken in any quarter of the globe.[2]

In this Victorian frame, designed for a national gaze, Johnson's deathless prose is forgotten, while his flawed body and its evanescent utterance are reanimated as uncanny familiars from the childhood of a nation.

In the context of the autopsy and the tradition of medical Johnsonians it inspires, the literary labor of preserving Johnson's immortality as an anecdotally embodied character exemplifies Luke Wilson's characterization of the practice of anatomy as a romance that recuperates, reunifies, and reanimates dissection's violent reduction of the body to its parts. While "the demonstration that the body is dead—its dissection—preserves by contrast the animation of spectator and anatomist alike," Wilson argues that "in the anatomy *as anatomy,* the figurative reanimation of the body—its reconstitution as physiologically actual or functional—results in a fantasy in which the body is felt to be alive as some sort of agent controlling the anatomist and the performance as a whole—a body giving an account of itself."[3] For Wilson, anatomy's "inversion" of dissection's destruction casts it "as romance; its trope is the reanimation of the dead; and while the loss here of the contrast preserved in the anatomy of dissection occasions a morbid anxiety, the simultaneous identification with the reanimated cadaver produces a certain elation in the apparent reversibility of mortality."[4] The anatomist in this scenario—medical and literary Johnsonian alike in this context—identifies with both the corpse and the seemingly divine power that revivifies him by restoring him to knowledge.

Johnson's autopsy took place at a time of increasing medical professionalism, when scientific anatomy strove (not always successfully) to become the province of private instruction rather than of public performance. While public anatomies had been traditionally performed on criminal corpses four times a year since 1540, by the time of Johnson's autopsy in December 1784, private anatomies—which included a small group of medical men and often involved a semi-public lecture accompanying the dissection and a feast afterward—had largely supplanted this carnivalesque brand of public instruction.[5] The history of Johnson's autopsy is one of a progressive removal of the corpse from public view, a removal that accompanies the autopsy's disembodied dissemination in print: occurring on December 15, 1784, "at his house in Bolt-Court, in the presence of some gentlemen of the faculty,"[6] its results were summarized for the public in that month's issue of the *Gentleman's Magazine*.[7]

But just as public anatomy still thrived and indeed overlapped with professional anatomy in eighteenth-century England, this particular autopsy still remains a spectacle that haunts anatomy's adherence to abstract laws and adds a particular personal urgency to its trope of reanimation: viewers and readers of Johnson's body for centuries afterward have transposed anatomy's suspension between life and death onto a complementary set of opposing poles: individuality and anonymity. Johnson, as we shall see, can never fully die, nor can his particular example be completely subsumed to medicine's general rules.[8]

The "romance" of my title begins with the autopsy of Samuel Johnson and unfolds in a series of questions. This romance is queer, it must be noted at the outset, precisely because the desires and identifications it sets in motion cannot be reduced to or exposed as a particular kind of sexuality. The strangest difference here is historical: the particular, pronounced, anything-but-latent queerness of an anatomical tableau that undercuts familiar distinctions—between heterosexual and homosexual desire, between disembodied gaze and embodied object—making them unstable and even interchangeable.[9] Fascination and identification with Johnson's corpse creates a new subject, one that we would find familiar—the English male professional individual—in an uneasy relation to emergent social and national totalities.[10] This singular subject of knowledge emerges as the culmination of a series of substitutions of parts for wholes, a chain that exchanges the eccentric body of the author, itself divided into collectible parts, for the inchoate body of the nation.[11] But this exchange is haunted by differences—whether in the case of Johnson, his seemingly meaningless gesticulations, or death itself, or in the case of Britain, the ungovernable differences—social, racial, sexual, and economic, between and within nation and empire—that must be disavowed. Desire and disavowal haunt us to this day when we summon the English canon in Johnson's image.

How does Johnson's dead body inspire and inform a professional male community and a national consciousness? How do Johnsonians, both men of letters and medical doctors, from the eighteenth century to the present, engage in a mutual effort—the shared fleshly origins of which have been denied as literature and medicine have diverged—to reanimate Johnson by preserving their author's body in all its particularity? What sort of desire drives the Johnsonian romance plot? What bonds are forged, what divisions bridged, by bringing the dead Johnson to life?

## Anatomy's Primal Scene

The romance of anatomy is a story that prevents itself from being told. In an analysis of a nineteenth-century painting of a male anatomist contemplating a female corpse, Elisabeth Bronfen in *Over Her Dead Body* characterizes autopsy as unstable tableau: "The security that preservation of this object of sight [the corpse] might afford is subverted by the opposite realisation that suspending temporality and with it narrative also leads to death in the form of stasis."[12] As the figure of the corpse vacillates between inanimate object of metaphor and allegorical subject of memento mori, Bronfen's romance of anatomy is distinctly gendered and the object of the male anatomist's gaze is in her terms necessarily female. Samuel Richardson's rake hero Robert Lovelace provides an eighteenth-century example of the overdetermined and (often) gendered cultural link between narrative desire, visual stasis, and death that Bronfen elaborates. The libertine's rage to penetrate the virtuous facade of the heroine Clarissa propels the novel's narrative and results in his final demand:

> I think it absolutely right that my ever-dear and beloved lady should be opened and embalmed. It must be done out of hand—this very afternoon. . . . Everything that can be done to preserve the charmer from decay shall also be done. And when she *will* descend to her original dust, or cannot be kept longer, I will then have her laid in my family vault between my own father and mother. Myself, as I am in my soul, so in person, chief mourner. But her heart, to which I have such unquestionable pretensions, I *will* have. I will keep it in spirits. It shall never be out of my sight.[13]

Lovelace's paradoxical desire, dependent upon and continually thwarted by the mortal body, is for an end to narrative and the defeat of death. With what Bronfen terms the "auto-icon" of Clarissa's heart, both part and symbol of her embodied being, forever preserved and contained in spirits (both the heart and its spirits grounding their metaphorical essences in literal puns), the rake's plotting to possess her can come to an end; death can be at once acknowledged and, in this fetishistic defeat of its temporality and decay, denied. Desire is fulfilled and sustained at the level of the visual by the embalmed heart that

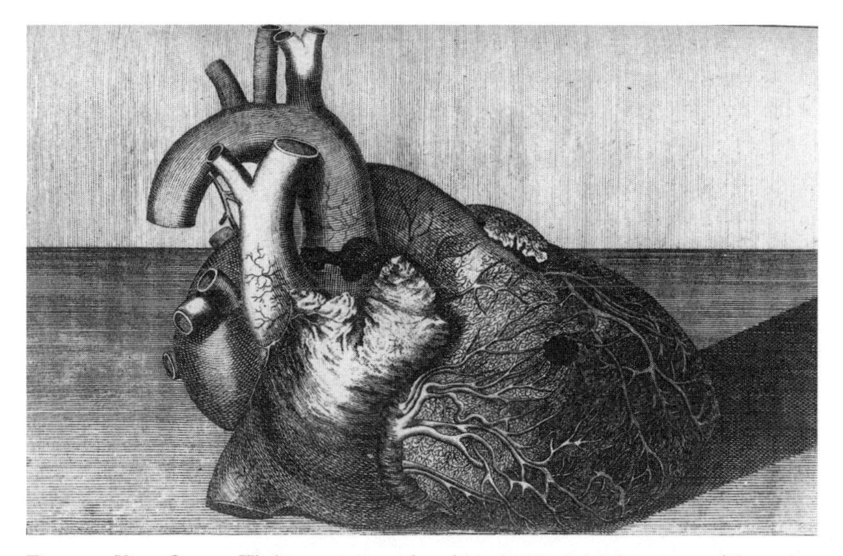

Figure 5. King George II's heart, as reproduced in *Gentleman's Magazine* 32 (November 1762). This image originally appeared in color. Reproduced by permission of the Huntington Library, San Marino, California.

cannot be touched.[14] Lovelace's emphasis on "will" seems to reanimate Clarissa's corpse with her own will, as if the perverse charmer resists him till the end even in inexorable decay. Such will can only be defeated within the medium of the flesh: Clarissa's heart, both trophy and relic, possessed and "in spirits," enlivened and envisioned forever. (For a royal equivalent and precedent, see fig. 5.)

In Bronfen's account, the aesthetized cadaver's femininity renders it doubly other, allowing the anatomist to fix his ambivalent identification with the corpse and anxieties about death and castration upon the fetishized female body. Yet as Abigail Solomon-Godeau has argued about the genre of the nude more generally in the nineteenth century, the femininity of the objectified body is a historical contingency rather than a psychic necessity; throughout the eighteenth century in Europe, the traditional "nude" was assumed to be male.[15] To ignore such contingency is automatically to equate death with the "lack" of female castration. Such an equation is belied by Freud's brief digression in his essay on fetishism from a primary explanatory discussion of the male denial of the sight of maternal castration to a counterexample of two young men each of whom "had refused to acknowledge the death of his father." While Freud's essay is concerned with making lack a strictly female characteristic, it nevertheless reveals a structural difference between female castration as visible impression and male death as unnarratable event. The former comes to serve as a kind of fetish for the latter.[16] Similarly for Bronfen, the visual scenario of sexual difference replaces specific narrative complicities of desire and death that compromise all distinctions, including those between the sexes, as well as the clear division between anatomist and corpse.[17] Such complicities are revealed by Lovelace's crazed fantasy, in which Clarissa's bottled heart, the part Lovelace himself so often was accused of lacking, replaces the female body itself, both memento mori and the libertine's longed-for self-reflection, his story's continuing end.

If we separate the historical particularities of an individual death from the generic spectacle of castration, the unstable identifications that inform the scene of Johnson's autopsy come into clearer focus. In the static narrative of the autopsy, Samuel Johnson similarly is made to give his "final account," reanimating the living men around him through anatomy's romance. But in the case of Dr. Johnson, this romance is fueled by homosocial identification with a male celebrity who cannot be subsumed into femininity's faceless alterity. Endowed with a monumental name and a recognizable face, Johnson's body is at once monstrously unique and representative of a nation, a body whose lack and whose death are too close to home, and thus all the more passionately denied.

## Embodied Ghosts

The Johnsonian romance that I trace to its origins in the autopsy ends with the profession of eighteenth-century English literature as it is practiced today, a profession still haunted by Johnson's ghost. Age of Johnson courses are taught in classrooms around the country where the great man's mode of speech is sometimes imitated (most famously by Walter Jackson Bate in his Harvard lectures, who according to legend left his own deathbed to perform the death of Johnson) and his presence always resurrected in anecdotes, kicking the stone to refute Berkeley thus. If we turn to the anecdote of the stone, we can see the autopsy as a kind of double for Boswell's printed page—both endow an author's ghost with the material density of certainty:

> After we came out of the church, we stood talking for some time together of Bishop Berkeley's ingenious sophistry to prove the non-existence of matter, and that every thing in the universe is merely ideal. I observed, that though we are satisfied his doctrine is not true, it is impossible to refute it. I never shall forget the alacrity with which Johnson answered, striking his foot with mighty force against a large stone, till he rebounded from it, 'I refute it *thus*.'[18]

This anecdote has been analyzed as evidence of Johnson's philosophical sagacity, but I am interested in reading it as a device that inspires our faith in Johnson's "solidity." I am influenced here by Elaine Scarry's analysis of the literary text as a source of instruction for the imagination and vivification of object-worlds in *Dreaming by the Book.* Interestingly, Scarry claims that "being 'under the sway' of another person is critical to the production of vivacity," evoking the power of the author of the text that inspires imagining. But Boswell stages his biography as a representation of his own falling under Johnson's sway; it is Johnson—the most consummately achieved and vivacious illusion in the history of biography—to whom the reader of the *Life of Johnson* must therefore submit. In juxtaposing Johnson's body, rebounding with the strength of the "mighty force" of his kick, with the "large stone," Boswell accomplishes something similar to what Scarry describes in her description of Proust's achievement of the solidity of his room at Combray through the depiction of the images of a magic lantern passing over the walls. Scarry, citing J. J. Gibson's analysis of perception, describes a feat of the mind called "kinetic occlusion," a "'wiping-out' or 'shearing-away' of what lies behind, followed by its restoration," which is structurally similar to disavowal, to the fort-da game in which the absence and restoration of the lost object create a kind of faith, a renewed "depth at an edge in the world."[19]

In a rehearsal of such a fort-da game on an existential level, Johnson's kicking the stone, in its demonstration of the force of the collision of man and object, enacts the solidity of both, a solidity bolstered by the anecdote's unique proximity to historical reality. We are left with a sense that the man is as immutable and enduring as the stone, even though Boswell's imaginative vivification is always shadowed by his hero's death, a death evoked not only in the preface to the completed work, but by Johnson's first appearance in the *Life* as the ghost of old Hamlet.[20]

Such solidity has been summoned so effectively that it seems to link Johnson's physical to literary corpus in a heaviness that, for a romantic like Edgar Allan Poe, was too hard to bear:

> Give me, I demanded of a scholar some time ago, give me a definition of poetry? "*Tres-volontiers,*"—and he proceeded to his library, brought me a Dr. Johnson, and overwhelmed me with a definition. Shade of the immortal Shakspeare! I imagined to myself the scowl of your spiritual eye upon the profanity of that scurrilous Ursa Major. Think of poetry, dear B—, think of poetry, and then think of—Dr. Samuel Johnson! Think of all that is airy and fairy-like, and then of all that is hideous and unwieldy; think of his huge bulk, the Elephant! and then—

> and then think of the Tempest—the Midsummer Night's Dream—Prospero—Oberon—and Titania![21]

The bounds of definition—what Boswell found comforting about Johnson's kicking of the stone—are linked here to the prison of the flesh; Johnson's weighty *Dictionary* (and we might remember it here being tossed out the window of Becky Sharpe's coach in another act of defiance) and "huge bulk" give ballast to Poe's championing of Shakespeare's ability to abandon his own body through the evanescence and ethereal population of dramatic form. We'll return to this opposition between two differently embodied authors and to the connection between embodiment and style in the next chapter. For now it is enough to consider this enduring association between Johnson and the defining bounds of a flesh that is all too solid.

The Johnson Society of Lichfield's 1999 printed *Transactions* is bookended by two acts of Johnsonian memorial that replicate this dialectic of disembodied memory and burdensome corporeality, Alexander and the bunghole. The first, the novelist Beryl Bainbridge's "Presidential Address," is an excerpt from a novel then in progress and now in print, *According to Queeney*, which narrates the friendship of Samuel Johnson and Hester Thrale from a variety of perspectives, most notably the critical eye of Hester's daughter Queeney of the title. Bainbridge chose to read what she thought at the time was the novel's opening, an account of Johnson's mental breakdown and Hester and Henry Thrale's lifesaving intervention, the subject of the anecdote that began this book. This harrowing portrait of Johnson in a state of near-bestial madness is narrated from the perspective of Mrs. Desmoulins, a friend of Johnson's dead wife, native of Lichfield, and member of the odd, fractious, self-made family that constituted his London household. Her account is punctuated by the memory of innocent dalliances with the great man when he was denied his sick wife's bed. Bainbridge concludes her talk with "I don't know what I'm going to write next. All I can say to you this evening is, *God bless Samuel Johnson.*"[22] The novel Bainbridge went on to complete, while privileging the viewpoint of the title character, begins and ends with a different frame, that of desire. One could say that Bainbridge's main subject is Johnsonian desire in its multiplicity and ambiguity—the desire for Johnson that motivated both the surgeons and the biographers (the narrative is punctuated by Queeney's reticent responses to the requests of Letitia Hawkins—John Hawkins's daughter and also a writer of Johnsoniana—for intimate information about Johnson and Thrale, and intimations of Johnson's own desire, a desire that must be edited out of the traditional picture of paternal authority summoned to reassure generations of male audiences). Her story begins with a depiction of the autopsy and with the solitary Mrs. Desmoulins, who bids farewell to the body at the book's first and final moment, alone in Bolt Court, with her memories and her longing.[23]

The same volume of the *Transactions* ends with a paper by Philip Spinks—"currently a Paramedic with Warwickshire Ambulance" who was "introduced to Johnson by a friend"—titled "The Post Mortem Examination and Death Mask of Samuel Johnson" and prefaced with apologies for the morbid subject matter from both the author and the editor. The risk is worth it, both conclude, for the "patience, calmness and spirituality" revealed in Johnson's final days and for the author's conclusion that "even after his death, Johnson was still of importance to his friends and acquaintances: for some, a medical specimen—for others, a work of art."[24] Bainbridge's novelistic account of the taking of Johnson's death mask reveals the self-reflexive nature of this importance: "When the wax had cooled and he pulled away the cast, the eyelids were dragged open; he was too engrossed in scrutinising the imprint of the face to notice the staring aspect of the original."[25] Even in the presence of the corpse itself, the original matters less than the enduring image.

Both Bainbridge's speech and Spinks's research are inspired by desire—whether that of the novelist exposing the frailty of Johnson's body and mind to a twenty-first-century audience, of the amateur Johnsonian "introduced to Johnson by a friend,"[26] of the abandoned and lovelorn Mrs. Desmoulins of Boswell's *Life* and Bainbridge's imagination, of the surgeons who performed the autopsy, or of the artist who took the death mask and immortalized it as a bust—to know and thus to claim a man revealed as fully embodied, fully mortal. Around the figure of Johnson's corpse, surgeons and artists unite in the need to delineate and display an uncommon individual and to anchor his origins in the flesh.

Johnson's body thus inspired a poetics of collection common to scientific and aesthetic sensibilities alike. While the surgeon William Cruikshank ordered Johnson's autopsy, it was the painter and fellow-author Joshua Reynolds who visited the great man's deathbed late at night to take a mold of his face for a death mask.[27] Transformed into a bust by the sculptor James Hoskins (fig. 6), one of the few known representations to display the scars of scrofula on Johnson's neck turns the autopsy's paradoxical balance of personal particularity with clinical objectivity into a curious blend of neoclassical type and aberrant detail. Reproduced in the first English translation of Lavater's *Essays on Physiognomy* (1789–98), when art editor Thomas Holloway chose to supplement the original two portraits of Johnson "exactly copied from the French Edition"—one "*a general idea* of the character; the other *a careful copy* after a well-known portrait"—with an image "engraved after a cast taken from nature,

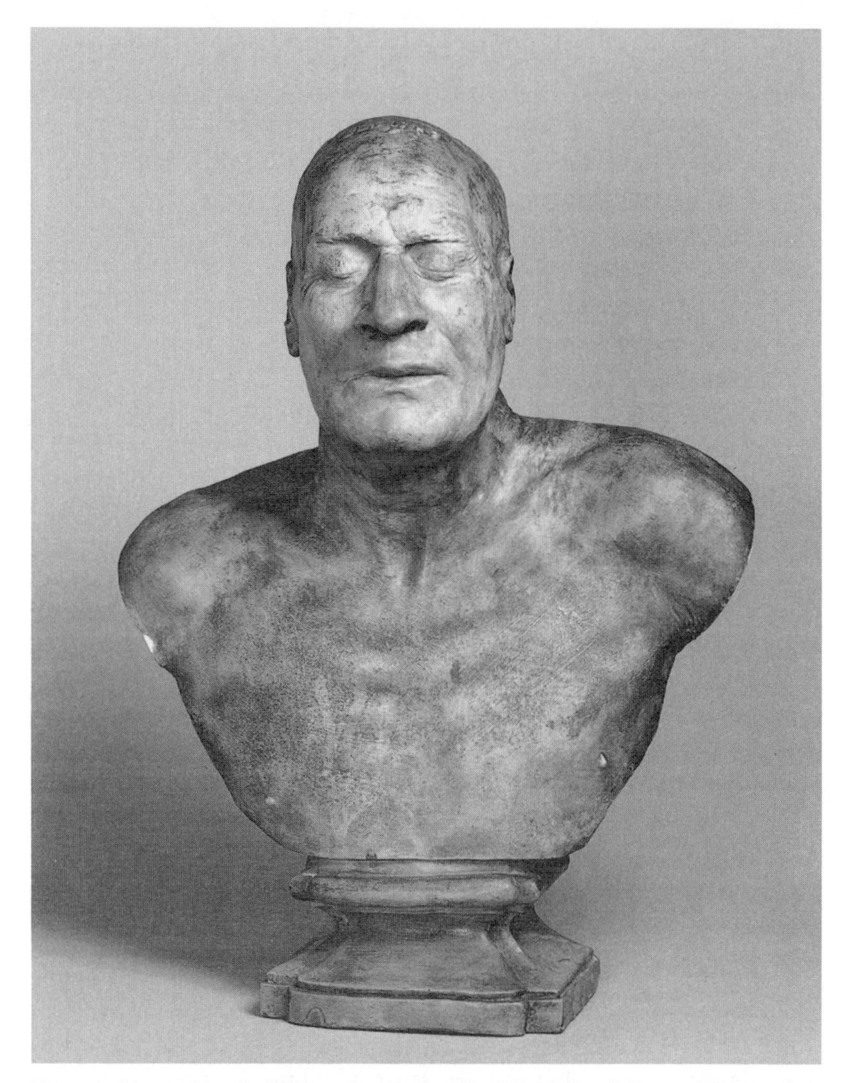

Figure 6. *Samuel Johnson* by William Cumberland Cruikshank (1784). National Portrait Gallery, London. Plaster cast of the bust incorporating Samuel Johnson's death mask. While the surgeon Cruikshank supervised the taking of the death mask, the sculptor James Hoskins transformed it into a bust.

as a proof of Mr. Lavater's Physiognomical Sagacity, and a confirmation of his doctrine," this collaboration of art and science outdid both idealized and realistic portraiture in its truth to life (fig. 7).[28]

Continuing the tradition of corporeally based realism, physician and Johnsonian Lawrence McHenry, in an essay for the Journal of the American

*Of the Heads of* Johnſon *presented on this Plate, those marked 1, 2, are exactly copied from the French Edition. The former seems to be* a general idea *of the character; the other* a careful copy *after a well-known portrait. The Editor has taken the liberty to introduce between them one engraved after a cast taken from nature, as a proof of Mr. Lavater's Physiognomical Sagacity, and a confirmation of his doctrine.*

Figure 7. Thomas Holloway's illustration to Johann Caspar Lavater, *Essays on Physiognomy, Designed to Promote the Knowledge and Love of Mankind,* trans. Henry Hunter (London, 1789–98), 1:194. Special Collections, UCLA Arts Library. The image in the center, based on Johnson's death mask, was added to this first English edition as "a proof of Mr. Lavater's Physiognomical Sagacity, and a confirmation of his doctrine."

Medical Association entitled "Art and Medicine: Dr. Johnson's Dropsy," bases his dating of Irish wax sculptor and royal medallion modeler Samuel Percy's miniature portrait group of a tavern scene of Johnson and his friends as 1783 or later "entirely on medical evidence, namely the time when Dr. Johnson developed dropsy" (fig. 8).[29] Noting the swelled legs of the "lifelike" figure, McHenry groundlessly speculates that since "the wax face of Johnson by Percy in many ways resembles" the only known bust of Johnson taken from life by Joseph Nollekens, "hence, it too may have been taken from life."[30] Percy, renowned for the vividness of his portraits made from molds taken from the sitter, then tinted and ornamented with "little touches such as lace, flowers, comb, ring or jewels," also had a "lucrative side-line," advertised as "Masks taken from the dead on the shortest notice and likenesses made from them." Percy's model of Johnson, in the opinion of recent experts, turns out to be

Figure 8. Samuel Percy, miniature wax tavern scene of Samuel Johnson and the Club. © Museum of London.

neither, but McHenry's desire to imagine his embodied presence is powerful enough to martial medical evidence in the service of fantasy.[31] To gaze at that wax diorama, its detail so lovingly rendered in a medium so close to the flesh, is to experience the miniature's nostalgia for a self-enclosed deathless world, a world—like that of Macaulay's vision—close to that of childhood memory. In all these examples, clinical attention to the subject's mortal body unites with the desire to reanimate a Johnson "taken from life," an imaginary Johnson, an uncanny familiar, for whom there is no original.[32]

## Untold Stories

Wednesday, December 15, 1784: Opened the body of Dr. Samuel Johnson for Mr. Cruikshank, in the presence of Drs. Heberden, Brocklesby, Butter, Mr. C. and Mr. White. He died on the Monday evening preceding. About a week before his death Mr. C. by desire of his physicians scarified his legs and scrotum, to let out the water which collected in the cellular membrane of those parts, Dr. Johnson being very impatient to have the water entirely gone, the morning of the day on which he died repeated the operation himself, and, cutting very deep lost about ten ounces of blood[;] he used a lancet for this purpose—he was in too weak a state to survive such an apparently trifling loss. For several years past he had been troubled with

> asthma for which he commonly used to take opium, and found that nothing else was of any service to him, he had discontinued this practice some years before he died.[33]

The surgeon James Wilson thus begins his record of Samuel Johnson's autopsy. Wilson, the senior pupil of Johnson's surgeon William Cruikshank, who along with Matthew Baillie inherited the mantle of lecturer on anatomy at William Hunter's school, entitled his account "Asthma" and inserted it in its proper alphabetical place after "aneurism" and "apoplexy" in a bound vellum folio volume, the cover emblazoned in capital letters with the word DISSECTIONS. Beneath this title is written "of Morbid Parts By my father James Wilson [Surgeon] and Myself, his loving Son! James Arthur! To be kept, and carefully by one who will know its value. J.A.W." James Wilson Jr., like Johnsonians heedful of paternal memory, not only bound in his father's book his own records of postmortem examinations, more regularly codified and purged of random detail, but also dined out on the volume's historical interest, lecturing to the Royal College of Surgeons on selected extracts of his paternal record.

Wilson Sr. labeled his account with a pathological rubric, yet his opening paragraph eludes such classification by gesturing toward a pathetic story it neither begins nor properly ends, as the next paragraph shifts to a detailed description of the opened body. While this initial narrative turn is not atypical of Wilson Sr.'s style, which usually frames the static visual record of the body's interior with the circumstances that brought the corpse to the anatomist's table, the story it tells, fraught with Johnson's desperate gesture, exceeds the autopsy's narrative scope. In effect, Wilson's account defeats its own ostensible purpose, answering what we might consider the central autoptic question of cause of death without knowledge of the corpse's interior by pointing to the living body's action. In this truncated narrative, doctor and patient compete for the use of the lancet in a joint and uncanny rendering of the great man's body as living corpse. Johnson's act of self-scarification, perilously close to self-castration and even self-dissection, renders him "too weak . . . to survive such a trifling loss."

While the story's end thus seems to be clear before the autopsy begins, the obvious question of why the autopsy was performed is thus left unanswered. One medical commentator has suggested that the surgeons intended to dispel rumors of suicide, rumors that Johnson's biographer John Hawkins fueled by denying and that Wilson's record itself fails to dispel completely.[34] Johnson did indeed precipitate his own death by his self-scarification, and I will return to this incident repeatedly over the course of this book. The narrative submerged in these rumors, one of death and its denial, is of agency gone awry in a house of mirrors.

In the *Gentleman's Magazine* of February 1785, lawyer and amateur man of letters Thomas Tyers offers a more compelling motive than concern about Johnson's reputation as a good Christian:

> At the request of Mr. Cruikshank, the executors permitted the body to be opened, on the suggestion, that his internals might be uncommonly affected, which was the case on inspection. The dead may sometimes give instruction to the living. The Cyrus of Xenophon ordered his breathless body to fertilize the earth that had given it nourishment. Johnson's inside had not the soundness of that of old Parr (as related by Harvey), not far from whom he is now deposited. . . . Perhaps, "of no disease he died," like the character in the Tragedian: for who can tell wherein vitality consists?[35]

The anecdotal allusions used to justify the scientific impulse reveal a less rational curiosity: the unanswerable question of cause of death gives way to the need to penetrate the interior of an uncommon individual. What uncommon mark might an exceptional mind make on the body? Yet the onus of intention for such a violation is placed rhetorically on Johnson himself as the great moralist continues to "give instruction to the living" after death. Cyrus's dying words to his sons, as he contemplates death with equanimity—"whether I be with the divine nature, or be reduced to nothing"—give literal fleshly meaning to Johnson's moral project before and after death in a pagan version of Christian communion:

> When I am dead, my children, do not enshrine my body in gold, or in silver, or in any other substance; but restore it to the earth as soon as possible; for what can be more desirable than to be mixed with the earth, which gives birth and nourishment to everything excellent and good? I have always hitherto borne an affection to men, and I feel that I should now gladly be incorporated with that which is most beneficial to men.[36]

The comparison with the heroic Persian monarch, whose will seems to reanimate a "breathless body" in this telling, further bolsters the uncanny agency of what becomes for its own time a national corpse, an exemplary and collectible body that can speak to future generations.[37]

With the mention of Old Parr, Johnson's body is explicitly defined as a curiosity and placed in a British tradition of exceptional corpses that are monumentalized, along with the bodies that gave birth to the literary canon, in Westminster Abbey.[38] Thomas Parr, a poor country laborer and national prodigy for his exceptional longevity, "having been visited by the illustrious Earl of Arundel . . . was brought by him from the country to London; and . . . was presented as a remarkable sight to his Majesty the King." "After having lived to one hundred and fifty-two years and nine months, and survived nine princes," Parr was examined by seventeenth-century physician and anatomist William Harvey "by command of his Majesty."[39] Parr's value as a curiosity rests particularly in his reputation for exceptional sexual prowess, and Harvey's 1635 "anatomical examination" pays special initial attention to the healthiness of "the organs of generation":

> So that it seemed not improbable that the common report was true, viz. that he did public penance under a conviction for incontinence, after he had passed his hundredth year; and his wife, whom he had married as a widow in his hundred-and-twentieth year, did not deny that he had intercourse with her after the manner of other husbands with their wives, nor until about twelve years back had he ceased to embrace her frequently.[40]

Having reified Parr's exceptional masculinity, the record continues with a distinctly pastoral air. Discussing the exceptional soundness of the intestine, Harvey marvels at the ability of Parr's body to digest an exceptionally coarse diet: "On this sorry fare, but living in his home, free from care, did this poor man attain to such length of days." Social mobility, in fact, is determined to be the cause of death: "the chief mischief being connected with the change of air, which through the whole course of life had been inhaled of perfect purity," and with the shift to the rich foods and strong drink of aristocratic tables.[41] Parr, the embodiment of the healthy essence of rural old England, takes his place in Tyers's account next to the great man of letters, both memorable and upwardly mobile British corpses. Like Johnson, Parr endures as a doubled body: when William Blake depicts him in one of his "Visionary Portraits," he imagines Old Parr when young (fig. 9), displaying this prodigy of longevity as the epitome of eternal youth. As Blake's spiritual portrait of Parr reminds us, there is a dimension to Johnson's autopsy that echoes the practice of sacred anatomy, in which the corpse remained a particularized object of adoration and source of potential divinity. Dissection of saints revealed not the evidence of anatomy's general truths but rather the unique marks of God's hand on the flesh.[42]

But while the burial of the humble Parr, exemplary for his sexual vigor, in Westminster Abbey seems governed by a nostalgia for an earlier and more fixed model of class difference, the ongoing exhumation of Johnson, exemplary

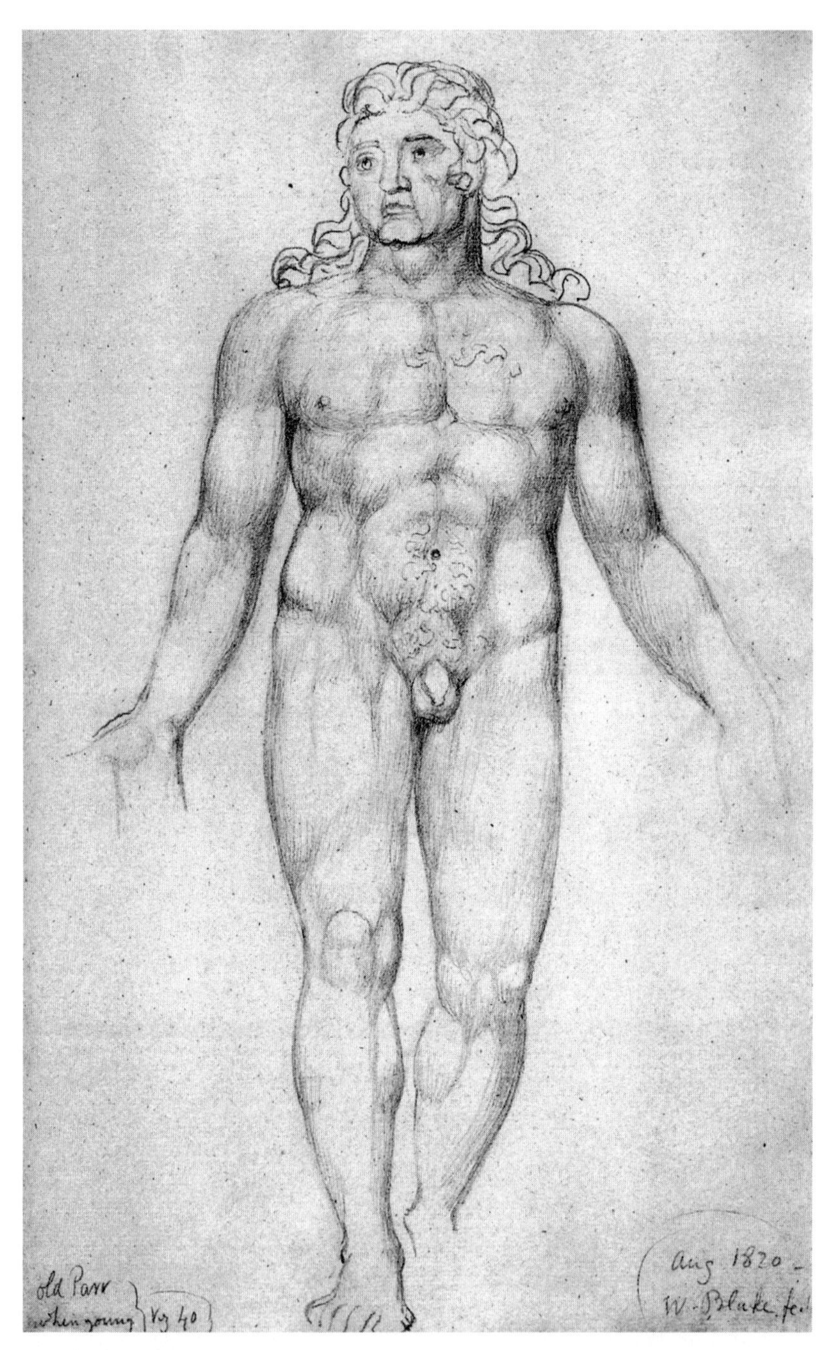


Figure 9. William Blake, *Old Parr When Young* (1820). Reproduced by permission of the Huntington Library, San Marino, California.

for his moral and intellectual strength, denies his sexuality in exchange for successful social self-fashioning. The national moralist must endure as either nursery companion (Macaulay) or unsexed father (Boswell, to a great degree);[43] the Johnsonian view at the time of his death is a child's-eye perspective that must partially disavow the sexual and therefore mortal habitation of the great man's intellect.[44]

The reward for such a bargain is an uneasy immortality—Johnson is never fully laid to rest. The obituary record in the *Gentleman's Magazine* registers this ambivalence about the body as it shifts abruptly from a clinical report on the body's interior to Lovelace-like metaphorical present-tense praise of the enduring moral qualities of the man: "Together with the ablest Head, he seems possessed of the very best Heart at present existing . . . while his writings point out to us what a good man ought to be, his own conduct sets us example of what he is."[45]

Johnson's corpse vacillates between life and death as the surgeons—and centuries of readers of the autopsy narrative in print—alternate between an imaginative sympathy that reanimates the dead body in all its particularity and depth, and a fetishistic curiosity to see that body's secrets, to literalize that depth as physical interior. This instability of identification has long characterized anatomy in the realm of literature. In Edward Ravenscroft's *The Anatomist; or, The Sham-Doctor* (1697), for example, a comedy that was popular throughout the eighteenth century, a resourceful male servant aiding his master in a love affair with the anatomist's daughter resourcefully performs the roles of both surgeon and corpse as expediency demands: "I would rather," he declares, "act the Dr than the dead Body," but the play's structure demands performance of both roles as part of his trickster repertoire. Underlying the play's comic surface is a darker fantasy that allows the audience to witness two different situations in which characters masquerading as corpses barely escape the anatomist's knife and live to tell about the suspension between life and death: "I had rather be a Sot than an Anatomy, I will not have my Flesh scrap'd from my Bones. I will not be hung up for a Skeleton in Barber-Surgeon's-Hall."[46] Such living death, in the case of Dr. Johnson, is the pattern of individuality itself.

The allusions that suffuse Johnson's autopsy and the commentary on it haunt the event with the uncanny awareness that Johnson himself might not be fully dead. The autopsy record and its printed responses try unsuccessfully to conclude a story of death that has multiple endings dictated by conflicting desires. What the autopsy cannot narrate is the confrontation with the idea of death itself that motivated Johnson's act of self-scarification. The autopsy is written in the interest of the living and for the sake of the desire, on the part of both medical doctors and men of letters, to live forever through a collective knowledge that is experienced as intimate individual conversation with a (living) dead author.

## Embodied Afterlife

When James Wilson Jr. lectured to his colleagues on his father's record of Johnson's case, he chose to recount it along with the record of the autopsy of the popular actor John Henderson (unnamed in the senior Wilson's text) as examples without great pathological novelty, but "not altogether without interest in a meeting like the present, as they relate to men of literature and genius." The actor whom Wilson calls "the only Falstaff that has lived and breathed," and Johnson, "the first of English critics" who was so often likened to Falstaff and who addressed Falstaff as a living being in his commentary on Shakespeare's text—"But Falstaff unimitated, unimitable Falstaff, how shall I describe thee?"—are united in death by the gentlemanly curiosity that constitutes a professional community of surgeons through the display of good taste.[47] Henderson, it is also worth remembering, was one of the last actors to be buried in Westminster Abbey.[48]

Such an affirmation of a cultured literary heritage was particularly important for surgeons, whose professional association was only to be granted the prestige of a Royal College in 1800 and whose class status was compromised by surgery's definition as manual labor and its direct contact with the body's materiality and mortality.[49] Wilson's after-dinner examination of Johnson's body provides us with a sublimated repetition of the banquets and performances that accompanied dissections in Renaissance anatomy theaters and that followed them in eighteenth-century private dissecting rooms; his inheritance of the autopsy record and his rehearsal of the contents in this context transform the dissection of Johnson's corpse into a kind of civilized professional sacrament. The medical society lecture in this context is the ultimate sublimation of and collective identification with the spectacle of anatomy.

The community that results, so this fortuitous linking of Johnson and Henderson reveals, cannot leave the material realm altogether behind; it is also a kind of contagion. Henderson was also immortalized in a footnote of Boswell's *Life* on Johnson's "bow-wow way" of speaking, in which Boswell deplores the proliferation of "second-hand copies" and "overcharged imitations or caricatures" inspired by the actor's incorrect mimicry of Johnson's manner, false and proliferating imitation that the biographer made it his business to correct, paradoxically, in print. Without the wished-for alternative that Johnson's "mode of speaking . . . could be preserved as musick is written," Boswell's record of Johnson's conversation remains haunted by his subject's elusive voice, as he beseeches his readers "to endeavour to keep in mind his deliberate and strong utterance."[50]

Wilson's oral recitation of the autopsy record for his medical colleagues provides an uncanny mirror to the *Life*'s project of accurate reanimation of Johnson's spoken tones in print. In the most successful apotheosis of what John Wolcot termed the national disease of "Johnso-mania," Boswell boasts that he has "*Johnsonised* the land; and I trust they will not only *talk,* but *think* Johnson."[51] And while Johnson's peculiarities of manner continue to inspire contagious imitation in classrooms and Johnson societies around the country, the autopsy manuscript itself ends with an addendum of physical contagion: "N.B. Mr. White assisting me to sew up the body pricked his finger with the needle; next morning he had red lines running up the arm, and a slight attack of fever." The imaginative sympathy between surgeon and corpse leaves the trace of its contagion on the body.

Wilson's manuscript volume remains in the archives of the Royal College of Physicians, but Johnson's autopsy has fascinated doctors enough to have an afterlife in print. Excerpted in the *Gentleman's Magazine* (as we have seen) shortly after his death in December 1784, as well as in Hawkins's 1787 *Life* among others, from its first verbatim appearance in the first volume of the *London Medical Journal* in 1849—the transcript of another medical oration by one George James Squibb[52]—to the present day, Johnson's opened body has been puzzled over and diagnosed (conclusions range from high blood pressure and kidney disease to chronic heart failure linked to both aortic valvular disease and to emphysema) in a capacious and tenacious medical literature, itself a kind of embodied double to the critical literature in which an immortal Johnson with an equally scrutinized living body continues to inspire today's Johnsonians. This opened cadaver has endured not only in the printed records of after-dinner lectures often delivered by those indirectly related to the doctors involved in the autopsy, not only in the "Xhumation" section of the *Lancet* and other historical sections of medical journals, but quite literally (we might even say anecdotally) in preserved parts.[53]

In a footnote to Wilson's observation in the autopsy record that "the air cells on the surface of the lungs were also very much enlarged," Squibb writes, "Dr. Latham, sen., averred that the plate of *emphysema* of the lungs in *Baillie's Morbid Anatomy,* was taken from the lungs of Dr. Johnson."[54] A great deal of medical ink has since been spilled over the true identity of the lung that appears in Baillie's 1799 *Series of Engravings* . . . intended to accompany his 1793 *Morbid Anatomy of Some of the Most Important Parts of the Human Body.* Like the autopsy record itself, these debates, as they translate the body into text and back again into images, involve questions of property, textuality, and the attribution of names—both to persons and diseases. In the first edition of his *Anatomy*, in a subsection entitled "Cells of the Lungs Very Large," Baillie notes: "The only specimen of this sort of disease which I am acquainted with, is in the collection of Mr. Cruikshank; and the person in whom it was found, had been very long subject to difficulty of breathing."[55] Cruikshank's ownership of the specimen convinces even skeptical medical readers that the lung Baillie describes in print in 1793 belonged to Johnson, but the debate about the identity of the corresponding image in Baillie's atlas, attributed by Baillie to a specimen in his own collection, persists, and with it the question of the proper diagnosis to be drawn from Johnson's autopsy. The image of the lung (as we have already seen), reproduced often in twentieth-century medical textbooks, has its own afterlife, not unlike the relic of an unnamed saint.[56]

Such proliferation of detail generated by the autopsy record barely begins to indicate the fascination, at once clinically objective and downright fetishistic, with which doctors struggle to reconstitute the words and images that dismantled Johnson's body. Metaphors used by Wilson and Baillie in describing the corpse, standard usage in anatomy texts of that day, become crucial and curiously cannibalistic indicators of that body's individuality—the cells of Johnson's lung "most of them of the size of a common garden pea, and some few . . . so large as to be able to contain a small gooseberry," the gallstone described in the postmortem as "about the size of a pigeon's egg," misprinted in Squibb as "pigeon's head" and termed in John Hawkins's *Life* "the size of a common gooseberry."[57]

We might relate this curiously flexible and closely observed language to the hyper-realistic pathological poetry of a body become object that Foucault marveled over in *The Birth of the Clinic*, in which "the grain of things [is] the first face of truth, with their colours, their spots, their hardness, their adherence."[58] But it makes more sense to see such language as characteristic of the discourse of English anatomy in its rhetorical reconstitution of dissection's fragmentation. Luke Wilson's account of Harvey's performance of that reconstitution in his anatomy lectures is germane to the rehearsal of Johnson's autopsy in print:

> Where the direct description of the body calls attention to the particularity of the body at hand as an absolute in itself, the analogical mode refers *away from* the cadaver and toward other particular bodies; the notion of a particular body subsists not in any one of these bodies but in the differences between them. There is sameness in these analogies, to be sure, but they are felt to be instructive exactly to the

> degree that they offer some *alternative* of the part or topic in question. The analogical movement away from the body, however, is reabsorbed in the oral and visual circuits that make up the performative reconstitution of the body. Difference is recuperated as relation, and the reconstituted body appears in the anatomy theater as a relation subsisting between anatomist and cadaver.[59]

Harvey's project in his use of analogy is to create an idea of a typical body subject to anatomy's general laws through detailed observation of a particular corpse; his method is not unlike Johnson's own as literary critic and biographer in its wariness of the deployment of unruly detail for its own sake, its ambivalent insistence on "just representations of general nature."[60] Johnson's corpse, not just anyone's, is reconstituted, by contrast, as unique and individual. Nevertheless, the surgeons' use of analogy emphasizes that uniqueness as relational and above all, as a rhetorical and visual bond between the corpse and the anatomists' perceiving eyes. The autopsy rehearses the ways in which, as David Le Breton has argued, the Cartesian separation of body from mind cuts off *cogito* from *cogitamus:* opening Johnson's corpse allows for the construction of a professional collective whose members vacillate between personal embodied identification and disembodied group anonymity.[61] The order that is constructed in place of the sacred order that linked nature to the divine and one body to another is a social secular order of masculine medical knowledge.

Katharine Young, relying on Foucault, characterizes modern anatomy's objectification of the body as an epistemic break from a medical narrative semiotics, to one in which "medicine is now merely spectacular, a reflection of and upon the visible. Limiting medical knowledge to the visible protects it from and in the same gesture prevents speculation. . . . Interiority, which used to be mystery, concealment, opacity, is split open and revealed as more surface."[62] The desire manifested by Johnson's autopsy is thus produced and thwarted by the paradox of the individual named body. This particular patient was a national icon of letters whose embodied presence was preserved in literary parts that served as anatomy's textual equivalent, a plethora of anecdotes. Originating at the crossroads of history and medical diagnosis (as Joel Fineman has argued), troubling the relationship between aesthetic immediacy and clinical objectivity, the anecdote haunts the surgeon's efforts at anatomical anonymity with its living particularity, just as the autopsy haunts the surgeon's "I" with a vision of himself as the bodily remnants of a famous intellect.[63]

Like Lovelace preserving Clarissa's heart, the doctors need to see, to know, the very interiority, the essence of that "uncommon man," which is by definition ineffable and invisible. To dissect Dr. Johnson, in other words, it is crucial to preserve the visual "objecthood" of his corpse, while yearning to penetrate such bodily anonymity, to create a narrative depth.[64] The genre that accomplishes this ambivalent desire is the anecdote.

## Living Deaths

> The anecdote, let us provisionally remark, as the narration of a singular event, is the literary form or genre that uniquely refers to the real.[65]

Joel Fineman's characterization of the anecdote in this section's epigraph links the literary and medical men of our inquiry through their common desire to grasp and emulate the singular. While literary Johnsonians—from actors Henderson, Foote, and Garrick to biographers and drawing-room and classroom performers Boswell and Walter Jackson Bate—are known for their penchant to imitate the living Johnson, it would seem that medical Johnsonians prefer to collect him and fix him in death. Yet these two groups of gentlemen, medical and literary men of letters whose professional identities were very much in flux and only beginning to separate at the end of the eighteenth century, have a great deal at stake in their common interest in Johnson's body.[66] A brief survey of essay titles from the medical and literary realms is revealing in this regard: from the medical side, "Dr. Johnson, My Patient"; "A Postmortem on Dr. Johnson"[67]—which refers both to the text of the autopsy and the essay itself; from the literary side, "Dr. Johnson in the Flesh."[68] Whether preserved in individual organs and illustrating distinct categories of disease or immortalized in personal anecdotes, Johnson's body poses a continuing problem of the relation of part to whole, of the organ to the body, the author to the nation, the individual life to the communal denial of death. The Johnsonian phenomenon unites the literary and medical professions in a common disavowal of their union and its object, a common desire to reanimate the body of a dead male genius and thus to live together forever.

Both literary and medical Johnsonians enact, in a kind of unconscious mutual mirroring, the paradox of a modern individuality for which the body has become "nothing more than a leftover," but which nevertheless cannot abandon the body as a figure for the self. This phrase, meant to describe the end result of Descartes's cogito, has also been used by Stephen Greenblatt to evoke the representational dilemma of a post-Reformation England confronted with a demystified Host that no longer possesses transubstantiative power, a figure for the body that is no longer divine, only the material object of metaphor.[69] Descartes in the *Meditations* cannot imagine himself embodied without confronting himself as corpse: "In the first place, then I considered myself as having a face, hands, arms, and all that system of members composed of bones and flesh, as seen in a corpse which I designated by the name of body."[70] Johnson's individual distinction, the contagious physically particular imitability that changed the speech of a nation, is his autopsy's mirror image.[71] Alive or dead, Johnson's body figures paradoxes of representative individuality that are haunted by what they deny: the doctors by the particular person whose dissected flesh they reorder into anatomy's textual body yet whose name they cannot abandon; the literary men by the death and anonymity that the mortal body of their immortal genius cannot help but signify.

Fascination with Johnson's body unites aesthetic and medical knowledge in what Jonathan Sawday has termed an autoptic vision that penetrates the other in order to see itself and that is dependent on "a worship of parts."[72] John Bender has linked the "technical practices of the eighteenth-century novel to anatomical science," revealing a violence that undergirds the eighteenth-century narrative construction of sympathy and self-regulating subjectivity. For both novelist and clinician in Bender's account, the murderous omniscience of this penetrating realist gaze gains its authority from death and is dependent upon the fictional (and as his essay discusses, feminized) anonymity of its object.[73] The spectacle of Johnson's autopsy, by contrast, provokes a less totalizing, more narcissistic brand of attention to embodied male particularity. The self-regarding fetishism of medical and anecdotal collectors of Johnsonian detail denies death and narrative alike in the preservation of presence. It is important to note in this regard that out of all the autopsies recorded by James Wilson Sr., only Dr. Samuel Johnson is fully named. The body of this national individual can be displayed, paradoxically, only in literal and anecdotal parts.

For those who deplored the proliferation of Johnson's peculiarities in Johnsoniana, the anecdote's exposure of private and petty details rewrites autopsy as murderous fragmentation. In his mock-eclogue *Bozzy and Piozzi,* John Wolcot, in the persona of the judge of their anecdote contest—his future target Sir John Hawkins—addresses Boswell and Thrale:

Just like *two Mohawks* on the man you fall—
*No murd'rer,* is worse serv'd at SURGEON'S HALL.
Instead of adding *splendor* to his name,
Your books are downright *gibbets* to his fame.
Of those, your anecdotes—may I be *curst,*
If I can tell you, *which* of them, is *worst.*
You never, with *posterity* can *thrive*—
'Tis by the *Rambler's death alone,* you *live*—

Like *wrens,* (that in some volume, I have read)
Hatch'd by strange fortune, in a *horse's head.*
Poor Sam was rather *fainting* in his *glory,*
But lo! his fame, lies *foully dead* before ye.
*Thus,* to some dying man, (a frequent case)
Two doctors come, and give the *coup de grace.*[74]

Rather than successfully conveying the particularity of Johnson's life through anecdotal detail, such murderous exposure cuts Johnson into pieces. The biographers become at once savages—bloodthirsty Mohawks—and surgeons, reducing Johnson to the fate of a common murderer whose punishment, as Hogarth's fourth stage of cruelty reminds us was the case in the eighteenth century, goes beyond hanging to include public dissection or, worse, the grisly display of the gibbet.[75] This murder of Johnson's fame makes Thrale and Boswell monstrous natural anomalies who live on death or, even more disgraceful, doctors who profit by it. Just as the threat of dissection carried punishment into the afterlife by depriving criminals and, later, the poor of decent burial, so in Wolcot's words, Johnson's fame, his life after death, is brutally murdered by anecdotal indiscretion.[76] Neither the anonymous body of the criminal denied a grave (an exemplary body whose public abjection and suffering linked it in Renaissance anatomy theaters to that of Christ) nor the once-sacred and perpetually scrutinized body of the monarch, Johnson's curious body takes a liminal and markedly particular place in the English imagination.

An anonymous biographer in the same year, 1786, similarly observes:

> There is, perhaps, no instance of an individual that has sustained so much injury to his posthumous reputation from the ill-directed zeal of injudicious friendship, as Doctor Samuel Johnson. A life marked only by the common sterility of inactive literature, has been expanded into a source of voluminous anecdote—been tortured into becoming the vehicle of almost general history—and what is still worse, of no small portion of individual calumny. . . . With a new and base species of intellectual anatomy, they have laid open his heart after his death, and have produced to the observation of mankind parts of its constituent materials not always creditable to the dead, but what is perhaps of still more consequence, in the highest degree, afflicting to the living.[77]

Anecdotes, in this account through indecent exposure, make something out of nothing, turning the common sterility of a literary life into diseased, detailed fertility, inactive retirement into excessive interest; like the gibbet itself, they torture individuality into "almost general history," displaying dirty laundry in an anatomy theater.[78] The very flaws, the particulars that in Boswell's account have allowed him to Johnsonise the land, in the view of anecdote's detractors, embalm the great man in a violated state of living death.

## Anecdotal Immortality

> The detail, we might say, is as small as Napoleon.[79]

Whether Boswell's and Thrale's anecdotal monuments murder Johnson to dissect or immortalize him, their details have evoked a true Johnson for Johnsonians in a romance that elides historical differences, particularly inspiring medical readers of various stripes. Lecturing to the New York Psychoanalytic Society in 1942, Edward Hitschmann, MD, in an analysis of Johnson's character that is also an attempt to define the character of the English nation and ultimately of the "Anglo-Saxon" race, proposes "to show here once again that psychoanalysis is the best available method for interpreting and constructing coherent pictures of the development of a personality . . . by a study of the 'most unforgettable character I have met,' the famous Englishman of the eighteenth century, *Samuel Johnson.*" Describing Johnson in familiar Rabelaisian terms that push him to the borders of the human—"his laugh should remind one of a rhinocerous [*sic*] and his mouth of that of Gargantua, his spirit of the far reaching trunk of an elephant," Hitschmann nevertheless comments on Johnson's complete centrality to English culture and self-image:

> He was called, "the most national man of English letters"; "the embodiment of the essential features of the English character", "an awful majestic philosopher" and, "a tremendous converser.". . . Carlyle acknowledged him as "the largest soul that was in all England" and as the "ruler of the English nation for some time, not over them but in them." In the parliament Johnson was called, "the pattern of morality." People saw in him, "their magnified and glorified selves." In spite of his lifelong suffering from a compulsive neurosis, the distorting convulsions of his severe tic and his attacks of depression, one biographer even declares: "Johnson embodies the healthiest instincts of the Anglo-Saxon race."[80]

What is most remarkable about this passage is its bricolage of Johnsonian quotations from across the centuries, few of them attributed, so that the "most unforgettable character I have met" articulates at once Hitschmann's personal "I" and the generic Johnsonian moment of identification.[81] In a similar move in plural form, the esteemed English physician and Johnsonian Sir Humphrey Rolleston, in a 1924 address, opines even more definitively, "It might naturally be thought that, apart from Napoleon, few if any famous men would have attracted more speculation about their medical ailments than Samuel Johnson, both because Boswell's *Life* provides such a minute and intimate account, and because 'he embodies all that we most admire in ourselves.'"[82]

Be he mythical monster or national dictator, Johnson still endures not despite, but because of his bodily failings. The nineteenth-century surgeon George Squibb accounts for this paradox by attributing the moral power of Johnson's legacy to his physical ailments:

> It was therefore to the corporeal condition of Johnson, and to the train of thinking more immediately produced by that condition, that we are so much indebted. . . . His mind, tinctured by morbid feelings even from earliest childhood, was of the most imaginative cast, and had it not indeed been for the consciousness of a bodily infirmity which continually possessed him, his mind might not have had that bias which led to such deep powers of reflection and a knowledge of moralities. Had he been blessed, therefore, with "mens sana in corpore sano," he might only have been known to us as the poet, dramatist, and historian (and not moralist).[83]

Such "embodiment" in these medical appraisals is at once literal—the basis for a diagnosis of Johnson's particular ailments or neuroses—and metaphorical—the basis for a sympathetic identification with, for Hitschmann, "the embodiment of the essential features of the English character,"[84] or for Rolleston, "all that we most admire in ourselves." Both the medical and the literary reader of Johnson share an autoptic vision (as we have seen) that is not just a seeing by oneself (the word's literal meaning), but a seeing of oneself, an identification dependent on the disavowal of difference and of death.

David Simpson has described the rage for anecdotes in eighteenth-century England as "a compulsive mode of representation," a socializing and subjectifying genre giving temporary printed closure to conversation's presence while avoiding more divisive endings to larger stories called history, culminating in twentieth-century New Historicism's romance with the local and the personal. What Isaac D'Israeli in 1793 characterized as anecdote's empirical "science of human nature," its rejection of "vague theory" for "certain experiment," allows us, in Simpson's words, indeed "to speak with the dead, and in so doing bring ourselves to life" and to imaginary community.[85] One can see similarly structured denials of difference in the history of medical knowledge:

the Hunters' (both William and his surgeon brother, John) as well as Baillie's emphasis on the anatomical classification and separate examination of individual organs modeled on natural history rather than the analysis of tissues based on a systemic understanding of the body with which Bichat revolutionized pathology at the end of the eighteenth century in France is a similar privileging of the part for the whole, an inclination toward the individual inflected by social divisions in the eighteenth-century English medical profession not unlike those larger social divisions that enabled and were effaced by the Johnsonization of the nation.[86]

By denying larger systemic wholes for individual parts, medical and literary Johnsonians alike deny their own deaths in a romance of personal immortality. Johnsonian desire vacillates between the dream of a professional community that will create a collective body of universal knowledge that transcends time, and the personal romance of an individual Johnsonian "I" that, as in Hitschmann's lecture, is spoken through a collective identification with a singularly embodied hero.

I close with the anecdote from John Hawkins's *Life of Johnson,* which brings us back to our beginning and which ends Hawkins's narrative:

> He had often reproached his physicians and surgeon with cowardice; and, when Mr. Cruikshank scarified his leg, he cried out—"Deeper, deeper;—I will abide the consequence: you are afraid of your reputation, but that is nothing to me."—To those about him, he said,—"You all pretend to love me, but you do not love me so well as I myself do."[87]

In this account, the living Johnson, who had declined opiates several days before, struggles against impending death and a dropsical body turned "bloated carcase." This accusatory culmination of what Hawkins describes as an embattled death, omitted from Boswell's version in accordance with the prevalent late eighteenth-century ideal of death as painless acquiescence, is perhaps most striking and most disturbing for Johnson's denial of sympathy. Cruikshank's fear of causing pain is, in Johnson's angry words, fear of loss of reputation—a cynical version of Adam Smith's impartial spectator who governs his own behavior by imagining the reactions of others.[88] Not moral sympathy but rather Johnson's own self-love and desire for life are shown in this anecdote of his stoic endurance of pain and what Boswell called the "bold experiment" of self-scarification, which according to Hawkins and the autopsy record precipitated his death. Johnson's ruthless act of self-dissection exposes the underside of the sympathy that had stayed Cruikshank's hand: the doctor's identification with his learned patient is undermined when the patient, about to be (like the death mask) "taken from life" and preserved in parts forever, attempts to take life back, taking his own life in the process.

Yet the surgeons and (in his own anecdotal forum) Boswell had the final word, opening Johnson's body at Cruikshank's behest in order, so the rumor goes, to disprove suspicions of suicide. This rumor, preserving Johnson's heroic mastery to the last, conceals—even reverses—what his desperate self-scarification—the act of a body contemplating itself as corpse—had revealed: not the will to die but the individuating and self-disintegrating fear of death itself. The autopsy reclaims and resurrects Johnson in an affirmation of a moral and masculine community of self-contained individuals: through biographers' and surgeons' founding acts of anecdotal preservation, the Johnsonian story might never end, Johnson's conversation might go on forever, and the Age of Johnson might always bear his name.

"To trace continuous connections across distinct parts is to know the local actions of the hand that manifest the motions of the soul," writes Katherine Rowe in her analysis of the role of the hand as both subject and object of anatomy.[89] The anatomist's hand can serve as the missing link between body and soul, between dead matter and divine handiwork, between Alexander and the bunghole, between holes and wholes. When we turn to the local actions of Johnson's hands, to his body in motion, that link disappears, leaving us with living mystery.

CHAPTER TWO

# STYLE'S BODY: THE CASE OF DR. JOHNSON

Here, as before, never, so help you mercy,
How strange or odd soe'er I bear myself—
As I perchance hereafter shall think meet
To put an antic disposition on—
That you, at such times seeing me, never shall,
With arms encumber'd thus, or this head-shake,
Or by pronouncing of some doubtful phrase,
As "Well, well, we know," or "We could, an if we would,"
Or "If we list to speak," or "There be, an if they might,"
Or such ambiguous giving out, to note
That you know aught of me

*Hamlet,* 1.5.177–87

In the eyes of his contemporaries, Samuel Johnson was both a monument and a monster. He was, in other words, the image of an author. At once subject and object, printed voice and aberrant spectacle, the Dr. Johnson who longed for what he praised in his elegy to his friend—the unlettered surgeon who ministered to the poor, Robert Levet—as "the power of art without the show"[1] embodied literary authority as a paradox for his age. The immortal Johnson whose distinctive style of writing and speaking lives on in English literature classrooms across the country as a signature for certainty enacted in his mortal person a series of apparently compulsive movements, mutterings, and rituals that in different ways to different viewers compromised agency itself. Dubbed by contemporaries as both Great Cham (tartar monarch) and Caliban of Literature, by recent critics as both "colonial master" and "fetish object of the literary world," the living Johnson provided a spectacle of bodily defect in perpetual motion that was inextricably bound during his own life to a distinctively fixed literary style.[2] This chapter works to uncover the historical logic of a link that Johnsonians of our own day have tended to ignore, a link that in its own time took the form of the vanity of authorial wishes.

For eighteenth-century England, the case of Johnson—and by "case" I mean at once "body" and "example"—united style with substance, text with body, universal truth with the individual writer, posing paradoxes that attempted to make personal agency and interiority legible on the body of the author. My purpose here is neither to reinforce nor to undermine an enduring impression of Johnson's immeasurable greatness, nor to write a familiar narrative (one critiqued by Lennard Davis) of a powerful mind heroically overcoming a defective body.[3] Rather, I want to contemplate the ways in which both Johnson's monumentality and monstrosity mirror each other at a particular historical moment. I read Johnson's singularity as representative of a late eighteenth-century chapter in the Western formulation of the mind/body problem during which mind and body, through the complex workings of sensibility, are both inextricably interconnected and inscribed in each other's image; ineffable interiority, not fully separated from the body, becomes an uncanny thing in itself.

Neither the once-sacred body of the king, nor the incorruptible body of the saint, nor the malleable body of the actor, the unique body of the eighteenth-century author, preserved in anecdotal form, figures the ineffable and embattled substance of individuality and intention. I begin with a meditation on the relation between the seemingly meaningless repetition of Johnson's tics and the meaningful repetition of his literary style, paying special attention to his use of that most condensed of vehicles for paradox, the heroic couplet, inherited from another authorial monster, Alexander Pope, in one of his most thematically representative texts, *The Vanity of Human Wishes.* We will return to this pairing in the form of an opposition in chapter 5. I then turn to Johnson as an embodied figure of such "vanity," by pondering eighteenth-century definitions of, speculations about, and depictions of those "convulsive starts and odd gesticulations"[4] that so fascinated audiences in polite drawing rooms and that have recently been diagnosed as a twentieth-century medical version of mind/body interconnection, Tourette's syndrome. And I conclude with a reflection on the relationship between a form of Johnsonian criticism that seems compelled to imagine the author in the image of his monstrous style and Johnson's own compulsion to repeat. What follows, in short, reflects on the ways in which Johnson's physical particularity turned authorship into a performance, an ambiguous enactment of agency by a body in motion that made monstrosity exemplary. If the autopsy exemplified the anecdote's position at the borders of life and death, to consider Johnson's living body and its relationship to interiority is to locate the anecdote—the narrative equivalent of a single gesture—between mind and body, intention and its lack.

## The Ends of Style

> Years following Years, steal something ev'ry day,
> At last they steal us from our selves away;
> In one our Frolicks, one Amusements end,
> In one a Mistress drops, in one a Friend:
> This subtle Thief of Life, this paltry Time,
> What will it leave me, if it snatch my Rhime?
> If ev'ry Wheel of that unweary'd Mill
> That turn'd ten thousand Verses, now stands still.
>
> ALEXANDER POPE, *The Second Epistle of the Second Book of Horace Imitated,* lines 72–79

One of the first readers to link Alexander Pope's body of work to his physical corpus was Samuel Johnson, whose mercilessly detailed description of Pope's shrunken form in his *Life of Pope,* a life criticized for the brutality of its anecdotal attention to its subject, seems determined to re-anchor Pope's polished art in the writer's aberrant person.[5] There Johnson's account of Pope's "petty peculiarities" reduces his most important predecessor as a professional author to a helplessly feminized "person well known not to have been formed by the nicest model," a person deformed, in fact, by excessive literary effort, a fleshly allegory for the vanity of authorial wishes.[6] Johnson's scrutiny of Pope's body points us toward an eighteenth-century reading of authorship as visibly and aberrantly embodied personification, a figure that Pope himself exploited to his own advantage. Pope's literary career was modeled on deformity's ambiguity; shaping, and shaped by, the "monstrous contingency" of modern authorship. Throughout this poet's life of literary imitation, deformity guaranteed him a marked and marred originality. As I have argued in *Resemblance and Disgrace: Alexander Pope and the Deformation of Culture,* for Pope, deformity and form made the ultimate couplet, a couplet that, as this section tries to show, Johnson both resisted and rewrote.[7]

The brief excerpt from Pope's imitation of Horace's second epistle of the second book with which this section begins condenses a literary tradition into a metronome of loss and finally into a contemplation of death. The evocation of time's predation through the couplet's hierarchy of pauses—all but the second and penultimate line relying heavily on caesurae, the second couplet's emphasis on "one" like the relentless ticking of a cosmic clock—culminates in a semi-apostrophe to time's force that echoes a host of predecessors from Pope himself, to Dryden, to Milton, to Montaigne, before turning to a startlingly original moment of emotional self-revelation: "What will it leave me, if it snatch my Rhime?" That blunt monosyllabic question with its emphasis on the violence of "snatch," as stark as the line that preceded it is musically overdetermined, is then rephrased, paradoxically, as the poet's contemplation of himself as machine devoid of affect and of agency, "that unweary'd Mill / That turn'd ten thousand Verses," and, finally, devoid of life imagined as the power to write: "Now stands still."

What can these lines, distinguished by their exceptional lyrical power from a poetry more often satiric or didactic than plaintive, reveal about Pope's control over and submission to the couplet form? What can they begin to tell us about the relationship of literary style to the body of the author? How does poetic form possess, even figure, those who create by its means? How, to invert the question, is style a kind of embodiment of meaning at once inimitable and imitated? How then, does Samuel Johnson, in part through his own idiosyncratic use of the couplet, inherit and transform his predecessor's work as monarch of letters into his own ambivalent brand of embodied literary authority? The curious fact that the two most prominent men of letters in eighteenth-century England were so strikingly physically visible in such different ways was at once coincidental—who could have predicted Pope's tuberculosis of the spine or Johnson's nervous tics?—and overdetermined.

These two authorial bodies were as much framed by cultural modes of vision as they were arbiters of cultural production; for both writers, style depended on embodied particulars, and like the body itself, at once signified and limited individuality and agency. For Pope at this moment, style at once subsumes and threatens to erase the self: the silent answer to the multivalent question "What will it leave me?"—as subject and object waver in the balance—is "nothing."

Pope's seamless paradoxes of couplet form embody for Johnson a prospect of art without "end," without closure or moral purpose. Unable to stomach Pope's ability to pause from life in order to extend the domain of letters, Johnson bases his own authority on his instructions to the aspiring scholar in his own satiric masterpiece and achievement in classical imitation, *The Vanity of Human Wishes,* to "pause a while from Letters, to be wise."[8] When Johnson fully appropriates the couplet in the *Vanity,* he rejects the monstrosity of Pope's spectacularly visible self-authorization, his refusal to leave the stage of writing, through a transformation of form itself. The *Vanity*'s form demonstrates how the author of perfect art must be brought back to the body and reminds every author that he is authored by another.[9]

If Pope's indelibly marked deformity allowed him to embody imitative originality for a public newly obsessed with the author's person and literary property, Johnson's "tricks," "antics," "convulsions," and "gesticulations," as we shall see, put authority on display by turning it to perpetual motion without apparent end, as if determined by the rhythms of the couplet itself. So distinctive is this style's analogical linkage of gesture to thought to conversation to printed word that it becomes an icon for imitators from Garrick to Boswell to contemporary Johnsonians. Boswell wished that Johnson's "bow-wow way" could be "preserved as music is written," and Hannah More, "umpire in a trial of skill between Garrick and Boswell, which could most nearly imitate Dr. Johnson's manner, . . . gave it for Boswell in familiar conversation, and for Garrick in reciting poetry," which the latter did, according to Boswell's description of one couplet performance, "with ludicrous exaggeration . . . with pauses and half-whistlings interjected, looking downwards all the time, and, while pronouncing the four last words, absolutely touching the ground with a kind of contorted gesticulation." Even Johnson's seesawing—"his head," one female observer remarked, "swung seconds"—seemed a kind of bodily mimicry of the couplet; while William Cooke's observation of his "rolling about his head, as if snuffing up his recollection," before breaking out into twenty lines of Juvenal's tenth satire in the original Latin marks a similar purpose to a seemingly meaningless act.[10] And his mode of composition of the *Vanity*, ordering the rhythms of potentially obsessive thoughts into the regularity of the couplet form almost without writing's mediation, elucidates Johnson's characteristic opening of the couplet into larger blocks than Pope's: Boswell marvels at the "fervid rapidity" with which the poem was written, remarking that Johnson "composed seventy lines of it in one day, without putting one of them upon paper till they were finished."[11]

The *Vanity*'s rejection of satire, which Johnson's choice of Juvenal as model after Pope's Horace would seem to belie; its enactment of what Walter Jackson Bate called satire manqué, the continually "active balance" with which it undoes and expands its own impulses to judge (its brutality to that epitome of the genre, Swift, who "expires a Driv'ler and a Show" [318], is as much a result of that impulse as of its failure)[12]; its near-obsessive reliance upon personification and abstraction; its resistance to narrative in the form of endless repetition of the same end to every attempt at self-signification; its disembodied preoccupation with the visible and with the destructive power of objectification; and its final abandonment of agency and closure in the form of a prayer—all demonstrate the same Johnsonian couplet of meaningful and meaningless repetition. From the perspective of the *Vanity*'s melancholic reduction of all human attempts at distinction to the same meaningless end, the perpetual motions of Johnson's tics, much like Pope's image of himself as an unwearied couplet mill, render the great author a kind of personification of self-destructive intention. In all his oddity, peculiarity, and particularity, Johnson the man becomes an exemplary character precisely because of his resistance to perfect form. The detritus, the scattered waste of the synecdochal signs of fame that fill the *Vanity,* and the satirist's thwarted and at times angry voice ("hear his death, ye blockheads, hear and sleep" [174]), all evidence the death-ridden inevitability of particular embodiment, which Johnson sees through and mourns in the *Vanity* and which brought him to life.[13] The personifications of the *Vanity,* in this regard, have a curiously contagious effect, threatening to reduce even the disembodied author to a rhetorical and physical figure.[14]

The Johnsonian penchant for personification, while extreme and noted by many of his critics as characteristic of his style,[15] was far from unique. The proliferation of personification in the eighteenth century, as Steven Knapp has shown, signaled a larger cultural anxiety about the unruly animating power of enthusiastic agency; the trope's rhetorical contagion threatened absurdity or, worse, solipsistic fanaticism.[16] When Johnson criticizes Milton's licentious use of allegory in the personification of Sin and Death in *Paradise Lost,* for example, he is also implicitly censuring a potentially revolutionary authority nearly satanic in its excess.

Johnson's personifications in the *Vanity* are by contrast strictly rhetorical. They govern a poem about the futility of action by the power of their own limitations. In reading them, Chester Chapin claims, "the impulse toward visualization . . . is rather a hindrance than a help." But by the sheer weight of their stasis, these figures become curiously animate, possessing as much "metaphorical force" as the verbs that they rule, bringing even grammatical relations to life: "Where then shall Hope and Fear their Objects find?" (343). Chapin characterizes these "vivified abstractions" as couplets in miniature, balancing moral truth and individual examples in "verbal reflection of a particular moral paradox."[17]

From this perspective we can read the opening lines of the *Vanity,* a poem in which the human agents are as empty of meaning as the rhetorical ones, as ticlike in its compulsion to repeat—"Let Observation with extensive View, / Survey Mankind from China to Peru" (1). At once controlled and excessive, the grandeur of this evocation of a universality at once abstract and spatial, its magisterial surrender of agency by an imperative to Observation, which erases any particular observer, and its elaboration of a simple action into a series of ornate repetitions, all typify the Johnsonian style in the simultaneous undermining and evocation of meaningful purpose. Coleridge—whose patience for Johnson was as limited as his emulation of him was intense—paraphrased the line with acute annoyance, "Mere bombast and tautology as if to say 'Let observation with extensive observation observe mankind extensively.'"[18]

Similarly, just as nervous tics have been described as the bodily hieroglyphics of a personal narrative whose origins have atrophied into mere mechanics,[19] so the hyper-condensed careers in the *Vanity* serve only to point morals and adorn tales. Just as the dynamic fluidity of fictional agents in Milton threatens contagion to ostensibly "real" agents, so the Johnsonian personification's yoking of "live" metaphors to dead abstractions[20] infects the poem's historical subjects and renders them personified characters, "moral paradoxes" in a plotless, redundant story. It is the personifications who act, "Hope and Fear, Desire and Hate" who "O'erspread with Snares the clouded Maze of Fate" (5–6), while the poem's persons, no matter how active, seem by contrast curiously passive. Wolsey and Marlborough are particular examples no more or less vivid than the unnamed "young Enthusiast" (136) whose quest for scholarly fame moved Johnson to tears in the reading of it. Names in the poem often serve as encapsulated history, the narrative impulse truncated to a single word. This poem, to which only a prayer can put an arbitrary end,[21] by "pierc[ing] each Scene with Philosophic Eye" (64), reframes both history and its authors as spectacle.

The *Vanity* makes us aware of the inadequacy of its own attempts at order, despite how such attempts are as insatiable in their own way as those of the ambitious, greedy subjects it portrays. Even the anonymous bearer of a reasonable wish in the poem's context—the wish for a life without narrative, the progression of an age "that melts with unperceived decay, / And glides in modest innocence away" (293–94), a potentially endless life as unconscious subject and invisible object—is consumed by the poem's relentless need to close:

> Year chases Year, Decay pursues Decay,
> Still drops some Joy from with'ring Life away;
> New Forms arise, and diff'rent Views engage,
> Superfluous lags the Vet'ran on the Stage,
> Till pitying Nature signs the last Release,
> And bids afflicted Worth retire to Peace. (305–10)

Johnson rewrites even the humble life that replaces ambitious solipsism with what he would similarly praise in his elegy to Robert Levet as the "narrow round" (25) of social virtue, the submission to an inhumanly regular process of inevitable loss that makes its end—Nature's signing of a "last Release" from life refigured as debt—like the end of the poem itself, a devoutly to be wished escape from selfhood. The passage with which we began, Pope's poignantly personal identification with and refusal to abandon the couplet form, is directly alluded to here, but Pope's satiric end is collapsed into its beginning, and individual authorship recuperated into the inevitable progression of general art and shifting perspective: "New Forms arise, and diff'rent Views engage." Johnson transforms Pope's heroic refusal to stand still into a universal desire for an end to life's infinite gradations of loss.[22] Both Pope's original appropriation of imitation's stage with deformity's trademark and Johnson's impersonal subjection of individuality on that same stage to a divine author's ends form particular couplets of mind and body, of the disembodied power of art with the embodied particulars of spectacle and show. Both authors are monstrous characters and national monuments in the eighteenth-century theater of authorship.

## Couplets

> Whatever its sophistication, style has always something crude about it: it is a form with no clear destination, the product of a thrust, not an intention, and, as it were, a vertical and lonely dimension of thought. Its frame of reference is biological or biographical, not historical: it is the writer's "thing," his glory and his prison, it is his solitude.
>
> ROLAND BARTHES, *Writing Degree Zero*

What can Johnson's particular use of the couplet form in poetry, a form in which he was distinguished but not prolific, tell us about a style that was predominantly defined in prose? If we frame the paradoxes of *The Vanity of Human Wishes* with contemporary criticisms of Johnson's style at the end of his career, his highly compressed couplets seem to bear the same relation to his prose that his prose—in its excessive artfulness, its ornate Latinity, its elaborate balance of abstractions, and its (to some) monstrous hybridity—bears to the plain style that his critics favored and that Johnson himself bears to the general idea of authorship. Making explicit the thematic preoccupations that haunt Johnson's work throughout his career, the *Vanity* demonstrates in exceptionally concentrated form how style for Johnson is itself an empty performance, a defamiliarization of mimesis that advertises its own artifice, rendering itself a vain spectacle. In Barthes's terms, style—bearing the trace of the body, a matter more of biological reflex than intention—is Johnson's distinguishing mark, the unique "thing" rendering him at once subject and object, his glory and his prison.

Johnson's vivification of abstraction in his prose is both a formal problem and a philosophical and epistemological project. Formally the question concerns the utility of particular detail for conveying general truth: how to make meaning most effective, persuasive, and emphatic without, as the poet Imlac says in *Rasselas,* numbering the streaks of the tulip? Philosophically (and thematically) the dilemma involves the effectiveness of exemplarity itself: how to distinguish one's own sentiments from general truth, how to cut a figure in the world, without coming to the same vain end?[23] The author's desire to delight a large audience through instruction thus is undermined by an awareness of that or any desire's futility. If we begin with an awareness of how the genre of literary imitation intensified the questions of authorship and originality that preoccupied both Johnson and his audience, the *Vanity* can be read as the exemplary form of Johnson's public prose, the mournful strain that haunts its sonorous distinction.

By uncovering the structural homology that links Johnson's poetic style to his nervous tics, we make visible his paradoxical relationship to originality, to authorship, and to agency. By taking on the couplet form and the imitation of a Roman original (specifically Juvenal's tenth satire), Johnson lets himself be partially determined by his most formidable satiric neoclassical predecessor, Alexander Pope, in the act of writing his own distinction. Style thus appears to be both a kind of compulsion and a singularity vulnerable to imitation, suspended, like tics and habits, between volition and its lack. Such paradigmatic authority invites readers to disavow its particular origins in ambivalence about self-assertion: this most celebrated of talkers, whose "language was so accurate, and his sentences so neatly constructed, that his conversation might have been all printed without any correction," never spoke in public until he was spoken to; this most prolific of authors famously insisted that "no man but a blockhead ever wrote, except for money."[24] In all its emulable substance, distinctive literary style generates itself out of the void that the *Vanity* repetitively evokes in commemorative erasure of all attempts at personal distinction, out of an ambivalent response to the poet's own authorial self-consciousness and ambition.

In the potentially self-effacing poetry of classical imitation, Johnson's labors were unmistakably his own and irrefutably opaque. David Garrick echoed many criticisms of Johnson's difficult prose when he complained that the *Vanity* was "as hard as Greek."[25] The first part of this chapter contrasted Johnson with Pope in order to consider what was distinctive about Johnson's use of the couplet form. I would now like to consider how Garrick's jest resonates with other critical accounts of Johnson's prose that saw him as monstrous in his distinction. Archibald Campbell's Lexiphanes and Charles Churchill's Pomposo[26] begin a series of parodies and criticisms of Johnson himself, in the guise of "hard words," ornate and excessively Latinate diction that persist after his death. William Cadogan, in his marginalia to Boswell's *Life,* remarks, "His Brutality & thy Folly, Bozzy, were ever, I believe, very prominent & *con*spicuous

& his Style i.e. the construction & form of his sentences was easy & *per*spicuous—they were without difficulty analysed, but without his Dictionary not to be understood."[27] Cadogan's contrast of the clear visibility of Johnson's "brutality" with the easy opacity of Johnson's prose links embodied presence to the printed word by emptying style of substance.[28]

Robert Burrowes astutely identifies Johnson's penchant for Latin phrases as a kind of poetification of prose and similarly focuses on the seductively audible, rather than visual or clearly comprehensible, nature of Johnson's distinction in print: "The first of these reasons for substituting, in place of a received familiar English word, a remote philosophical one, such as are most of Johnson's abstract Latin substantives, is its being more pleasing to the ear. But this can only be deemed sufficient by those who would submit sense to sound, and for the sake of being admired by some, would be content not to be understood by others."[29] For Burrowes, Johnson's catering to the ear is a cultivation of deliberate obscurity; his style thus becomes a an accumulation of "confirmed and prevailing habits" and a dangerous imperial project, "new-raised colonies disdaining an association with the natives, and threatening the final destruction of our language."[30] Johnson's style in this account is monstrous, "confound[ing] the distinction between poetry and prose," overly singular in its willful exclusion of women and unlearned readers in its attempts at erudition[31] and, like any great style based on "peculiarities," contagious: "Many . . . from admiration of his general excellence are led at last involuntarily to resemble him," just as Burrowes often seems, either voluntarily or involuntarily, to be imitating Johnson himself.[32] Stylistic ambition makes Johnson not manly but deviant and indecipherable; just as, "determined to deviate from the English language," Johnson in his "antipathy to the French" becomes not more English but rather a Latinate hybrid.[33] In his refusal of Frenchness but incorporation of the foreign matter that informs Frenchness—Latinity—into English, Johnson resists while promoting the coherence of the English national body and lays bare the process of exchange the English language ought to aestheticize and conceal. Elizabeth Montagu, intent upon rivaling Johnson as critic of Shakespeare and defender of English literary honor against the attacks of Voltaire, finds herself driven beyond the bounds of English in characterizing Johnson's use of it: "We have no word in our language to express the fault I have to find with Mr j—ns Style, it is trop recherché."[34]

These responses to Johnson's prose take on new meaning in light of Neil Saccamano's reading of Joseph Addison's use of the concept of false wit in the *Spectator* to delineate proper agency as the proper embodiment of language.[35] False wit, in this account, made visible when language's agency, the authority of the verbal frame itself, becomes apparent, tortures language figured as body. The figure of the body is a kind of narcissistic projection, as I read it, of cultural fantasies of agency, wholeness, coherent limits, individuality, and iterability, which false wit disrupts. The paradox that Addison in particular and Augustan writers more generally continually confront is the interrelation of nature and art, which this chapter has figured as Johnson's particular exemplification of the mind/body problem. Both Addison's fantasy of self-regulating nature and Johnson's efforts at self-evident exemplary truths are haunted by an awareness that the singular is difficult to distinguish from artifice, mimesis, and monstrosity.

The critical language about Johnson's stylistic deformity, in this context, renders the author not as an *agent* of false wit but somehow an object of it. Johnson's prose displays itself as a body deformed by art (deformed bodies may be unnatural—made not born), a monstrous blend of poetry and prose, art and artlessness. As an author, Johnson embodies the paradox of repeatable singularity, almost as though he stands for the untranslatable power of language itself and the ways in which language compromises any author's control.

James Thomson Callender's *Deformities of Dr. Samuel Johnson* (1782) is one of many texts to demonstrate style's basis in personal embodied peculiarity and habit. In Johnson's particular case, style's link to the body has a degrading force. Callender's title-page epigraph from *Rambler* 176 aptly quotes Johnson's judgment that "baiting an *AUTHOR* . . . is more lawful than the sport of teizing other *animals,* because for the most part HE comes voluntarily to the stake."[36] Here that which elevates an author above an animal, his will, is that which renders him less than human. John Courtenay, attempting to distinguish the literary from the moral when considering Johnson as an example (a project with which Johnson himself, for whom character was always exemplary, allegorical, and enlivened by its peculiarities, would have been in sympathy),[37] creates in couplets a kind of meta-couplet that balances Johnson's human imperfection with his stylistic uniqueness:

> In solemn pomp, with pedantry combin'd,
> He vents the morbid sadness of his mind;
> In scientifick phrase affects to smile,
> Formed on Brown's turgid Latin-English style[38]

While laboring to distinguish the man from the author, Courtenay's couplet logic unites moral foible with prose peculiarity in the language of defect. Most shocking is a comparison of Johnson's fear of death to that of Maecenas, which balances pagan superstition with failed Christian heroism:

A coward wish, long stigmatiz'd by fame,
Devotes Maecenas to eternal shame;
Religious Johnson, future life to gain,
Would ev'n submit to everlasting pain:
How clear, how strong, such kindred colours paint
The Roman epicure and Christian saint!

Courtenay goes on to comment that Maecenas, "had he liv'd in more enlighten'd times," would have trembled at the 1680 comet or Mary Toft's mock-births. In a note to this passage, Maecenas's words haunt Johnson's deathbed declaration with a vision of deformity:

Let me but live, the fam'd Maecenas cries,
Lame of both hands, and lame in feet and thighs;
Hump-back'd, and toothless;—all convuls'd with pain,
Ev'n on the cross,—so precious life remain.[39]

The eponymous patron of the arts is united with the exemplary author who defied aristocratic patronage in the fear of death and the language of deformity that makes the self a spectacle. Courtenay's comparison unites not only pagan and Christian but also monster and man, nobleman and slave (for only slaves die on the cross), and finally subject and object, moralizing and internalizing the language of deformity as it articulates an inner fear. Defect is no longer associated explicitly with Johnson's body but rather with what is implied to be, as the Roman and the Englishman blur, his superstition and credulity, his unenlightened affinity for monsters and divine signs. Just such a hybrid being, in its Christianization of a Roman text, and in its fear of an end, is Johnson's *Vanity of Human Wishes;* just such a monster is Johnson himself.

## Re-viewing and Repeating

To talk about Dr. Johnson has become a confirmed habit of the British race. . . . Such sayings as "Hervey was a vicious man, but he was very kind to me; if you call a dog Hervey I should love him," throb through the centuries and excite in the mind a devotion akin to, but different from, religious feeling.

In his 1898 address (quoted above), "On the Transmission of Dr. Johnson's Personality," politician and essayist Augustine Birrell, who had also authored a pithy mosaic of Johnsonian sayings titled "The Gospel According to Dr. Johnson," describes an experience of identification at once common enough to be part of the national unconscious—a "confirmed habit of the British race"—and personal enough to create individual converts—"throb through the centuries and excite in the mind a devotion akin to, but different from, religious feeling."[40] "The transmission of Dr. Johnson's personality," in other words, comes to be figured not as contagion but rather as a vehicle of communal rhetorical repetition in the service of individual ritual. In their juxtaposition of near-unconscious national "habit" with passionately original acts of individual identification, and in their summoning of Johnson's living voice—a voice that is simultaneously repeated and reinhabited—from print's dead letter, Johnsonians from Johnson's own day to the present make of literary communion a secular religion. On the border between the dead and the living, as Freud reminds us, and at the root of the "unheimlich," is the familiarity of home,[41] figured by the homely appeal of habitually recollected aphorism and embodied anecdotal detail.

While Boswell's biography was perhaps the most unstinting and precise in its use of personal detail,[42] it was far from unique in its close attention to Johnson's bodily particularity. John Wolcot, among many others, complained of those biographers, particularly Boswell and Hester Thrale, who, following Johnson's own example in *The Lives of the Poets:*

> *Blest!* . . . his philosophic phiz can *take,*
> *Catch* ev'n his *weaknesses,* his NODDLE's *Shake*[43]

Yet the enduring afterlife of the anecdotal Johnson in "confirmed habit" documents the durable transience the *Life* most successfully constructed from such careful scrutiny of Johnson's peculiarities. "I have *Johnsonised* the land," Boswell declares; "and I trust they will not only *talk,* but *think* Johnson."[44]

Boswell's hope that his biography will transform the theatrical mimicry of "talking" Johnson into the internalized identification of "thinking" Johnson hints at a tension between the unique oddity of Johnson's physical presence and the universal wisdom and imitable expression of his words. "It is certain," Boswell writes of the *Rambler* in particular and Johnson's style in general, "that his example has given a general elevation to the language of his country, for many of our best writers have approached very near to him; and, from the influence which he has had upon our composition, scarcely any thing is written now that is not better expressed than was usual before he appeared to lead the national taste."[45] Or as Sir James Mackintosh put it in a private journal in 1811: "As a writer, he is memorable as one of those who effect a change in the general style of a nation, and have vigour enough to leave the stamp of their own peculiarities upon their language."[46] This tension between exemplary authoritative style and singular defect is also a long-disavowed connection, a connection that is a constant subtext, we might even say the subconscious, of the *Life*.

"Johnson, indeed, stands as one of England's greatest characters, as well as one of her greatest caricatures," wrote one Johnsonian doctor;[47] he articulates beautifully the tenuous line between internal depth and visible excess, between imitable exemplar and contagious monstrosity, that epitomized the history of the word "character" in eighteenth-century England,[48] a line that Johnson, in both his printed and personal guises, both challenged and helped to define. In its ideal form, the relationship between Johnson's eccentricity and the national character is one of mutual reflection and coherence: Johnson, a human concrete universal as articulated, for example, in the elegantly balanced, couplet-like oppositions of the final "character" of Boswell's *Life* (itself Boswell's nod to Johnson's own biographical structure and style) epitomizes the morally exemplary, heroically masculine regulation of difference that distinguishes the British nation, itself at once universal/imperial/British and particular/local/English.

Eccentricity has long been a part of how the British have distinguished themselves from other nations. We might begin by pointing to William Temple's late seventeenth-century essay "Of Poetry" in which he argues for a greater "variety of Humor in the Picture [of the British stage than of the Roman], because there is a greater variety in the Life." Attributing the English abundance of "Originals" to "the Native Plenty of our Soyl, the unequalness of our Clymat, as well as the Ease of our Government, and the Liberty of Professing Opinions and Factions," Temple concludes: "We have more Humour, because every Man follows his own, and takes a Pleasure, perhaps a Pride, to shew it."[49] Considering the variety of climate in particular, Temple continues, "We are not only more unlike one another than any Nation I know, but we are more unlike our selves too at several times." A nation of unruly difference, Temple's Britain (like George Cheyne's in *The English Malady* some fifty years later) is proud to be called "The Region of Spleen."[50] When Laurence Sterne goes to France on his *Sentimental Journey* (1768), he links the polished uniformity of French manners to the feminization of commerce and to the domination of commerce by women:

> The genius of a people where nothing but the monarchy is *salique* [forbidden to women], having ceded this department, with sundry others, totally to the women—by a continual higgling with customers of all ranks and sizes from morning to night, like so many rough pebbles shook long together in a bag, by amicable col-

> lisions, they have worn down their asperities and sharp angles, and not only become round and smooth, but will receive, some of them, a polish like a brilliant—Monsieur *le Mari* is little better than the stone under your foot—[51]

By the end of the century, the originality Temple championed—the definition of character that resonates with singularity and eccentricity—strains against new norms of masculine comportment, which, more in line with the ironic acquisitiveness of Sterne's curious Yorick, who inhabits his bodily sensation while hiding his body from view, demand restrained uniformity rather than singular excess. Put another way, in the immediate aftermath of his own historical moment, Johnson's monument is suspended between two possible distinctions: as Samuel Taylor Coleridge paraphrased Joshua Reynolds, "The greatest man is he who forms the taste of a nation, and . . . the next greatest is he who corrupts it."[52]

While Boswell saw Johnson's style as the transformation of the dead lumber of commonplace into living wisdom, rivals for the position of leader of the national taste disagreed. Johnson's friend, fellow citizen of Lichfield, and frequent critic Anna Seward implied that Johnson failed to achieve the highest ranks of "superlative genius" because he lacked "exhaustless variety of style;—the Proteus ability of speaking the sentiments and language of every character, whether belonging to real or to imaginary existence; and that so naturally, as to make the reader feel that so must have spoken every man or woman, angel or fiend, fairy or monster, whose shape is assumed."[53] The implicit standard of comparison underlying this criticism, is of course, Shakespeare, who similarly provided woman of letters and bluestocking salonier Elizabeth Montagu with an example of a truly English because "particularly unJohnsonian" style.[54]

Curiously presaging Boswell's conflation of frontispiece and biographical object (when he first sees Johnson in the *Life,* he describes him as resembling the portrait that graces the finished book), and echoing other criticisms of Johnson that equate social, physical, and stylistic aberration, frustrated at Johnson's disruptions of her polite gatherings, and by turns contemptuous of and intimidated by his literary prowess,[55] Montagu inveighs against

> the envy, & the malice, & the railing, of such wretches as Dr Johnson, who bear in their hearts the secret hatred of Hypocrites to genuine virtue, & the contempt of Pedants for real genius. But no more at present on so odious a subject as the Doctor & his malicious falsehoods. I wish his figure was put as a frontispiece to his works, his squinting look & monstrous form w$^{d}$ well explain his character. Those

> disgraces which make a good mind humble & complacent, even render a bad one envious & ferocious. Lady M. Wortley Montagu says of a deformd Person who had satirized her,
>
> 'Twas in the uniformity of fate,
> That one so hateful sh$^{d}$ be born to hate.[56]

Montagu locates her fantasy of a correlation between Johnson's body and mind in a tradition that begins with Lady Mary Wortley Montagu's similar attack on Alexander Pope (whom she had branded, in her *Verses Address'd to the Imitator of . . . Horace,* at once "resemblance and disgrace" of humanity's "noble race").[57] Elizabeth Montagu's genteel femininity (which confines her criticism to a private letter) relies on the earlier aristocratic Mary Wortley Montagu's published precedent to silence the author by advertising his body's clear message. Like Pope, too much of a monster and a "character," almost, we might say, too much of a self, Johnson's marked singularity is his greatest lack, distinguishing him from aesthetic, social, masculine,[58] and human norms.

Horace Walpole in 1779 argues for a similar and equally paradoxical danger: Johnson's style is monstrous not just in its inability to imitate, but also in that, like its author, it is too easily imitated:

> A marked manner, when It runs thro all the compositions of any Master, is a defect in itself, & indicates a deviation from Nature. The Writer betrays his having been struck by some predominant Tint, & his Ignorance of Nature's Variety. . . . He approaches the nearest to universality, whose Works do not put it in the power of our Quickness or depth of Sagacity to observe certain characteristic touches that ascertain the specific Author.[59]

The end result of such identifiable distinction is, to name one particularly powerful example, Thomas Macaulay's condemnation of the late style of Frances Burney, which he claims is more "broken Johnsonese" than English. Burney in Macaulay's account seems to exemplify one extreme of female Johnsonian, writing a caricatured version of English in order to produce a gallery of unnatural caricatures.[60] The other extreme, as natural and easy as Burney is grotesque, would be D. A. Miller's version of Jane Austen, producing sparklingly Johnsonian disembodied maxims that write herself out of any kind of embodied authority or sexual possibility in her own texts.[61] The female Johnsonian can therefore either give birth to monstrous bodies or have no body at all.

But while Macaulay allows Johnson himself to have excelled at his own style, Walpole's criticism excludes him from the realm of proper expression, noting that Johnson's words, while not quite meaningless,

> form a hardness of diction, & a muscular toughness that destroy all ease & simplicity . . . He destroys more Enemies with the Weight of his Shield than with his spear, & had rather make 3 mortal wounds in the same part than one [referring to Johnson's habit of tripling phrases]. . . . [His Humor] is the clumsy gambol of a lettered Elephant. We wonder that so grave an Animal should have strayed into the province of the Ape, yet admire that practice should have given the bulky Quadruped so much agility.[62]

We are reminded here of Johnson's infamous comment likening women preachers to dogs dancing on their hind legs—both remarks serve to mark the limits of proper comportment, proper ambition, and proper humanity.

Consider by contrast Cuthbert Shaw's 1766 satiric poem *The Race,* in the mode of Pope's *Dunciad*:

> Here Johnson comes—unblest with outward grace,
> His rigid morals stamp'd upon his face.
> While strong conceptions struggle in his brain;
> (For even Wit is brought to-bed with Pain)
> To view him, porters with their loads would rest,
> And babes cling frighted to the nurse's breast.
> With looks convuls'd, he roars in pompous strain,
> And, like an angry lion, shakes his mane.
> The Nine, with terror struck, who ne'er had seen
> Aught human with so horrible a mien,
> Debating whether they should stay or run—
> Virtue steps forth and claims him for her son.[63]

Whether critics patronize or praise him, Johnson personifies his style in a gracelessness that marks him as at once hyper-masculine (warrior, elephant, or lion) and subhuman. In all accounts, gentlemanly ease is forfeited for the rigor of style's rigid distinction. The author's lack of social refinement renders him a beast; in Shaw's terms, at once a monster and a monster-breeding mother whose imagination is possessed with "strong conceptions." The social monstrosity of such marked productivity may dismay the aristocratic Walpole, but for the hack writer Shaw, it also signals the triumph of middle-class masculine morality.

Yet in Johnson's case, particularity, rather than universality, is what haunts even the most devoted critics in the form of a ghost summoned by reading. Take, for example, Walter Scott, who is revisited, it seems, by Montagu's vision of a man in the image of his frontispiece:

> Of all the men distinguished in this or any other age, Dr. Johnson has left upon posterity the strongest and most vivid impression, so far as person, manners, disposition, and conversation are concerned. We do but name him, or open a book which he has written, and the sound and action recall to the imagination at once his form, his merits, his peculiarities, nay, the very uncouthness of his gestures, and the deep impressive tone of his voice. . . . It was said of a noted wag that his *bon-mots* did not give full satisfaction when published because he could not print his face. But with respect to Dr. Johnson this has been in some degree accomplished; and although the greater part of the present generation never saw him, yet he is, in our mind's eye, a personification as lively as that of Siddons in *Lady Macbeth* or Kemble in *Cardinal Wolsey.*[64]

Scott's Johnson is both a personification and performance of authority so successful that like Sarah Siddons in *Macbeth* or J. P. Kemble in *Henry VIII,* he turns a supporting role into a starring one, thus retitling the play (on the topic we might parenthetically note of the vanity of human wishes) in which he is nevertheless a character rather than an author.

However contested their vision, both Boswell and his followers assert a national "we" based on a vision at once intimately personal and collective, an author turned character's immortal fame, remembered as long as the English language is spoken, built from eccentrically embodied trifles. Called brute, bear, unlick'd cub, savage, monster, idiot, laborer, or robber; "Irish chairman, London porter, or one of Swift's Brobdingnaggians"; effeminate, Polyphemus, madman, Oddity;[65] possessed of a style irresistibly imitable yet incapable of impersonation;[66] of a body at once super- and subhuman marked by its excesses and lacks;[67] neither completely manly nor properly gentlemanlike—Johnson comes to mirror the nation's imagination of itself as language in an unruly balance of universal transparency and aberrant detail.[68]

Moral frailties and stylistic faults thus come to form the basis for a common vision; take, for example, John Courtenay's remarks in *A Poetical Review of the Literary and Moral Character of the Late Samuel Johnson* (1786):

> But who to blaze his frailties feels delight,
> When the great author rises to our sight?
> When the pure tenour of his life we view,
> Himself the bright exemplar that he drew?

> Whose works console the good, instruct the wise,
> And teach the soul to claim her kindred skies.[69]

The vision of the author supplants stylistic and moral flaws alike. Once both have been reviewed, they are rejected so that the author is completely fused with the man, the "tenour of his life" made identical to "his works." "Johnson" like Longinus in Pope's *Essay on Criticism,* "*is himself* that great *Sublime* he draws" (681). For Johnsonians to read him is to envision the author himself as exemplary.[70]

When Leslie Stephen, having completed a narrative of Johnson's life for the 1878 lead volume of the English Men of Letters series, turns to his subject's "peculiar" prose style, describing it as "mannerism" and adherence to "blind habit," he concludes with a kind of hallucination of both a prose and a body deformed by lack of agency, reduced to pure repetition: "Johnson's sentences seem to be contorted, as his gigantic limbs used to twitch, by a kind of mechanical spasmodic action. . . ." Though granting his prose "the merits of masculine directness" and transparency, Stephen opines that Johnson's conversation was both livelier and pithier than his writing:

> His face when in repose, we are told, appeared to be almost imbecile; he was constantly sunk in reveries, from which he was only roused by a challenge to conversation. In his writings, for the most part, we seem to be listening to the reverie rather than the talk[.] . . . We seem to see a man, heavy-eyed, ponderous in his gestures, like some huge mechanism which grinds out a ponderous tissue of verbiage as heavy as it is certainly solid.[71]

Stephen's reverie of Johnson as maxim machine, the heir to Pope's vision of himself as couplet machine, is the uncanny complement and tenacious heir to Courtenay's vision of a moral exemplar. It is to the bodily origins of that reverie of mechanical repetition that I now turn.

## Defining

> Since the eighteenth century the word *tic* has faced the perils of definition many a time, and has as often all but succumbed.
>
> HENRY MEIGE AND E. FEINDEL, *Tics and Their Treatment,* xiii

The word "tic," a locus classicus for the mind/body link,[72] is a French import to English, brought into medical discourse (its first usage in the OED is from an 1824 medical journal) from the lower reaches of "popular expression" in order to resolve "the perils of definition" of bodily motions without proper names. The preface to Henry Meige and E. Feindel's classic and groundbreaking 1902 study, *Tics and Their Treatment,* the first in the wake of Charcot and Gilles de la Tourette to theorize extensively the nature of tics, pays special attention to the "justice" the two have done to the word "tic" itself, remarking:

> If popular expression sometimes confounds where experts distinguish, in revenge it is frequently so apt that it forces itself into the vocabulary of the scientist. In the case under consideration, Greek and Latin are at fault. The meaning of the word tic is so precise that a better adaptation of a name to an idea, or of an idea to a name, is scarcely conceivable, while the fact of its occurrence in so many languages points to a certain specificity in its definition.[73]

From the perspective of this chapter's preoccupation with style and its link to embodiment, the simple word "tic" thus can seem to be a kind of lexicographical cure, a mark of scientific objective precision, in contrast to the deliberate and ornate Latinity of the Johnsonian style of the *Rambler.* These twentieth-century medical efforts to define it properly have their roots in eighteenth-century literary and medical ambiguities.

The opening sentence of *Tics and Their Treatment* is definitive in its understanding of the history of a set of nervous symptoms as primarily a linguistic problem. These doctors rescue the word "tic" from etymological ambiguity and almost inevitable misunderstanding. Such salvation from "the perils of definition," a phrase almost Johnsonian in its near personification of an abstraction, would set proper limits for the word "tic" itself and the body in motion it delineates: "Our work will not be superfluous," their preface concludes, "if we succeed in allotting to the word a definite position in medical terminology, or if any information we have amassed prove of service to future observers." For these authors, tics, unlike the involuntary spasms with which they are often confused, manifest a degenerate disease of the will and can be treated by discipline: "The man who tics has both the debility and impulsiveness of the child. . . . He does not know how to will; he wills too much or too little, too quickly, too restrictedly." If "tic was a psychical disease in a physical guise,"[74] proper diagnosis became a form of seeing through the body, an unmasking of physical gesture that relied upon proper linguistic terminology. Clearly defined words would therefore lead to properly observed, properly diagnosed, and ultimately properly controlled bodies.

In Johnson's own time, his erratic movements, when not diagnosed as wholly involuntary convulsions, spasms, or a "devil's jig," are described in words that echo or submerge the French term: "gesticulation," "antic," or "trick."[75] If we turn to those definitions in Johnson's *Dictionary*, and in physician Robert James's *Medicinal Dictionary* (1743–45), to which Johnson was an important contributor,[76] we can see how over a century before Meige and Feindel's fixing of proper verbal categories, the body in motion posed problems of naming, interpretation, and etymology that collapsed words with what they were intended to represent. Observed in action, Johnson's movements jeopardized the limits between body and mind, and between intention and its lack. In their definition—in their translation into representative or exemplary status—such motions threatened linguistic categories, collapsing definitional or diagnostic objectivity with subjective ambiguities of agency, the linguistic equivalent of Meige and Feindel's "disease of the will."

Johnson's dictionary defines "gesticulate" as "to play antick tricks; to shew postures," and thus as intentional, theatrical (a "shewing" of the body in display), and deceptive. If we turn to the adjective "antick"—etymologically linked to "antient, as things out of use appear old" and thus to social constructions of novelty and fashion, and defined as "odd; ridiculously wild; buffoon in gesticulation"—Johnson's choice of examples from Shakespeare emphasizes the word's sense of masquerade, and with it the undermining of authority inherent in such excessive performance:

> What! dares the slave
> Come hither cover'd with an *antick* face,
> And fleer and scorn at our solemnity?

The example for the noun form of "antick" internalizes such theatrical ambiguity:

> within the hollow crown,
> That rounds the mortal temples of a king,
> Keeps death his court; and there the *antick* sits
> Scoffing his state.

And under the verb form:

> Mine own tongue
> Splits what it speaks; the wild disguise hath almost
> *Antickt* us all.[77]

With a fluidity of agency that makes noun and verb forms difficult to distinguish, whether metaphorically linked to death (a connection that will bear further investigation) and parodying royal authority, or splitting a speaking self deceived by the "wild disguise" of its own words, "antick" in Johnson's illustrations undermines authenticity, authority, and uniform identity.[78] The ambiguities of "trick"—first defined as "sly fraud," moving through "dexterous artifice," "vicious practice," to "a juggle; an antick; any thing done to cheat jocosely, or to divert," and ending with the seemingly innocuous "a practice; a manner; a habit"—further demonstrate the potential moral "perils" underlying what in this definition has become the very stuff of personal distinction and effective representation.[79] As the language of moral disapprobation diminishes, the language of intention narrows (culminating in the exculpatory "to cheat jocosely, or to divert") and finally, with "habit," virtually disappears. Haunting all these Shakespearean uses of "antick" is of course Hamlet's decision "to put an antic disposition on," an expression of intention to perform inauthenticity that renders all intention ambiguous, and that is accompanied, as the opening epigraph to this chapter reminds us, with his injunction to the reader to remain still while he moves, to give nothing away, to appear to know nothing.

If we turn to James's *Medicinal Dictionary* and first examine the definition for Boswell's diagnosis of Johnson, St. Vitus's dance, the similarities are striking:

> *Sydenham* says that St. Vitus's Dance is a kind of Convulsion, which principally attacks Children of both Sexes, from ten to fourteen Years of Age. It first shews itself by a certain Lameness, or rather Unsteadiness of one of the Legs, which the Patient draws after him like an Idiot; and afterwards affects the Hand on the same Side, which, being brought to the Breast, or any other Part, can by no means be held in the same Posture for a Moment, but is distorted, or snatched by a kind of Convulsion, into a different Posture and Place, notwithstanding all possible Efforts to the contrary. If a Glass of Liquor be put into the Hand to drink, before the Patient can get it to his Mouth, he uses a thousand odd Gestures; for, not being able to carry it in a straight Line thereto, because his Hand is drawn different Ways by the Convulsion, as soon as it has reached his Lips, he throws it suddenly into his Mouth, and drinks it very hastily, *as if he only meant to divert the Spectators.*[80] [emphasis mine]

Here "odd" gestures are voluntary efforts to battle involuntary convulsions, which are misread and perhaps determined by "the spectators" as anticks (in Johnson's definition), as comic performance. Even when the body in question is defined as unable to control its gestures, the metaphor of performance, here consciously employed in an effort of description, seems unavoidable.

The term in James's dictionary that bridges the discourses of objective description and theatrical metaphor is "gesticulatio," defined as "a species of gymnastics, consisting in a spontaneous Agitation of the Parts, and throwing the Body into different Postures, much like Actors on the Stage. Gesticulation, says *Oribasius,* is a middle kind of Exercise, between dancing and Mock-fighting, but more like the latter, and is useful for the same Intentions [see UMBRATILIS PUGNA]; but it is more adapted to Children, Women, and aged Persons, and those of Weak and thin Bodies." The association with effeminacy inherent in the St. Vitus's dance definition is clarified here,[81] while the theatrical ambiguity of agency remains in "spontaneous," which could mean either willingly (its literal sense) or without obvious cause.

But theatricality in this case becomes not only metaphor but also cure. The definition of "umbratilis pugna" further elucidates the "use" of such posturing, "a species of Gymnastics, in which the Patient fights with Head and Heels, or boxes and wrestles with a Shadow." Here the parodic element in gesticulation is not linked, as it was in Johnson's dictionary, to the subversive power of illusion and deceit but rather to the therapeutic effects of a self-consciously theatrical performance that advertises its own futility. Fredric Bogel, quoting Roland Barthes's essay on wrestling in explication of Bertrand H. Bronson's famous characterization of Johnson Agonistes, has used just such a metaphor to describe Johnson's version of ironically assumed, performative authority: "[He gives] to his manner of fighting the kind of vehemence and precision found in a great scholastic disputation, in which *what is at stake is at once the triumph of pride and the formal concern with truth.*"[82] The connections Bogel draws in his discussion between Johnson's self-asserting pride and self-effacing concern with truth, between self-evident artless content and pure artificial form, are given new resonance when we consider the gestures, at once theatrical and unintentional, that characterized and divided Johnson in the flesh.

## Diagnosing

> Between a tic and a spasm there is an abyss.
>
> CHARCOT[83]

"What could have induced him to practise such extraordinary gestures who can divine! his head, his hands and his feet often in motion at the same time."[84] Frances Reynolds was not alone in asking the question; Johnson's tics put both body and mind on display, creating an audience of interpreters. Frances Burney was one of many to describe the spectacle Johnson made in company:

> . . . he has naturally a noble figure; tall, stout, grand and authoritative: but he stoops horribly, his back is quite round: his mouth is continually opening and shutting, as if he were chewing something; he has a singular method of twirling his fingers, and twisting his hands: his vast body is in constant agitation, see-sawing backwards and forwards: his feet are never a moment quiet; and his whole great person looked often as if it were going to roll itself, quite voluntarily, from his chair to the floor.[85]

Burney juxtaposed what seems to be an involuntary distortion of a "naturally" noble figure by stooping and "constant agitation" with the vision of a body with a will of its own, leaving the great mind at its mercy.

The case of Johnson transforms a centuries-old imagination of the deformity of genius into a parodic body, a body in motion whose shape and gestures challenge the limits of social intelligibility. Lord Chesterfield, in his oft-quoted description of a "respectable Hottentot"—according to Boswell, "generally understood to be meant for Johnson"[86]—comments less sympathetically:

> There is a man, whose moral character, deep learning, and superior parts, I acknowledge, admire, and respect; but whom it is so impossible for me to love that I am almost in a fever whenever I am in his company. His figure (without being deformed) seems made to disgrace or ridicule the common structure of the human body. His legs and arms are never in the position which, according to the situation of his body, they ought to be in, but constantly employed in committing acts of hostility upon the Graces. He throws any where, but down his throat, whatever he means to drink; and only mangles what he means to carve. Inattentive to all the regards of social life, he mistimes or misplaces every thing. He disputes with heat and indiscriminately, mindless of the rank, character, and situation of those with whom he disputes; absolutely ignorant of the several gradations of familiarity and respect, he is exactly the same to his superiors, his equals, and his inferiors; and therefore, by a necessary consequence, absurd to two of the three. Is it possible to love such a man? No. The utmost I can do for him is, to consider him a respectable Hottentot.[87]

Johnson is a monster, then, at once corporeal *and* social. If his figure seems designed to "disgrace or ridicule the human body," the movements of that body, while not clearly intentional, actively deform social ritual and destroy proper social distinctions. Just as early modern biology imagined the female body as the monstrous imitation of the male, so Johnson's ill breeding epitomizes by its aberration the pattern of the eighteenth-century gentleman. Johnson's tics in Chesterfield's account become "anticks" that parody proper norms of comportment, and his penchant for "disput[ing] with heat and indiscriminately, mindless of the rank" of his opponents ridicules the rules of social rank and deference.

If we consider the relation of Johnson's tics to rhetorical gesticulation, we bring to the fore norms of late eighteenth-century masculine comportment that Johnson himself—despite his apparent flagrant disregard for them—endorsed. Boswell observes in the *Life:*

> He had a great aversion to gesticulating in company. He called once to a gentleman who offended him in that point, *'Don't attitudenise.'* And when another gentleman thought he was giving additional force to what he uttered, by expressive movements of his hands, Johnson fairly seized them, and held them down.[88]

When he speaks against gesticulation, Johnson echoes a distrust of populist rhetoric that ironically resonates, as we shall see, not only with a general ambivalence about classical oratory in the period, but also with contemporary critiques of his ornamental style:

> At Mr. Thrale's, in the evening, he repeated his usual paradoxical declamation against action in publick speaking. 'Action can have no effect upon reasonable minds. It may augment noise, but it never can enforce argument. If you speak to a dog, you use action; you hold up your hand thus, because he is a brute; and in proportion as men are removed from brutes, action will have the less influence upon them.' MRS. THRALE. 'What then, Sir, becomes of Demosthenes's saying? "Action, action, action!"['] JOHNSON. 'Demosthenes, Madam, spoke to an assembly of brutes; to a barbarous people.'[89]

Similarly commenting on the difference in "the state of the Judicial eloquence of England" from that of Greece or Rome, Adam Smith remarks, "The eloquence which is now in greatest esteem is a plain, distinct, and perspicuous Stile without any of the Floridity or other ornamentall parts of the Old Eloquence." Smith continues: "The behaviour which is reckoned polite in England is a calm, composed, unpassionate serenity noways ruffled by passion. Foreigners observe that there is no nation in the world which use so little gesticulation in their conversation as the English." Only the "Rabble . . . express all the various passions by their gesture and behaviour."[90] If corporeal composure and restraint mark proper British masculinity, and if rhetorical gesticulation is a sign of lower-class barbarity, what are we to make of Johnson's own movements, not quite intelligible as gestures, but, like gestures, not fully intentional?

Boswell's account, a passage uneasily sectioned off from the flow of his narrative, enacts a similar series of contradictory impulses toward interpretation, moving from minute and random singularities to the positing of meaning and intention through a metaphor that reinforces Johnson's superiority in conversation:

> That the most minute singularities which belonged to him, and made very observable parts of his appearance and manner, may not be omitted, it is requisite to mention, that while talking or even musing as he sat in his chair, he commonly held his head to one side towards his right shoulder [see fig. 10], and shook it in a tremulous manner, moving his body backwards and forwards, and rubbing his left knee in the same direction, with the palm of his hand. In the intervals of articulating he made various sounds with his mouth, sometimes as if ruminating, or what is called chewing the cud, sometimes giving a half whistle, sometimes making his tongue play backwards from the roof of his mouth, as if clucking like a hen, and sometimes protruding it against his upper gums in front, as if pronouncing quickly under his breath, *too, too, too:* all this accompanied sometimes with a thoughtful look, but more frequently with a smile. Generally when he had concluded a period, in the course of a dispute, by which time he was a good deal exhausted by violence and vociferation, he used to blow out his breath like a Whale. This I suppose was a relief to his lungs; and seemed in him to be a contemptuous mode of expression, as if he had made the arguments of his opponent fly like chaff before the wind.

Boswell ends his description of Johnson's movements with an aesthetic justification:

> I am fully aware how very obvious an occasion I here give for the sneering jocularity of such as have no relish of an exact likeness; which, to render complete, he who draws it must not disdain the slightest strokes. But if witlings should be inclined to attack this account, let them have the candour to quote what I have offered in my defence.[91]

Here the curious gaze is not an end in itself but a means toward effective mimesis, and the opacity of bodily detail in Boswell leads to a transparent understanding of Johnson's monumental greatness.

Johnson himself attributed his repetitive movements to "bad habit" and coupled the self-diagnosis with an admonition to the young woman bold

Figure 10. James Heath, *Samuel Johnson,* engraved after the 1756 portrait by Sir Joshua Reynolds, used as frontispiece to the first edition of Boswell's *Life of Johnson* (1791). Note that Johnson is shown with his head tilted toward the right shoulder, a pose that is visible in the earliest known portrait of Johnson (believed to be painted around 1736) and that is termed in medical discourse a "tonic" or attitudinal tic based on habitual pose rather than motion. Reproduced by permission of the Huntington Library, San Marino, California.

enough to inquire about their cause, "Do you, my dear, take care to guard against bad habits."[92] Meige and Feindel posit the need for such mindfulness by defining tics as thoughts that imperceptibly turn to gesture, the line between them as difficult to draw as it is to define the domain of "habit" in mind or body: "As one thinks, so does one tic."[93] The paradoxes with which doctors and patients alike in the past two centuries have described nervous tics are structured like a series of couplets: physical enshrinements of psychic detail, grotesque caricatures of everyday acts, voluntary capitulations to an almost irresistible force, immediate habits.[94] It is significant too, in this regard, that medical scholarship on Tourette's syndrome, unable to resolve the question of whether or not tics were voluntary, began with the Rousseauian "Confessions of a Victim to Tic" and has come full circle to reconsider patients' first-person accounts of their experiences.[95] This disease, with which Johnson has been recently and frequently diagnosed, through its challenging of clear limits between body and mind, and through its affinities with mimesis (tics often begin with an irresistible desire to mimic someone else's strange gestures), displays an abyss between intention and its lack.[96] Such an abyss at once posits and undermines intention, origin, and order.

The embattled history of scholarship on Tourette's reinforces the power of Johnson's example to compromise our ability to distinguish between tics, symptoms, habits, and compulsions. What difference is there between a physical and psychological tic, between a bodily compulsion and a mental obsession? Johnson's public evincing of psychological turmoil by praying in public in silent mutters and audible whispers; his recitation at moments of conversational absence of poetic contemplations of death: "Ay, but to die and go we know not where"; his perpetual resolves to overcome his compulsion to resolve, that is, his "scruples"—all make an answer impossible.[97] How does a tic differ from a symptom? Here what his contemporaries referred to as convulsions, gesticulations, the devil's jig, his head shakings, and seesawings, which made him roll rather than walk, come to mind. But such purposelessness can have its purpose: seesawing, like a bodily mimicry of the couplet, appeared to aid him, we'll recall, in the memorization of poetry.[98] How does a nervous tic differ from a habit? We might think here of Johnson's need to touch every post he passed on the road, or his elaborate whirlings and twistings at doorways, his "gigantic straddles" on the street, his rapt attention to the precise positioning of his feet in polite drawing rooms.[99] Johnson's "air of great satisfaction"[100] at the successful completion of such rituals might also lead us to ask, along with James's dictionary, whether a tic is a performance, and if so whether that performance is voluntary or involuntary and for an audience, real or imagined? If, as one early medical authority states, "the victims of tic are thankful for solitude as the only way of escaping observation," what are we to make of Johnson's need for that audience, his lifelong fear of solitude, his perpetual desire to "escape from himself,"[101] or his authorial persona's magisterial command of impersonal observation?

In the accounts of contemporaries, Johnson's tics at once warded off and figured anxieties about repetition, representation, mastery, and meaning that preoccupied and shaped his habits of mind. In a more general historical sense, such questions of agency and intention preoccupied both eighteenth-century authors negotiating their role in a newly professionalized and commodified print culture, and eighteenth-century doctors imagining a nervous system that would prove the body to be more than a machine. If Robert Whytt's "sentient principle" enlivened earlier mechanistic models of the body with divine intention, how to make sense of Johnson's seemingly purposeless tics?[102] Michel Foucault invokes a vision of corporeal intention gone chaotically indecipherable when he describes Whytt's pathologization of the mysterious female body, a body at once physically and morally weaker than the male, a body in excessive sympathy with itself.[103] When we consider the gendered equivalence in the taxonomy of nervous disease of male hypochondria and female hysteria, Johnson's case in particular, and that of the man of letters in general, takes on added mystery. How might the public figure of the moral arbiter Johnson be aligned in contemporary imaginations with such a pathologically private body? And how might such a body, debilitated by its nervous secrets, be made heroically representative of a literary culture? Indeed, "nervous" in late eighteenth-century usage has ambiguously gendered meanings, conveying both physical effeminacy and, when applied to a literary style such as Johnson's, sinewy, muscular masculinity. One answer to such a question relies on contemporary accounts of Johnson's conversation, one of the most salient testing grounds for what we might call the embodiment of style.

If a diagnosis of Tourette's allows us to see tics as purposeful and purposeless, habitual and symptomatic, mental and physical, it also brings us surprisingly close to Joshua Reynolds's penetrating view that such gestures were "not involuntary" in Johnson's case, but rather were the product of solitary reverie turned to spectacle in company. Johnson's tics, Reynolds argues, "proceeded from a habit which he had indulged himself in, of accompanying his thoughts with certain untoward actions, and those actions always appeared to me as if they were meant to reprobate some part of his past conduct." These actions occurred in moments of absence from conversation and were cured by a return to it.[104] Reynolds was one of many to puzzle over the contrasts between Johnson in person and in public, Johnson in private and Johnson in his published work, and in so doing to remark the striking opposition of his passionate need to talk for victory in public and his earnest submission to the rules of virtue and truth in intimate tête-à-tête and, especially, in print. "In mixed company," Reynolds observes, "he fought on every occasion as if his whole reputation depended upon the victory of the minute, and he fought with all the weapons. If he was foiled in argument he had recourse to abuse and rudeness."[105] The tension that Boswell feels compelled to excuse in his closing commemoration of Johnson would be for Reynolds the central point of any biography he might produce: "What appears extraordinary," writes Reynolds of the author of *The Vanity of Human Wishes,* "is that a man who so well saw, himself, the folly of this ambition of shining, of speaking, or of acting always according to the character [he] imagined [he] possessed in the world, should produce himself the greatest example of a contrary conduct."[106]

In person, Johnson neglects to watch himself by watching an idea of himself too fixedly: by acting "always according to [his] character," he becomes a spectacle of an uncontrollable rage for authority. While in print his authority is based on a skeptical questioning of personal agency, his conversational style, like mock-epic, transforms the "loose sparkles of thoughtless wit" into a performance of aggressive self-assertion.[107] In print such a style bases its originality upon the iterability of the Latinate paraphrase, asserts what it undermines in controlled repetition, wards off chaos by degrees, and turns toward the agency of others for resolution—in the form of imagined sympathetic and judgmental readers at the end of the *Life of Savage,* or public opinion in the ambivalent acceptance of Nahum Tate's happy ending to *King Lear,* or celestial wisdom at the close of *The Vanity of Human Wishes,* a comfort that *Rasselas* abandons in a "conclusion, in which nothing is concluded." In person, by contrast, print's solitary struggle against singularity in the eye of posterity becomes an agonistic drama for a polite company, an enactment that, like the tics that accompanied it, is repetition with a difference, not authority but its antick staging.

Johnson's conversation, as Reynolds describes it, transforms ambitious intention into something automatic: the "continual practice" of shining in conversation "made that a habit which was at first an exertion." In this account, the "habit" of "speaking his best"[108] cures the unconscious repetition of the nervous tic with the conscious repetition, what Boswell and Reynolds call the labor, of the Johnsonian style.[109] We can see this process in action when Boswell observes Johnson, who "seemed to take a pleasure in speaking his own style," rephrasing himself into Johnsonese: "It has not wit enough to keep it sweet" thus becomes "it has not vitality enough to preserve it from putrefaction."[110] Or consider Johnson's unintentional pun on "the woman had a bottom of good sense," revised with prideful resolve at the mirth of the company, as Boswell describes it:

> To assume and exercise despotick power, [he] glanced sternly around, and called out in a strong tone, 'Where's the merriment?' Then collecting himself, and looking aweful, to make us feel how he could impose restraint, and as it were searching his mind for a still more ludicrous word, he slowly pronounced, 'I say the *woman* was *fundamentally* sensible;' as if he had said, hear this now, and laugh if you dare. We all sat composed as at a funeral.[111]

Here the sheer force of authority on display in the translation into Johnsonese of a pun that remains a pun silences the group's amusement at the master's inability to control his language. Such authority asserts itself through self-conscious rehearsal and makes itself strange through repetition.[112] Conversation thus brings a melancholy mind otherwise painfully preoccupied with itself, and from which "the great business of his life . . . was to escape," into safety by bringing it into view.[113] Conversation is convulsion transformed by the "cure" of company into the performance of distinction:

> For every person who knew him must have observed that the moment he was left out of the conversation, whether from his deafness or from whatever cause, but a few minutes without speaking or listening, his mind appeared to be preparing itself. He fell into a reverie accompanied with strange antic gestures; but this he never did when his mind was engaged by the conversation. [These were] therefore improperly called . . . convulsions, which imply involuntary contortions; whereas, a word addressed to him, his attention was recovered. Sometimes, indeed, it would be near a minute before he would give an answer, looking as if he laboured to bring his mind to bear on the question.[114]

In Reynolds's account, tics, such as Johnson's compulsive prayer and recitation, signal the "reverie" of self-absorption. But while such actions can signify an indecipherable inwardness that demands the audience's diagnosis, at a word from the company, they are made legible as "preparation" for reentering the conversational fray.[115] The performance of style in the mock-heroic battle of conversation, in which Johnson holds forth in perfection fit for print according to his character,[116] thus becomes a kind of salutary habit, a meaningful repetition that orders the potential incoherence of self-enclosure and alleviates both the fear—and the lack—of an end.

By excluding Johnson's nervous tics from a consideration of his conversation, we ally ourselves with Hogarth, who in the midst of a talk with Samuel Richardson

> perceived a person standing at a window in the room, shaking his head, and rolling himself about in a strange ridiculous manner. He concluded that he was an ideot, whom his relations had put under the care of Mr. Richardson, as a very good man. To his great surprize, however, this figure stalked forwards to where he and Mr. Richardson were sitting, and all at once took up the argument, . . . he displayed such a power of eloquence, that Hogarth looked at him with astonishment, and actually imagined that this ideot had been at the moment inspired.[117]

Johnson's verbal mastery here becomes the subject of "display" to the artist's eye, a shockingly visible figure against the ground of his bodily incongruity.

Yet Hogarth's vision has something in common, I uneasily admit, with my efforts to think body and mind together in Johnson's case. In the context of Boswell's *Life,* this vision of the great man is profoundly disorienting, since here the objectifying eye scrutinizes not the physical antics of a known genius but rather Johnson's defining and usually disembodied trait, his eloquence. To label Johnson an "inspired ideot" is to make his speech, rather than his tics, the product of another. (Think, too, of Leslie Stephen's hallucinatory vision of Johnson's "imbecilic face" as he attempts to describe his prose.) My point in thus shifting our perspective is not to argue for the body's agency, its essentialized determination of literary form; rather, such a reframing, like the uneasy balances that the couplet form and Johnson's concretized abstractions in both poetry and prose enact, reveals the illusory nature of any attempt to define and identify agency. If the aberrant body and disembodied voice are read as inextricably related, the fragility and theatricality of Johnson's authority is left visibly in the balance. Hogarth's vision of a Johnson made momentarily strange to us foregrounds how, in diagnosing Johnson with a difference, mind and body are always inseparable and the power of art always asserts itself against a background of spectacle and show.

Hamlet's words, with which this chapter began, also remind us of another uneasy balance—the unspoken complicity between Johnson's antic body and the readers and viewers who attempt to determine that body's meaning. Here we might consider an anecdote in which the question of agency is blurred by audience participation. Commenting on Johnson's reluctance to discuss his "movement disorder," T. J. Murray observes:

> Perhaps the only exception might be his response to a lady who jokingly put her foot in the line of Johnson's hand, which was moving back and forth as he sat at

> the dinner table. Her shoe was knocked off and to the tittering company who recognized the joke, he responded, "I know not that I have justly incurred your rebuke. The motion was involuntary, and the action not intentionally rude."[118]

This anecdote reverses the earlier "bottom of good sense" moment, in which the laughter of the company at Johnson's expense is quelled by his intentional repetition of the phrase in habitual Johnsonian style. Here Johnson's public admittance of lack of intention shames the company who laugh at his expense. In both cases, the laugh is on Johnson's audience, on their need to know and to best him. This moment, Johnson knocking off the lady's shoe, has been restaged as emblematic of Johnson's own desire—an involuntary action crossing a fine line into what we would call a compulsion—as he becomes something of a foot fetishist in both Julian Barnes's *England, England* (as the actor playing Johnson drops to his knees and, "with a heavy yet bearishly precise flick," removes the shoe of his female employer in mid-reprimand) and Beryl Bainbridge's *According to Queeney* (in which Hester Thrale's daughter spies with disapproval on Johnson and her mother at breakfast as he removes her slipper).[119] The desire revealed in the original anecdote, however, the curious intent on display, is the lady's, and it is also our own.

CHAPTER THREE

# "LOOK, MY LORD, IT COMES": UNCRITICAL READING AND JOHNSONIAN COMMUNION

Let me not burst in ignorance; but tell,
Why thy canoniz'd bones, hearsed in death,
Have burst their cerements? Why the sepulchre,
Wherein we saw thee quietly in-urn'd,
Hath op'd his ponderous and marble jaws,
To cast thee up again? What may this mean,
That thou, dead corse, again in complete steel,
Revisit'st thus the glimpses of the moon,
Making night hideous, and we fools of nature
So horridly to shake our disposition
With thoughts beyond the reaches of our souls?
Say, why is this? Wherefore? What should we do?

*Hamlet*, I.iv.46–57

[Johnson on William Warburton's emendation of the above:]
The critick, in his zeal for change, writes with so little consideration, as to say, that Hamlet cannot call his father "canonized," because "we are told he was murdered with all his sins fresh upon him." He was not then told it, and had so little the power of knowing it, that he was to be told it by an apparition. The long succession of reasons upon reasons prove nothing, but what every reader discovers, that the King had been buried, which is implied by so many adjuncts of burial, that the direct mention of "earth" is not necessary. Hamlet, amazed at an apparition, which, though in all ages credited, has in all ages been considered as the most wonderful and most dreadful operation of supernatural agency, enquires of the spectre, in the most emphatick terms, why he breaks the order of nature, by returning from the dead; this he asks in a very confused circumlocution, confounding in his fright the soul and body.

SAMUEL JOHNSON, editorial notes to *Hamlet*[1]

> The amiable Dr. Adams suggested that GOD was infinitely good. JOHNSON. 'That he is infinitely good, as far as the perfection of his nature will allow, I certainly believe; but it is necessary for good upon the whole, that individuals should be punished. As to an *individual,* therefore, he is not infinitely good; and as I cannot be *sure* that I have fulfilled the conditions on which salvation is granted, I am afraid I may be one of those who shall be damned.' (looking dismally.) DR. ADAMS. 'What do you mean by damned?' JOHNSON. (passionately and loudly) 'Sent to Hell, Sir, and punished everlastingly.' DR. ADAMS. 'I don't believe that doctrine.' JOHNSON. 'Hold, Sir; do you believe that some will be punished at all?' DR. ADAMS. 'Being excluded from Heaven will be a punishment; yet there may be no great positive suffering.' JOHNSON. 'Well, Sir; but, if you admit any degree of punishment, there is an end of your argument for infinite goodness simply considered; for, infinite goodness would inflict no punishment whatever. There is not infinite goodness physically considered; morally there is.' BOSWELL. 'But may not a man attain to such a degree of hope as not to be uneasy from the fear of death?' JOHNSON. 'A man may have such a degree of hope as to keep him quiet. You see I am not quiet, from the vehemence with which I talk; but I do not despair.' MRS. ADAMS. 'You seem, Sir, to forget the merits of our Redeemer.' JOHNSON. 'Madam, I do not forget the merits of my Redeemer; but my Redeemer has said that he will set some on his right hand and some on his left.' He was in gloomy agitation, and said, 'I'll have no more on't.'
>
> JAMES BOSWELL, *The Life of Samuel Johnson, LL.D.*, 4:299–300

My second epigraph shows Johnson the critic in confrontation with a profoundly uncritical, indeed superstitious vision of the ghost of Hamlet's father. More accurately, he is confronting that vision through Hamlet's terrified queries on the printed page, queries that are filtered through the notes of other editors (in the case of this note, most particularly William Warburton) who attempt to order and elucidate them. Johnson finds Warburton's note (which had recommended that "hearsed in death" be changed to "hearsed in earth" "on the grounds that 'hearsed' meant 'reposited' and that 'death' is not a 'place' but a privation") rather than Shakespeare's words, "reprehensible," because of its failure of faith in both the power of the ghost and in common sense. Hamlet's confused words, Johnson argues, make perfect sense of a situation that transports him out of his senses: "Had the change of the word," he argues, "removed any obscurity, or added any beauty, it might have been worth a struggle, but either reading leaves the sense the same."[2] That sense Johnson paraphrases as follows: "Why dost thou appear, whom we know to be dead?"

This is the question this book asks of Samuel Johnson, and the answer has everything to do with the appearance of his ghost at the crossroads of critical and uncritical reading, and with the relation of that excessively embodied ghost to the disembodied regularity of print. As in the case of the ghost of old Hamlet, Johnson's endurance as a ghost hints at the possibility of unknown sins that keep him from resting in peace. We might consider here the debate between Johnson and Warburton over the adjective "canonized." Warburton had defended his emendation by pointing out "that 'canoniz'd bones' meant 'bones to which the rites of sepulture have been performed,' rather than 'bones made holy or sainted,' for the king had been murdered in a state of sin and therefore could not be sainted." Johnson observes that Warburton's "peevishness" in this note might result from his umbrage at "the incivility shown" to another editor, "who is represented as supposing the ground 'canonis'd' by a funeral, when he only meant to say, that the 'body' was deposited in 'holy ground,' in ground consecrated according to the 'canon.'"[3] This slippage—between holy bones and sacred ground, between death as unfathomable privation and death as material prison, between, as Johnson puts it, "the soul and body"—characterizes the secular ghost of Samuel Johnson as well, along with the literary canon, a repetition with a difference of the Anglican canon, which he stands for.

If we ask Johnson's ghost the questions Hamlet asks the ghost of his father, we must confront again the mysteries of his self-wounding and ensuing death. Was the act that precipitated his death intentional? Was it in fact a suicide? Was the fear of death that motivated his final gesture, while perhaps eradicating suicide as a possibility, itself unchristian? Is Johnson's ghostly persistence some sort of punishment? We might remember here that attitudes toward suicide were deeply ambivalent during the eighteenth century, torn between sympathy and even idealization of a certain kind of Roman stoicism best personified by the suicide of Addison's tragic hero Cato (a play that underlay the period's debate about poetic justice and aesthetic pleasure) and a legacy of Christian condemnation (itself shadowed by pagan rituals such as the burying of suicides at the crossroads) of the ultimate act of prideful will.[4] The "psychological equivalent of apostasy," suicide was the ultimate triumph of personal despair over Christian faith. Self-murderers, one late seventeenth-century author declared, "cast the Talent of Life in the Face of their Creator."[5] For an author who was haunted by the New Testament parable of the talents and its promise of punishment for those who waste what is given them, and who envied the humble Dr. Levet's "single talent well employed" (28), suicide would seem anathema. But suicidal despair at the thought of damnation was also a necessary, perilous, and potentially final step on the narrative path toward Christian repentance that Johnson himself knew well. We will return to this ambiguity in the next chapter. For now I want only to observe that Johnson's ghost wanders precisely because his final confrontation with death and its aftermath has never been successfully reduced to a single narrative; rather it rises from the unresolved temporality of an anecdote that will not fit into a happy Christian ending.

Like Johnson's ghost, like Hamlet's father—revenant of a purgatory erased from official doctrine but lingering in popular belief—the suicide is doomed "to wander between the world of the dead and that of the living"; "it was the very ambiguity of the state of eternal transition that made it the most terrifying."[6] It is that ambiguity—articulated by Boswell's Johnson in my third epigraph as the gap between God's infinite moral and finite physical goodness—that torments Johnson, causing him to disrupt "passionately and loudly" polite conversation exposed as Christian complacence. Boswell attempts to ward off propensities to repeat Johnson's arguments on the part of "the enemies of Christianity, as if its influence on the mind were not benignant," by arguing that "Johnson's temperament was melancholy, of which such direful apprehensions of futurity are often a common effect." His final refutation is in his own record of Johnson's death: "We shall presently see that when he approached nearer to his aweful change, his mind became tranquil, and he exhibited as much fortitude as becomes a thinking man in that situation."[7] Yet this record is much contested, and the "fortitude" Johnson exhibited in his final days might have been closer to that of a stoic Cato controlling death in his determination to live than a resigned Christian submitting to his final sentence. Ultimately the unresolved ambiguity of Johnson's excessive fear of death and its aftermath—a fear that baffled and unsettled the polite Anglican culture of his day, and that caused Anna Seward to remark that he "worshipped God as Indians worship the devil,"[8] a fear that is the driving narrative force of Boswell's *Life*—makes Johnson a ghostly familiar masking a terrifying prospect of ending and endlessness.

## The Image of Genius and the Compulsion to Repeat

This ghost takes the form of the monstrous yet lovable spectacle we encountered in chapter 2. Let's look again, from a different perspective, at Thomas Macaulay's vision of Johnson, which we should note at this point is actually a vision enabled by reading J. W. Croker's edition of Boswell's *Life of Johnson:*

> As we close [Boswell's book], the clubroom is before us, and the table on which stands the omelet for Nugent, and the lemons for Johnson. There are assembled those heads which live for ever on the canvas of Reynolds. There are the spectacles of Burke and the tall thin form of Langton, the courtly sneer of Beauclerk and the

> beaming smile of Garrick, Gibbon tapping his snuff-box and Sir Joshua with his trumpet in his ear. In the foreground is that strange figure which is as familiar to us as the figures of those among whom we have been brought up, the gigantic body, the huge massy face, seamed with the scars of disease, the brown coat, the black worsted stockings, the grey wig with the scorched foretop, the dirty hands, the nails bitten and pared to the quick. We see the eyes and mouth moving with convulsive twitches; we see the heavy form rolling; we hear it puffing; and then comes the "Why, sir!" and the "What then, sir?" and the "No, sir"; and the "You don't see your way through the question, sir!"[9]

This is a still life reanimated as caricature. The metonymic details—omelets, lemons, spectacles, sneers, smiles—move from the inhuman to the human, as the bodies of the members of Johnson's famous club come into focus and to life. Dominating the perspective is the Great Man himself, a monumental form marked by disease, adorned with condensed anecdote (the scorched wig borrowed from Hester Thrale's account of the disastrous results of Johnson's reading in bed by candlelight) and propelled, in an uncanny blurring of animate and inanimate, by a series of compulsions. Twitches, rollings, puffings, and habitual sayings turn his uncontrolled body and conversational mannerisms into the sort of automaton Freud would later call uncanny. But such automation coexists with a compelling subjectivity. A brief look at a portrait by Joshua Reynolds will give us the visual equivalent of the contradictions that Johnson's image embodies (see fig. 11).

Reynolds's apprentice and biographer James Northcote says of Reynolds's painting: "Sir Joshua had given to Dr. Johnson a copy of that portrait . . . in which the Doctor is represented with his hands held up, and in his own short hair; it is nearly a profile."[10] Reynolds also portrayed Johnson's squint in a later portrait (see fig. 12), which drew from his subject the famous response: "He would not be known by posterity for his *defects* only, let Sir Joshua do his worst. . . . I will not be *blinking Sam*."[11] But this earlier depiction, on view in his hometown of Lichfield in 1770, Johnson "much admired." Indeed, the image's classical allusions remove the squint from the iconography of personal defect into the semiotics of individual depth. What are we to make of Johnson's gesture? Draped in a toga-like robe, head bare, is he posed in an expressive, stylized movement of classical oratory, in accordance with the portrait's nineteenth-century title, *Johnson Arguing*? (None of the early modern rhetorical handbooks I've looked through provide an exact equivalent.) When we learn that Reynolds modeled his portrayal of Johnson on that of Socrates in Raphael's *School of Athens*, and that he later translated it into the image of Tiresias in *The Birth of Hercules*, how might these evocative earlier and later

Figure 11. James Watson, *Samuel Johnson* (1770), mezzotint after the 1769 portrait by Sir Joshua Reynolds. Reproduced by permission of the Huntington Library, San Marino, California.

versions haunt our interpretation of this figure's individuality and of the artist's "originality"? How might our perspective change when we realize that this portrait accurately depicts one of Johnson's strange compulsive "motions or tricks," which Sir Joshua himself believed often accompanied guilty thoughts,[12] and which his sister, Frances Reynolds, described as follows:

Figure 12. John Hall, *Samuel Johnson,* after the portrait by Sir Joshua Reynolds, circa 1775; frontispiece to first collected edition of Samuel Johnson's *Works* (London, 1787–88). Reproduced by permission of the Huntington Library, San Marino, California.

> As for his gestures with his hands, they were equally as strange; sometimes he would hold them up with some of his fingers bent, as if he had been seized with the cramp, and sometimes at his Breast in motion like those of a jockey on full speed.[13]

For twentieth-century doctor, medical historian, and Johnsonian Lawrence McHenry, in a 1967 medical article, Reynolds's image provides proof for a successful diagnosis of Johnson with Tourette's syndrome.[14] For art historian Duncan Robinson in 1992, by contrast, the portrait's "stark realism can only be explained in terms of the profound respect both artist and sitter had for mind over matter." For Robinson, this "extraordinary image" depicts "the eighteenth-century's greatest master of the English language poised on the boundaries of definition."[15] As compulsive repetition is frozen into a pose at once neoclassical and modern, stylized and verisimilar, Reynolds's portrait itself converges "on the boundaries of definition," boundaries that uncannily compromise the authorial mastery they portray.

We might think of this portrait, in its erring vacillation between contradictory modes of realism and idealism, suspended as well between two different classical interpretations (Johnson as Socrates, Johnson as Tiresias), as the visual equivalent of Johnsonian anecdote. By capturing the body in mid-gesture, the image renders that body meaningful, subjecting it to unending interpretation. Or perhaps we could imagine this image as emblematic of Johnsonian efforts to vivify, while preserving, a dead hero. The portrait is, in a sense, a lifeless body put into and removed from motion. Frozen in this image, Johnson's squint and gesture become permanent indexes of his internal depth; animated they mark him with physical excess.[16]

Whether enlivened by bodily tics turned conversational signature or mechanized and silenced (as in Leslie Stephen's vision from the last chapter) by the habitual production of ornate prose, Johnson's familiar ghost haunts the national collective it constructs with a perpetual repetition that affirms and threatens individual identification and original agency. The same "blind habit" that renders Macaulay's Johnson familiar renders Stephen's version of a twitching, mechanical Johnson alien. Both Johnsonian visions are summoned by an experience of reading, and in both cases the ghost of Johnson merges living speech with dead text.[17]

## "Look, my Lord, it comes": Prosopopoeia as Haunting

Both visions exemplify that most uncanny of tropes, prosopopoeia, the trope by which, as Paul de Man puts it in his classic essay "Autobiography as De-

Facement," "one's name . . . is made as intelligible and memorable as a face." We do not need de Man's paradoxical claim that while "all texts are autobiographical, we should say that, by the same token, none of them is or can be" in order to discern the autobiographical nature of the Johnsonian imagination that creates a communal self in the author's image. Indeed, Johnson arises from the pages of his biographies or from his own sentences in what feels like a concrete instance of de Man's generic description of prosopopoeia's ghostly operation: "Voice assumes mouth, eye, and finally . . . face, a chain that is manifest in the name, *prosopon poein,* to confer a mask or a face (*prosopon*)."[18]

Prosopopoeia, in de Man's account, both veils and reveals language's murderous impersonality; it animates by disfiguring, simultaneously mirroring and threatening its viewer in its conflation of text and vision. "By making the death speak," de Man argues in a curiously abstract formulation, "the symmetrical structure of the trope implies, by the same token, that the living are struck dumb, frozen in their own death."[19]

In the pages of his early biographies, Johnson is portrayed similarly and strikingly as a monument come to life and disfigured by the process. In the opening pages of *Journal of a Tour to the Hebrides,* published two years after the hero's death, Boswell describes Johnson thus: "His countenance was naturally of the cast of an ancient statue, but somewhat disfigured by the scars of that *evil,* which, it was formerly imagined, the *royal touch* could cure."[20] Hester Thrale, contemplating Johnson's death at the close of her 1788 edition of his *Letters,* asks, "And now . . . what remains?" Johnson's enduring soul, she concludes, is "like some fine statue, the boast of Greece and Rome, plastered up into deformity, while casts are preparing from it to improve students, and diffuse the knowledge of its merit; but dazzling only with complete perfection, when the gross and awkward covering is removed."[21] With the loss of Johnson's embodied originality—as figured and compromised by the complex image of the statue's idealized "ancient" form, plastered over and defaced in order to be reproduced—comes compensation for his death in the endurance of his soul, that of literature itself.[22]

The controversy over whether or not to monumentalize Johnson with a full-length statue resonates with Thrale's metaphor, recasting it into contradictory accounts of Johnson's corporeality. "The objection . . . to a whole length statue is very silly," Edmund Malone wrote in 1789, "for Sir J[oshua] R[eynolds] says that Johnson's limbs were so far from being unsightly that they were uncommonly well formed, & in the most exact & true proportion."[23] John Bacon himself defended his muscular and melancholic 1796 sculpture (see fig. 13) as an embodiment not of the man but of his prose:

Figure 13. Engraving of John Bacon's monument to Samuel Johnson in St. Paul's Cathedral, by James Basire for the *European Magazine* (April 1, 1796). Reproduced by permission of the Huntington Library, San Marino, California.

> I have . . . aimed that a magnitude of parts, and grandeur of style, in the statue, should accord with the masculine sense with which his writings are so strongly impregnated, and the nervous style in which it is conveyed to mankind. . . . By making him lean against a column, I suggest his own firmness of mind, as well as the stability of his maxims.[24]

Stilled as statue or corpse, Johnson becomes a monument, in Kevin Hart's phrase, the "rallying point for a community . . . the centre of a network of imaginary relationships and real desires."[25] Conversely, readers of the *Life* encounter a Johnson reanimated by communal memory and repeated ritual. "They will find," one reader enthused to Boswell, "their great Friend and Master as it were embalmed in your Narrative; and may daily live over those scenes which are long since past."[26] If we return to de Man's work on prosopopoeia, we can more easily perceive the necessary incompleteness of such an identificatory process, the endless repetition demanded by a ritual that is haunted by the mortality it works to revive.

"The death" animated by prosopopoeia in de Man's account is Shakespeare's, and the author of the trope of paralysis as monumentalization evoked in his description of prosopopoeia's "latent threat" is Milton, whose sonnet in memory of Shakespeare is quoted briefly by Wordsworth in his *Essays upon Epitaphs* (1810), the text with which de Man is primarily concerned. For de Man, Wordsworth's omission of the six lines from Milton that contain this crucial couplet on the power of Shakespeare's text—"Then thou our fancy of itself bereaving / Dost make us marble with too much conceiving"—is indicative of "the threat of a deeper logical disturbance" in the argument. Wordsworth's inconsistency on the subject of prosopopoeia, de Man argues, arises from his confrontation with the paradoxical nature of the "linguistic predicament" that constitutes both death itself and autobiography's attempts to veil "a defacement of the mind of which it is itself the cause."[27]

Shadowing de Man's abstraction of death is the particular power of the death of an author. That prosopopoeia's threat of reciprocity should arise when Wordsworth remembers the epitaph of one of his greatest canonical predecessors for his other greatest canonical predecessor is something over which de Man barely pauses. At a key moment in his first essay, Wordsworth's quotation of Milton—"what need'st thou such weak witness of thy *name*" (italics de Man's)—renders the generic epitaph legible. "In the case of poets such as Shakespeare, Milton, or Wordsworth himself," de Man writes, "the epitaph can consist only of what [Wordsworth] calls 'the naked name.'"[28] Only at this point in Wordsworth's essay, de Man observes, when the name on the tombstone is identified as that of a great poet, does the stone begin to speak.

And only at this point in *de Man's* essay is prosopopoeia defined. Prosopopoeia, then, or so de Man's extrapolation from Wordsworth seems to imply, is a particularly literary affair, pertaining not to death in the abstract but to the individual (yet representative) deaths and undying personified names of authors.

In her reading of *Hamlet* in *Shakespeare's Ghostwriters,* Marjorie Garber turns the uncanniness of de Manian prosopopoeia more explicitly toward questions of authorship that animate my own thinking on the cult of Johnson. Linking psychoanalytic constructions of repetition with deconstruction, Garber meditates on Shakespeare himself (who was known for his portrayal of the ghost in *Hamlet*) as the ultimate ghost, the ultimate absent presence, of the figure of the father/author, of history, and of writing itself.[29] Writing the *Life* in the shadow of Johnson's death, Boswell, as he describes his first meeting with his hero, summons him, as we have already noted, in the form of the ghost of Hamlet's father. Here I quote the meeting in full:

> When I was sitting in Mr. Davies's back-parlour, after having drunk tea with him and Mrs. Davies, Johnson unexpectedly came into the shop; and Mr. Davies having perceived him through the glass-door in the room in which we were sitting, advancing towards us,—he announced his aweful approach to me, somewhat in the manner of an actor in the part of Horatio, when he addresses Hamlet on the appearance of his father's ghost, 'Look, my Lord, it comes.'[30]

Equally uncanny is the following sentence in which Johnson appears as the image of the *Life*'s frontispiece: "I found that I had a very perfect idea of Johnson's figure, from the portrait of him painted by Sir Joshua Reynolds soon after he had published his Dictionary . . . which Sir Joshua very kindly presented to me, and from which an engraving has been made for this work."[31] (See fig. 10 for the frontispiece itself.) In the text's portrayal of the moment of its conception, Boswell's *Life* brings its hero to life as personification of print and messenger from the grave. This meeting—to which I return throughout the book—crystallizes the Johnsonian desire, in its strange confluence of authorial reticence and readerly eagerness, to transcend mortality while obeying a paradoxical injunction to remember death. In the case of Johnson, history is both evoked and evaded by the figure of the author's body. Not the elusively disembodied Shakespeare praised by the likes of Anna Seward and Edgar Allan Poe for the protean agility, negative capability, and ghostly mystery that Johnson lacked, this ghost haunts us particularly, in the flesh.

## Envisioning Johnson: Tradition and Aberration

> Even as Johnson, huge, ungainly, and infirm, has been immortalized and glorified by the brush of Reynolds, so his wisdom and his wit, his roughness and his tenderness, have been depicted for us by Boswell.
>
> With these masterpieces of bodily and mental portraiture before us, we may often say, "No more! the picture is perfect, the biography complete, we care for no inferior touch!" But there are other moods in which we feel that different aspects of both body and mind might have been shown. We have his portrait in repose, thoughtful, almost sublime, but we sometimes feel, "Would that an artist's eye had seen him at Uttoxeter doing penance in passionate repentance! Would that some one had noted the tender pathos of the farewell look on his dying servant, Catherine Chambers; or the glee with which, when almost penniless himself, he hid pennies in the hands of sleeping children in the London streets, lest they should awake breakfastless!"[32]

This passage from Robina Napier's preface to her edition of *Johnsoniana* (1884) epitomizes the visual, anecdotal nature of Johnsonian desire that we also encountered in the fetishistic disavowal of the autopsy. In her justification for compiling the first collection of "minor" material on Samuel Johnson that left anecdotal matter intact in original texts, rather than fragmented under topic headings, Napier's Johnson becomes a unified whole who is always in need of supplementation. He endures in the imaginations of posterity, thanks to painter Joshua Reynolds and biographer James Boswell, in dual images of ungainly body and exceptional mind joined in a character whose uniqueness encompasses contradictory extremes. While it is my intent to question Napier's firm separation of the provinces of body and mind, a separation belied by her use of the visual term "portraiture" to describe both image and text, what is most important to me here is her desire—imagined to be common to all readers and viewers of her hero—to see more. That "more" consists in an imagined intimacy with Johnson revivified from "repose" to a state that Napier imagines as his most alive. Such a desire summons Johnson in his most private, most sentimental, and in this case his most Victorian moments, from the anecdote on the page into momentary presence. Johnsonian desire consists here in a progressive escape from the bounds of history, printed text, and even the most limited of narrative confines, into the contemplation of a revivified familiar turned friend.[33]

This version of Johnsonian desire is the product of an encounter between an author figure so familiar to subsequent generations that he becomes an icon that transcends text, and a solitary reader whose access to such knowledge takes the form of a need. Craving ever more intimate knowledge of the completely known, Napier fantasizes herself representative of a collective gaze through an "artist's eye" at what has remained invisible—or, rather, what has been accessible only through print. Here anecdotal vivification of Johnson, so effective in Boswell's text, serves as a prompt toward further visualization, toward the desire for a portrait. But whether we turn to its beginnings in the play between Johnson and his biographers, most notably Boswell, of the author's game of self-revelation and concealment for posterity, or to mid-twentieth-century psychoanalytic speculation about Johnson's sexual behavior and potential madness, or to the recent debate in Johnsonian critical circles over Johnson as Jacobite and politically representative of an age unjustly Whiggified by academic liberals, the structure of such desire remains the same. A single author, "huge, ungainly, and infirm," in all his particularity of body and mind, has been rendered a vehicle for a solitary curiosity that is the stuff of collective representation. As Napier's "we" indicates, Johnson's ever-elusive interiority has become English public and psychic property through its capacity to be envisioned.

Whatever else we might say Samuel Johnson looked like, we can be sure that he presented to his own age a recognizable image of a mind and body in potentially monstrous conflict. Around 1772 Sir Joshua Reynolds, one of Johnson's closest friends as well as the highest-ranked portraitist of his day, painted what many Johnsonians consider "the best known of all portraits" of the great man (fig. 14). One of Johnson's first biographers, Sir John Hawkins, captured the source of the image's appeal when he remarked: "There is in it that appearance of a labouring working mind, of an indolent reposing body, which he had to a very great degree." This picture, according to Johnson, "seem[ed] to please every body" and was copied frequently.[34] In 1778 Reynolds made a replica of the painting for Johnson's friend, the wit Topham Beauclerk—of whom Johnson once said: "Thy body is all vice, and thy mind all virtue"[35]—who had inscribed on the frame the following lines from Horace's Satire 1.3:

> Iracundior est paulo, minus aptus acutis
> naribus horum hominum; rideri possit eo, quod
> rusticus tonso toga defluit et male laxus
> in pede calceus haeret; at est bonus, ut melior vir
> non alius quisquam, at tibi amicus, at ingenium ingens
> inculto latet hoc sub corpore. (29–34)

> ["He is a little too hasty in temper, ill-suited to the keen noses of folks nowadays. He might awake a smile because his hair is cut in country style, his toga sits ill, and

Figure 14. William Holland, *Samuel Johnson,* engraved after portrait by Sir Joshua Reynolds, circa 1772. Reproduced by permission of the Huntington Library, San Marino, California.

> his loose shoe will hardly stay on his foot." But he's a good man, none better; but he's your friend; but under that uncouth frame are hidden great gifts.]

Beauclerk's inscription places Reynolds's portrait in a classical tradition of images of scholars and poets in which exceptional greatness of mind is signified and belied by bodily difference. The meaning of such difference ranges from social aberration (Horace's slovenliness) to physical defect (Socrates' ugliness). (It is interesting to note in this regard that classicists have construed these lines as either Horace's self-portrait or a depiction of his friend Virgil.)[36] This eighteenth-century use of Horace's verse typifies contradictions in contemporary views of Johnson's body; his physical aberration is implied to be at once part of his nature—"that uncouth frame"—and within his control, a result of his neglect of regular washing, decorum, and proper dress. As Horace's choice of "incultus"—the meanings of which range from "neglected" to "uncultivated" to "savage"—indicates, in the author's case, bodily and behavioral norms collapse. Learned gentleman that he was, by gesturing toward Horace, Beauclerk marks Johnson as a true original, a man of "ingenium ingens"—a phrase that might be translated "vast, remarkable, uncouth" (echoing "incultus") or (to quote its most literal meaning of "outside the species") "monstrous" genius—whose uniqueness is modeled upon, indeed embodies, a great humanist tradition of moral authority. He thus provides an example of the ways in which the most exceptional of eighteenth-century bodies was read through classical lenses.

As he so often does in his poems to Maecenas (see, for example, the close to his Epistle I.i, imitated with an eye to self-portraiture by Johnson's great predecessor as man of letters and physical spectacle, Alexander Pope), Horace turns this conflation of natural and social flaws to his own satiric benefit. If the viewer judges by appearance alone, the poem continues, the flaws he identifies will in fact, however unwittingly, expose his own invisible moral shortcomings. The complex interrelations of uncultivated body and exceptional mind, of visible and invisible signs of character, of physical and social forms of defect, of monstrosity and original genius, of inheritance and particularity, characterize this book's trajectory as it pursues Johnson's case from classical precedent to the walls of Beauclerk's study to nineteenth-century Johnsonian centenary celebrations, to the twentieth-century Age of Johnson classroom and after-dinner speech.[37]

We can thus read Robina Napier's desire for "more" of Johnson as a kind of sentimentalized descendant, the more linked to vision as the more nostalgic (think, too, of McHenry's reflections on Percy's wax diorama in chapter 1), of the late eighteenth-century rage for Johnson's physical particularity—for eccentricities at once bodily and social—detailed in Boswell's *Life* and a host of other volumes of Johnsoniana. Napier's hero of united contradictions descends from the barely human prodigy about whom Soame Jenyns (among many others) complained in his 1786 epitaph:

Here lies poor Johnson. Reader, have a care,
Tread lightly, lest you rouse a sleeping bear;

Religious—moral—gen'rous and humane
He was—but self-sufficient, rude and vain:
Ill-bred and over-bearing in dispute
A scholar and a Christian—yet a brute.
Would you know all his wisdom and his folly,
His actions—sayings—mirth and melancholy,
BOSWELL AND THRALE, retailers of his wit,
Will tell you how he wrote, and talk'd, and cough'd and spit.[38]

## Actual Death

This lively, curiously solid version of Johnson's authorial ghost vexes an easy appropriation of a de Manian reading of prosopopoeia that conceives of death as only absence, as merely "a linguistic predicament."[39] Ultimately my evocation of de Man in reading Johnsonian repetition is shadowed by what he leaves out and what the eighteenth-century poetry of death insists upon—the mortality of the flesh.[40] While Garber can build a compelling psychoanalytic/deconstructive reading of the author's ghostly presence by focusing on the disembodied figure of Shakespeare, the character of Johnson—more beloved by many readers for his pungent sayings, anecdotal exploits, physical oddities, and medical history than for his literary production, more beloved precisely because he is not Shakespeare and Milton—seems to gain universality the more particularly and locally embodied it remains. What haunts Johnsonians, as we saw in chapter 1, is the awareness of the real corpse beneath the rhetorical figure. In a secular version of religious Communion, they build their "monument more durable than brass,"[41] and with it their professional and national identities, in the disfigured shape of a flawed mortal body.

A secularized Christian paradox fuels this Johnsonian dialectic of repetition and originality, community and individuality, immortal texts and personal mortality: the immortal spirit of literature dwelling in the mortal body of an author, commemorated in polite conversation and ritual. I want to return to the initial confrontation with death and its aftermath with which I began (but with a difference), with another form of repetition and with Dr. Johnson himself, who when not engaged by such conversation, would repeat to himself the poetry of death. The anecdote is Arthur Murphy's:

> The contemplation of his own approaching end was constantly before his eyes; and the prospect of death, he declared, was terrible. For many years, when he was not disposed to enter into the conversation going forward, whoever sat near his chair, might hear him repeating, from Shakespeare,

> Ay, but to die and go we know not where;
> To lie in cold obstruction and to rot;
> This sensible warm motion to become
> A kneaded clod, and the delighted spirit
> To bathe in fiery floods.

And from Milton,

> Who would lose,
> For fear of pain, this intellectual being?[42]

Repeating excerpts from act 3, scene 1 of *Measure for Measure* and book 2 of *Paradise Lost,* Johnson is haunted by the same two authors who shadow Wordsworth's *Essays upon Epitaphs* and de Man's essay on Wordsworth. As is often the case with Johnson, Milton's lines are significantly misquoted: the original excerpt from Belial's speech, in which he reflects for his compatriots in hell on the "final hope," the "sad cure" of divine annihilation, reads "for who would lose / *Though full* of pain, this intellectual being?"[43] Johnson shifts Belial's pain, and with it the thought of hell, to the terrifying domain of (to use one of his favorite words) "futurity." He thus uncannily anticipates what his biographers would already have known—the hour of his own impending death, when he desperately fought, with a scalpel as his weapon, a dropsical body turned "bloated carcase." "How many men in a year die through the timidity of those whom they consult for health!" he chastised a reluctant physician during his final days. "I want length of life, and you fear giving me pain, which I care not for."[44]

Johnson's obsessive, habitual repetition of poetic confrontations with death without end, refracted through the drama of his final self-wounding, also evokes a scene from Addison's *Cato,* fraught with echoes from the same moments in Milton and Shakespeare. Cato, with a book of Plato in his lap, "a drawn Sword on the table by him," soliloquizing on the immortality of the soul and the "pleasing, dreadful thought" of eternity,[45] affords a model for Hawkins's rewriting of Johnson as Roman stoic hero, whose last words were those of a gladiator, "Jam moriturus."[46]

But Johnson's repetition of these lines is fraught more with fear than with resolve. What ultimately was Johnson so afraid of? The following passage from *Rambler* 78 can begin to answer this question:

> For surely, nothing can so much disturb the passions, or perplex the intellects of man, as the disruption of his union with visible nature; a separation from all that

> has hitherto delighted or engaged him; a change not only of the place, but the manner of his being; an entrance into a state not simply which he knows not, but which perhaps he has not faculties to know; an immediate and perceptible communication with the supreme Being, and, what is above all distressful and alarming, the final sentence, and unalterable allotment.[47]

His fear of an end, one of the most enduring topoi of Johnsonian studies, is denied and propelled by an inverse fear of what follows "the final sentence," of endlessness. When he repeats Milton and Shakespeare in a habit grown as automatic as his compulsive gestures and the "guilty thoughts" that accompanied them, he rehearses both the fear of death and the terrifying prospect of eternal torment, confronting both an end to consciousness and consciousness extended unbearably beyond human limits.[48] Repetition enables Johnson to remember and forget the equally unimaginable and equally terrifying prospects of death and death's impossibility.[49]

In his *Life of Gray,* we'll recall from the introduction, in a spectacular example of such remembering and forgetting, Johnson singles out the "Elegy Written in a Country Churchyard" for special praise:

> The *Church yard* abounds with images which find a mirrour in every mind, and with sentiments to which every bosom returns an echo. The four stanzas beginning "Yet even these bones," are to me original: I have never seen the notions in any other place; yet he that reads them here, persuades himself that he has always felt them. Had Gray written often thus, it had been vain to blame, and useless to praise him.[50]

He similarly opined to Boswell that the only two good stanzas in Gray's poetry were the second two of those four stanzas he would later praise in the "Elegy," going on to recite:

> For who to dumb Forgetfulness a prey,
> This pleasing anxious being e'er resigned,
> Left the warm precincts of the cheerful day,
> Nor cast one longing lingering look behind?

"The other stanza," Johnson concluded, "I forget."[51] Johnson's forgetfulness here is doubly striking. First, the forgotten stanza rewrites Johnson's dual fear of death and deathlessness, the fear that dictated his final denial of the doctors' compassion, as a moral sympathy that eases the passage from the living to the dead, and fulfills the enduring desire of the dead for remembrance. This is the stuff of Johnsonian community:

> On some fond breast the parting soul relies,
> Some pious drops the closing eye requires;
> Ev'n from the tomb the voice of nature cries,
> Ev'n in our ashes live their wonted fires. (89–92)[52]

More striking is what Johnson forgets when he praises these lines as original: precisely the passages from Shakespeare and Milton to which they allude, and that he had repeated to himself for years.

Neil Hertz, in an essay that haunts this chapter, describes Johnson's sensation upon reading Gray's lines in the *Life of Gray* as a kind of uncanny "déjà vu"—"the wobble between never having seen them and always having felt them." In what Hertz calls a "touchstone for the academic study of literature in England and America," Johnson praises Gray for "an originality held within the ample but nonetheless strong constraints of a humanist aesthetic. His stanzas may not echo other poems, but his images and sentiments find their mirror and echo in every mind and every bosom: there they are available for the inspection of the common sense of common readers and serve to underwrite Gray's claim to poetical honors."[53] Gray's stanzas, in other words, are so "original" as to be universally true. Their heartfelt uniqueness is earned by an act of forgetting of the poetry of death. (Interestingly enough in this regard, we might remember that the text that won Boswell's approval as the best imitation of Johnson's prose style was an anonymous continuation of Johnson's thoughts on the "Elegy.")[54]

Hertz's elegantly compressed and complex analysis goes beyond our scope in these pages. But we can note that his discussion of that uncanny déjà vu that informs Dr. Johnson's misrecognition of Gray leads via de Man on prosopopoeia to a consideration of the elegiac pathos that haunts Descartes' Second Meditation, a pathos that arises when Descartes contemplates what Hertz calls the "figurative corpse" of a piece of wax. These experiments with wax began with Descartes' contemplation of his own body. Resolving "to consider the thoughts which of themselves spring up in my mind, and which were not inspired by anything beyond my own nature alone when I applied myself to the consideration of my being," Descartes, in a shocking act of identification, first "considered myself as having a face, hands, arms, and all that system of members composed of bones and flesh as seen in a corpse which I designated by the name of body."[55] In a rehearsal of eucharistic logic, Hertz argues, Descartes elegiacally posits a mind/body economy that affirms the undying original power of mental cognition at the price of mortal death. The "residue" of such mental power, one that imagines itself as "nature" free of figuration, in Hertz's logic, is "a figure for figuration generally: a [dead] body with signs on it."

What then of Dr. Johnson? Here I quote Hertz at length:

> Consider this possibility: when Johnson, bored with the conversation going on around him, falls to repeating those lines, he is not impersonating Claudio or Belial or even distinguishing Shakespeare from Milton; he is, rather, communing with Literature, taking it into his mouth. Taking it into his mouth in a necessarily duplicitous fashion, we could say—as signifier and as signified, as cadenced language (*this sensible warm motion, this intellectual being*) and as just representations of general nature. We shouldn't be surprised then, if the long-term effect of such oral repetition is not the strengthening of the memory . . . but the obliteration of the signifiers, the forgetting of those lines as poetry—as plangent phrasing—and their internalization as nature or truth. Like its religious model, this secular version of Eucharistic incorporation works to consolidate a collective or mystical body of communicants—here it is the set of common readers with whom Johnson rejoices to concur. He can concur with them because he has also forgotten what differentiates him from them—for example, his knowing those lines from Shakespeare and Milton, among other refinements of subtilty and dogmatisms of learning.

In this secular religion, forgetting enables a faith in "general nature" and universal truth, free from yet indebted to the dead letter. Returning briefly to Descartes' wax figure, Hertz asks, "Is there an equivalent 'body' in Johnson, a corpse that marks his forgetting the signifiers echoing within him?"[56]

There is such a corpse, not figurative but real, and it is Johnson's own. Immortalized and disfigured by the printed word it seems to transcend, evoking that same uncanny feeling—one poised between remembering repetition and forgetting it, between common habit and individual volition, between death and life—it still informs our literary community and individual acts of reading. Hertz continues, "We could take this anecdote as a fable—one with some historical validity—of the origins of the Anglo-American institution called English Studies, whose representatives can still be observed in contemporary polemics brandishing the common reader."[57] At this crossroads of solitary critical and communal uncritical reading, of remembering and forgetting, of literature's ghostly immortality and its bodily remainder, Johnson's ghost still haunts us.

## "Thou art a scholar; speak to it, Horatio"

We have begun to see how Johnson brought a collective audience to life in the newly imagined form of "the common reader." This fictional figure, to whom Johnson so often deferred in crises of judgment, summons the author's image in a communal mirror, solitary yet befriended through the thriving medium of print. I am curious about how Johnson, even or especially in our postmodern moment, still haunts the profession of English letters not as a great writer, or even as a great reader, but as a "great man, writing."[58] However we as critics might try to demystify this vision of Johnson, to turn back to the printed page and away from the human image, his ghost still beckons.

If we return to the volume *Dr. Johnson and His Circle*, written for the Home University Library of Modern Knowledge, only the second in the series (after Shakespeare) to focus on a single author, we encounter the literary critic John Bailey repeating a familiar truism, in a chapter titled "A National Institution," as he praises Johnson as "the embodiment of the essential features of the English character." Distinguished for "a sort of central sanity . . . which Englishmen like to think of as a thing peculiarly English," Bailey's Johnson lacks "genius" but possesses something better, "something broadly and fundamentally human . . . which appeals to all and especially to the plain man." "We never think of the typical Englishman being like Shakespeare or Milton," he writes, but thanks to Johnson's very typicality, that "quiet and downright quality which Englishmen are apt to think the peculiar birthright of the people of this island," "we can all imagine that under other conditions, and with an added store of brains and character, we might each have been Doctor Johnson." Johnson, in other words, belongs not to the literary critic but to the common reader.[59]

By evoking the memory of this figure from the national past, "as familiar to us as those among whom we have been brought up," Bailey, Macaulay, and many others transform the reading community into children viewing the world of adults, replacing the reader's critical distance with common admiration. Untainted by academic prejudice, this nostalgia for a literary childhood renders the author as comforting as an idealized, lost parent. Devoted readers over the centuries have given this monstrously lovable creature a human voice and face. He can leave even the most rigorous of critics speechless.

. . .

In his famous essay "The Double Tradition of Dr. Johnson," Bertrand H. Bronson charts the endurance of Johnson's ghost in excess of his texts in order to exorcise the author's spirit in the service of clear-eyed critical vision. "After his death there springs up an eidolon of an author," Bronson begins, "and it is of this everchanging surrogate, not of the original, that we inevitably form our judgments, and that by so judging we further change." Literary tradition, he argues, is thus "double": Bronson therefore proposes to pay unconventional attention not to "the operative power of tradition which we denominate influence," not to the realm of texts, but rather to "something more akin to a transmitted recollection, to a song or ballad," namely, the popular folk tradition in which Johnson "exists for us also like a character in one of our older novels, and on the same level of objectivity and familiarity."[60]

By analyzing this eidolon, Bronson implies, we can escape its subtle influence, an influence that operates at an uncritical level and in the realm of unconscious knowledge, of "recollection" and "familiarity." Bronson devotes the bulk of his essay to charting the work of "devoted specialists" in eighteenth-century literature whose labors have succeeded (thanks, he claims, to a decline in the critical currency of both Romantic individualism and isolated naturalistic detail) in successfully reforming the Johnsonian imago. These critical efforts have revived Johnson's conservatism and eighteenth-century orthodoxy, not in his familiar fixed image, but as a creative energy "that vibrates like a taut wire." But despite such academic success, Bronson concludes, "it appears likely that the folk-image still persists on a far higher level of culture than the specialist would ever dream possible."[61] As if to rehearse the futility of his own critical efforts to remake Johnson's ghost in the image of his texts, Bronson concludes by invoking the uncanny connection between printed text and authorial eidolon in the form of another famous ghost:

> But how can we sufficiently admire the vitality of this folk-image? It captures the imagination of generation after generation; it takes possession of some minds to such an extent that they spend years reading about Johnson and his circle, and even publish their own books on him, and all the while before them looms the same imago, unabashed and incorrigible. It is a humbling spectacle and a chastening one to the specialist. Each of us brings his burnt offering to the altar of truth, and the figure we invoke becomes momentarily visible, obscurely forming and reforming in the smoke above us, never the same. But the folk-image moves irresistibly onward, almost unaffected by our puny efforts to arrest or divert it.
>
> We do it wrong, being so majestical,
> To offer it the show of violence;
> For it is, as the air, invulnerable,
> And our vain blows malicious mockery.[62]

Bronson refers to the play that haunts his own evocation of the Johnsonian eidolon, *Hamlet* (1.1.145–48), as he summons the armored ghost of Hamlet's father, the same ghost evoked by Boswell at the founding moment of and in the *Life* of his and Johnson's first meeting, and which confounds Hamlet at the beginning of this chapter. *Hamlet* begins, so this allusion reminds us, with a dramatic refutation of scholarly doubt. Horatio dismisses the guardsman Marcellus's report of an apparition as "fantasy," only to be confronted with ocular proof in the form of the ghost himself, onstage, in full armor. "Thou art a scholar; speak to it, Horatio," Marcellus demands (1.1.42). But Horatio is paralyzed by his learning. "Harrow[ed] with fear and wonder," caught between an enlightened skepticism that had doubted the ghost's existence and the overwhelming evidence of his own eyes, he trembles and looks pale, his questions and commands rebuffed (1.1.44). The ghost responds neither to the scholar's words nor the soldier's assault. It disappears, summoned by the crowing of the cock to its unknown "confine," perhaps its temporary hell. It speaks only to its son and heir in whom it will live on, who will perform his bidding, rescue him from purgatory.

Johnson's ghost inspires in Bronson, skeptical but awestruck like Horatio, a similar involuntary admiration. Like the ghost of old Hamlet, Johnson's spirit will not speak to skeptical critics, but only to true believers, "generation after generation," who perpetuate an image that dwarfs the literary efforts of "specialists." The labors of the critic, sacrificed at the altar of truth, produce only changeable figures as ephemeral as smoke. Substantive as stone, Johnson's gigantic ghost stalks away, untouched by time or "puny" critics, inspiring instead popular books by those "possessed" by its powers and thus with the desire to reproduce it in print.

The ghost of old Hamlet haunts Johnsonian biographers, critics, and parodists alike. By summoning Shakespeare's words to echo past his own, Bronson concludes his essay suspended, like Horatio, between lay belief and learned doubt. Johnson's ghost, like that of Hamlet's father, refuses to rest. Whether vengeful or victimized, his spirit leaves the reader disarmed, prey—as was Johnson himself—to superstition. Through his own tragic and impenetrable doubt—a doubt at once anti-Enlightenment in its Puritan religiosity, and post-Enlightenment in its almost existential confrontation with the possibility of annihilation—the figure of Johnson transforms the symbolic remains of Christianity into a vehicle of community through the preservation and consumption of the author in print. The paradigmatic critic of English letters inspires uncritical reading at its spiritual height.

## "The soul of that writing"

> What I read now elevated my mind wonderfully. I know not if I can explain what I have felt, but I think the high test of great writing is when we do not consider the writer, and say, "Here Mr. Johnson has done nobly"; but when what we read does so fill and expand our mind that the writer is admired by us instantaneously as a being directly impressing us, as the soul of that writing, so that for a while we forget his personality, and, by a reflex operation, perceive that it is Mr. Johnson who is speaking to us. I feel quite well what I have now written. I wish I could make it clear in words.
>
> JAMES BOSWELL[63]

In his certainty that he *feels* the truth of his response to Johnson's text, and in his inability to express that feeling in words, James Boswell, in his journal entry for March 17, 1775, sounds more like (what indeed he was) an ardent amateur, or to our professorial ears a besotted student grappling with the need for a clearly articulated thesis, than a literary critic. Perhaps Johnson's ghost endures because it allows, indeed demands, this return to a preprofessional and pleasurable certainty, a return that is a retreat from critical distance and doubt. Rendering even the living Johnson a disembodied spirit, smitten by the power of his text, Boswell, that literary fan of epic proportions, enacts uncritical reading as sublime communion, a submission to the text that "impresses" the reader with the author's speaking voice. Resonating with Horatio's "wonder" at things undreamt of by his critical philosophy, Boswell's surrender to a text brought to ghostly life enacts a devotion that transcends the love of literature. Not Horatio but self-anointed Hamlet, Boswell is inspired by an encounter with literature experienced as connection to the author. He enacts uncritical reading as author love, helping to inspire two centuries' worth of Johnsonians to remember their ghostly father.

Boswell's textual epiphany starts with his frisson of uncritical and self-congratulatory pride in private familiarity with the man behind the celebrity. Reading excerpts in the newspaper of "Mr. Johnson's new pamphlet, *Taxation No Tyranny,*" he is "new struck with admiration of [Johnson's] powers."[64] He thus reappraises Johnson's familiar public style in the context of intimate association, a context that lends him glory: "I was proud, and even wondered that the writer of this was my friendly correspondent." Pride by association with Johnson's printed prowess leads (in a manner typical of the Boswell of the journals) to personal fantasy: "I thought that he who thinks well of my abilities might recommend me to the Sovereign and get me highly advanced, and how should I delight to add riches and honour to my family in a Tory reign, by the recommendation of Mr. Samuel Johnson."[65] This is intellectual height turned to social climbing.

Yet from the petty glories and mercenary fantasies of particular acquaintance with a famous writer rises the pinnacle of intellectual elevation of the passage that follows. Here the author is absorbed completely into the text, becoming "a being directly impressing us, as the soul of that writing," his personality and person forgotten, the word replacing the flesh. "Impress" here paradoxically contains a hint of physicality, as if the writing makes a literal mark upon the receptive reader, who has become warm wax to the imprint of the spiritual "being" the text bodies forth. This readerly rapture can be felt "quite well," but transcends words to become a transparent experience of extraliterary communion. Feats of style, the province of proper critical evaluation from a distance, are forgotten: a living voice is "speaking to us." Boswell's selfish "I" thus is effaced, by force of recognition, into the "we" of literary community.

This is one pole of Johnsonian communion—the couplet pair to the authorial corpse that will cast its shadow over the writing of Boswell's *Life,* a corpse that is never fully forgotten and to which this chapter will return.[66] While the Johnsonian medical community scrutinizes the burden and mystery of Johnson's materiality, as we saw in chapter 1, the Johnsonian literary community combats the loneliness of mortality and of the critic's mortal work.

The uncritical reading that Boswell exemplifies thus effaces the text in the service of the author "himself." Boswell struggles to describe a "reflex operation" of knowing and not knowing that results, uncannily, in recognition of the author as a kindred spirit. Such devout disavowal inverts the proper critical approach to the text. But it also exposes the critic's attention to the text as itself a form of fetishism—a fruitless substitution of the word for the living writer. From the Johnsonian perspective, critics disavow the vital power of the author's presence in the text by embracing the dead letter.

In her recent study of marginalia, H. J. Jackson has traced such communion book by book. Boswell's *Life of Johnson* plays a unique role in her project because of her inability to generalize about the history of its marginalia; what endures is its status over centuries of lively individual response as "a book that has been taken for a man."[67] So effectively does Boswell animate Johnson, speaking in published writing, private letters, and dramatic scenes of conversation, that readers are compelled in their own notes to talk back. At the intersection of private and public modes of discourse, having only personal idiosyncrasy in common, generations of readers have responded directly on the pages of his *Life* to a dead man reanimated by living speech. Leigh Hunt, for example, recounts his own experience of melancholy in response to Boswell's account of Johnson's youthful suffering: "I had it myself at the age of 21, not with irritation & fretfulness, but pure gloom & ultra-thoughtfulness. . . . During both my illnesses, the mystery of the universe sorely perplexed me; but I had not one melancholy thought on religion." In a Harvard copy of an 1887 scholarly edition, one reader highlights Johnson's remark, "A man may have a strong reason not to drink wine; and that may be greater than the pleasure," noting "see Aristotle Eth. Nich. Book I." Another reader retorts in the same margin, "You don't have to brag about taking Phil. A. You aren't Samuel Johnson. L.S.K."[68] In the case of the great critic, so this snarky exchange shows us, uncritical reverence prevails.

## Art Redeemed

> There are some authors who exhaust themselves in the effort to endow posterity, and distil all their virtue in a book. Yet their masterpieces have something inhuman about them, like those jewelled idols, the work of men's hands, which are worshipped by the sacrifice of man's flesh and blood. There is more of comfort and dignity in the view of literature to which Johnson has given large utterance: "Books without the knowledge of life are useless; for what should books teach but the art of living?"
>
> WALTER RALEIGH[69]

I have begun to uncover the ways in which the peculiarly uncritical worship of Johnson's ghost exposes the traditional critical reverence for the literary text as a form of fetishism. My epigraph from the Johnsonian Walter Raleigh takes this claim one step further. If, as Bertrand Bronson discovered to his own chagrin, the sacrifice of critical labor to the altar of truth results in Johnson's case only in airy phantoms, the sacrifice of authors to the altar of art involves a bloodier and equally useless tribute. Literature in this passage is a man-made god who demands the "flesh and blood" of the author himself. The finished work of art is a "jewelled idol" masquerading as a living deity. Johnsonians prefer the stolid humanity of the author's ghost to the arid perfection of a masterpiece. Rejecting the love of literature for its own sake as pagan idolatry, they reenact Christian Communion with the author himself. In the case of Johnson, such communion can take the form of private marginalia or civic ritual. Whether individual or communal, the love of Johnson disavows literary labor and, with it, human mortality. Nowhere has this been better exemplified than in Johnson's birthplace, Lichfield, home of the original and most revered Johnson society.

Every September in the English city of Lichfield, a name that means "field of the dead," the town's polite society gathers to celebrate the birthday of their

Figure 15. Johnson Society Annual Supper, September 1928, Guildhall, Lichfield (219th birthday celebration). Note that the male servant standing in the back left corner is dressed in eighteenth-century wig and livery. Reproduced by kind permission of the trustees of the Samuel Johnson Birthplace.

most famous native, eighteenth-century man of letters and moral philosopher Samuel Johnson. At the Johnson Society Annual Supper, celebrants enjoy a hearty British repast fit for their Rabelaisian hero: haunches, saddles, or joints of meat (2000 was the first time since the society was founded in 1910 that a vegetarian option was offered), followed by apple pie and cream, ale, cheese, punch, and the smoking of long clay "churchwarden pipes," handed round by a servant dressed in full eighteenth-century livery (see fig. 15). Formal toasts punctuate the proceedings in traditional order: the queen (proposed by the mayor of Lichfield), followed by "The Immortal Memory of Dr. Samuel Johnson," proposed by the president of the society, followed by five minutes of silence.

On the morning before the annual Johnsonian feast, a different sort of ceremony takes place. Civic community supplants the dinner's exclusivity as town dignitaries—including the mayor and mayoress, the sheriff, the alderman and councillors, the dean and canon of Lichfield—join the senior boys of Johnson's old school who form the Cathedral Choir, along with members of the Johnson Society and the general public in the market square. Ascending a ladder, the mayor adorns Johnson's statue with a laurel wreath (stored during the rest of the year at the base of Johnson's bust in Lichfield cathedral). "From a platform erected on the steps of the Birthplace—an innovation which was much appreciated," the Cathedral Choir sings "with their customary charm, the 'anthem' and appropriate hymns."[70] Choirboys receive a special token from the mayor in memory of their participation.

The singing of the Johnson anthem combines literary with religious memory in a living evocation of the author's death. Set to music by Arthur B. Platt in 1909 for the bicentenary birthday celebration, the text, taken from Johnson's last prayer, composed on December 5, 1784, eight days before his death "previous to his receiving the Sacrament of the Lord's Supper," reads as follows:

> Almighty and most merciful Father grant that my hope and confidence may be in Jesus' merits and Thy mercy. [semi-chorus]: Confirm my faith, establish my hope, enlarge my charity, Pardon my offences, and receive me at my death to everlasting happiness for the sake of Jesus Christ.[71]

The entire prayer—first published in Arthur Strahan's 1785 edition of Johnson's *Prayers and Meditations* and on sale at the Samuel Johnson Birthplace Museum as a calligraphed text superimposed on James Barry's portrait of an elderly Johnson (fig. 16)—reads thus:

> Almighty and Most Merciful Father I am now, as to human eyes it seems, about to commemorate, for the last time, the death of Thy Son Jesus Christ our Saviour and Redeemer. Grant, O Lord, that my whole hope and confidence may be in His merits, and Thy mercy; enforce and accept my imperfect repentance; make this commemoration available to the confirmation of my faith, the establishment of my hope, and the enlargement of my charity; and make the death of Thy Son Jesus Christ effectual to my redemption. Have mercy upon me, and pardon the multitude of my offences. Bless my friends; have mercy upon all men. Support me, by the Grace of Thy Holy Spirit, in the days of weakness, and at the hour of death, and receive me, at my death, to everlasting happiness, for the sake of Jesus Christ. Amen.[72]

Platt's anthem omits Johnson's evocation of the perspective of "human eyes." Since Johnson's death was a public spectacle recorded by many, this reminder of an uncertain worldly gaze might self-consciously evoke the particular witnesses of his final Communion. More abstractly, Johnson destabilizes his personal confrontation with mortality, in all its "seeming," by addressing disembodied divine omniscience. To frame the uncertain view from "human eyes" is to ascend, obliquely and ironically, to the possibility of the God's-eye view that

Figure 16. Anker Smith, *Samuel Johnson,* engraved in 1808 after James Barry's portrait of an elderly Johnson (1778–80). Reproduced by permission of the Huntington Library, San Marino, California.

a poem like Johnson's *Vanity of Human Wishes* initially summons in its figure of all-encompassing "Observation" and ultimately prays for in the form of "Celestial Wisdom." In the context of Johnson's life and work, even this simple prayer shows, conventional devotion is fraught with contradiction and paradox. Impending death can never be known, is always a "seeming," both because the time of one's end is known only to God and because death, or so Johnson hopes and fears, is only corporeal, a prelude to everlasting happiness or everlasting punishment. To identify with Johnson at this unedited moment would be unbearable, because at this instant of imminent death—the founding moment of Johnsonian memory—hope and fear are indistinguishable and unending. "Where then shall Hope and Fear their Objects find?" (343), the speaker of the *Vanity* finally and desperately asks. As we saw in chapter 2, in the context of a poem glutted with personified agents and devoid of human control, this query is both grammatical and existential. The questions that follow haunt Johnson's life and work:

> Must dull Suspence corrupt the stagnant Mind?
> Must helpless Man, in Ignorance sedate,
> Roll darkling down the Torrent of his Fate? (344–46)[73]

Rather than answer, the poem commands silence. "Enquirer, cease" (349), interrupts an anonymous voice. We never learn that speaker's identity, nor are we told whether the "Petitions" to heaven for sanity, obedience, patience, and resignation that the voice prescribes will be granted.

Platt also omits the original text's self-abnegating references to "imperfect repentance" and the "multitude of my offences." Like the voices in the *Vanity*, these, too, are double: at once typically pious (only truly repentant Christians could receive the Eucharist) and, in Johnson's case, personally fraught reminders of his extreme fear of death and the threat of death's eternally painful aftermath. This particular Christian humility borders on an almost Calvinist conviction of guilt disturbing to the comfortable belief of an Anglican establishment.[74]

The tranquil and eminently didactic death of Joseph Addison, recorded by Edward Young twenty-five years earlier in his *Conjectures on Original Composition*, was much more to the public taste.

> Forcibly grasping [his stepson's] hand, he [Addison] softly said, "See in what peace a Christian can die." He spoke with difficulty and soon expired. Through grace divine, how great is man? Through divine mercy, how stingless is death? Who would not thus expire?[75]

"By undrawing the long closed curtain of his death-bed," Young turns the author of that pagan exemplar *Cato* into the exemplary Christian "actor of a part, which the great master of the drama has appointed us to perform to-morrow." Death puts all upon the stage while distinguishing the earlier master of print culture as a truly moral genius whose virtue sets his performance apart: "Have I not showed you," Young asks, "a stranger in him whom you knew so well? Is not this of your favorite author,—*Notâ maior imago?* VIRG. [a greater image than the well-known one (*Aeneid* 2, 773)]. His compositions are but a noble preface; the grand work is his death."[76] The "grand work"—not written word but live animation of authorial likeness, imago, or ghost—transfixes an audience who, it is implied, violates the privacy of the dying man, drawing the bed curtains to reveal a stage. The performance that ensues is that of an author, an invisible "spectator" previously known through his texts alone, whom death renders a "stranger" authored by another.[77]

Young domesticates the unsettlingly uncanny transformation of familiar to stranger in his revelation of the author turned actor by offering unequivocal evidence of Addison's Christian virtue and salvation. Despite the title of his essay, Young admits, his ultimate goal went beyond mere literary criticism. His "chief inducement for writing at all" was to bring to light this particular author's final hours.

> For this is the monumental marble there mentioned [at the beginning of the text], to which I promised to conduct you; this is the sepulchral lamp, the long-hidden lustre of our accomplished countryman, who now rises, as from his tomb, to receive the regard so greatly due to the dignity of his death; a death to be distinguished by tears of joy; a death which angels beheld with delight.[78]

The image of "monumental marbles" in a "wide pleasure garden" evoked to describe Young's "somewhat licentious" and digressive text at its outset is channeled into the open tomb of a resurrected Addison.[79] The prying light of Young's curiosity submerges itself in the self-illuminating "sepulchral lamp" of his hero's own "long-hidden lustre," as Addison takes his final bow to long overdue deathbed applause.

As the ambiguities and controversy generated by Johnson's fear of death in general and last prayer in particular indicate,[80] the later author's tortured life and ambiguous death make a Johnsonian rewriting of Christ's resurrection impossible. Addison's tomb is reassuringly empty—Johnson's corpse endures. Even in the composition of a prayer familiar enough to become a popular commodity, Johnson both invites and undermines the easy exemplarity that facilitates collective identification. As they sing, the innocent choirboys and the proud town officials perpetually repeat, inhabit, and disavow the ambivalent and irreducibly singular moment of imminent death. Their mass commemoration violates solitude and erases aberrant fear, evoking and assuaging both in the creation of a uniquely literary kind of secular saint, to whose words they must always return since mere words are never sufficient.[81]

From the death of Addison to the life and afterlife of Johnson, the author in eighteenth-century England becomes an increasingly ambiguous object of religious curiosity. For this monumental man of letters and mass moralist in a burgeoning print culture, literary talents are inexorably linked to an unstable doubt monumentalized at the moment of incomplete self-reckoning.[82] Ventriloquizing their countryman at his final Communion praying for hope and redemption, the good people of Lichfield pray for a confirmation of faith—not in literature but rather in the author himself—that will render an uncertain end one of everlasting happiness. They render the moment of impending doom—the hero's most solitary and dangerous, the biographers' most elusive—one of collective desire. They remember Johnson at the moment of death while willing him back to life.

. . .

Let's return for a minute to those five minutes of silence at the Johnson supper—a long time to impose on guests at the social event of the Lichfield season. What does that silence signify? Ritualized silence, a counterpoint to the singing of the Johnson anthem, marks Lichfield's participation in collective Johnsonian memory, while bringing that memory to its limit in an encounter with death.

At Samuel Johnson's old haunt, the Cock Tavern, Fleet Street, London, the club of belletrists, journalists, statesmen, and scholars founded in his name in 1884, "exactly one hundred years from Dr. Johnson's death," met quarterly for supper and a paper presented by one of the members. Lionel Johnson's poem "At the Cheshire Cheese," surely one of the most dramatic of such presentations, conjures the great man from the dead in the form of an eminently clubbable ghost, to grace the "Brethren's" proceedings. The fantasy ends with a return to reality and that same silent affirmation of melancholy community:

> If only it might be! . . . But, long as we may,
> We shall ne'er hear that laughter, *Gargantuan* and gay,
> Go pealing down *Fleet Street* and rolling away.
> In silence we drink to the silent, who rests
> In the warmth of the love of his true lovers' breasts.[83]

Death haunts this idyll of a literary Last Supper. Summoned and silenced by collective reverie, that Gargantuan laughter resounds (in a moment the poet must have had in mind) in Boswell's *Life of Johnson* in "peals so loud, that in the silence of the night his voice seemed to resound from Temple-bar to Fleet-ditch." Ironically, the subject in Boswell's anecdote that reduced Johnson "almost [to] a convulsion" of hilarity was news of a friend having made his will.

> He now laughed immoderately, without any reason that we could perceive . . . called him the *testator,* and added, 'I dare say, he thinks he has done a mighty thing. He won't stay till he gets home to his seat in the country, to produce this wonderful deed: he'll call up the landlord of the first inn on the road; and, after a suitable preface upon mortality and the uncertainty of life, will tell him that he should not delay making his will; and here, Sir, will he say, is my will, which I have just made, with the assistance of one of the ablest lawyers in the kingdom; and he will read it to him (laughing all the time). He believes he has made this will; but he did not make it: you, Chambers, made it for him. I trust you have had more conscience than to make him say, "being of sound understanding;" ha, ha, ha! I hope he has left me a legacy. I'd have his will turned into verse, like a ballad.'
>
> In this playful manner did he run on, exulting in his own pleasantry, which certainly was not such as might be expected from the authour of 'The Rambler,' but which is here preserved, that my readers may be acquainted with even the slightest occasional characteristicks of so eminent a man.[84]

Johnson's epic and spectral mirth, echoing down Fleet Street in the clubmen's ears over a hundred years later, mocks the futility of individual authorial attempts to assert the self beyond the grave. In his efforts to transform Johnson into a Christian exemplar, Boswell edits out his own raucous participation in the joke when he transforms his original 1773 journal entry into this episode in the 1791 *Life.*[85] In the process, he creates a collective defined by their puzzled yet faithful gaze at an embattled hero whose confrontation with death at once invites and resists identification.

Their gaze seems to animate the dead. Johnson's ghostly laughter at the futility of personal wills haunts Johnsonians from Boswell to the present as they bring their hero back to life by collective will. Perhaps more than any other English writer, Johnson makes it clear that rumors of the death of the author have been greatly exaggerated. Whether preserving his body in parts or preserving his presence in anecdotes, both literary and medical Johnsonians have turned uncritical reading into author love, and thus into a kind of national secular religion based on the necessary insufficiency and self-transcending power of the printed text. Johnsonian communities vary across time and place, yet each version shares a desire to transcend time, place, and, above all, mortality, turning the individual communion of reading (that which the critic must murder to dissect) into a communal conversation with the author's spirit.

In their haunted disavowal of Johnson's mortality and in their reinterment of Johnson in their individual breasts, Johnson's devotees both confront and avoid the double nature, material and immortal, of Johnson's body. In a literary version of the cult of the saints, Johnsonians accomplish Christian miracles, joining, in Peter Brown's phrase, "Heaven and Earth at the grave of a dead human being."[86] Just as the graves of saints provided physical sites for new forms of community that crossed social bounds, contaminating (from the pagan perspective) the "public life of the living city" with the corrupt bodies and relics of the dead, so the initial rage for published anecdotes that immortalized Johnson in intimate detail brought the dead back into uncanny contact with the living.[87]

Object of longing and fantasy, Johnson thus endures (anti-Boswellian critical efforts notwithstanding) not in the dead letter but in romance's eternal present.[88] While such resurrection evokes the romantic trope of bringing the dead to life, Johnsonian tradition also evokes the genre of romance in its blurring of the borders of history and fiction, and of secular and religious realms of meaning. From this perspective, a text like Boswell's *Life,* which in the view of many critics achieves *both* objective truth and aesthetic integrity through its impregnation with Johnsonian ether, resonates with Northrop Frye's definition of romance as "secular scripture."[89]

Like Hamlet in the graveyard, Johnsonians contemplate their Yorick's skull in anecdotal form, with the same Shakespearean mixture of humor and pathos, scatological comedy and devout tragedy that eighteenth-century readers deplored in Shakespeare's play. Melancholic comedy and tragedy inhere in Boswell's staging and staged disavowal in the *Life* of Johnson's grim laughter at the thought of life after death. In the critical response that ensues, some of it defending a less commercial and grossly material form of Johnsonian reverence, some of it denouncing the Johnsonian phenomenon altogether, we can hear tonal echoes of satiric pagan responses to Christianity, Protestant attacks on the Catholic Mass, and early modern skeptical interrogations of Western religious belief (for example, Montaigne and Swift).[90] The Johnsonian devotion of Boswell and a host of other anecdotal collectors, in short, inspires and contributes to a late Enlightenment rewriting of the pagan confrontation with the corporeal nature of Christian faith.[91]

Over their mugs of ale at the Cheshire Cheese, churchwarden pipes at the Lichfield town hall, or glasses of wine at the Dorothy Chandler Pavilion, Johnsonians thus raise, however distantly, however decorously, what Stephen

Greenblatt—in a series of essays on the eucharistic controversies that dominated thought on the nature of linguistic signs in the early modern period—has recently termed "the problem of the leftover." More a problem of matter than the words that transform it, more the province of literature than theology, the Eucharist's material remainder joins the holy sacrament to human waste, Christ's immortality to mortal filth. Speaking of the Protestant reinterpretation of the sacrament as metaphorical, Greenblatt describes "an uneasy meeting: the conjunction of gross physicality and pure, abstracted spirituality, of Body and Word, of corruptible flesh and invulnerable ghost, of rotting corpse and majestical ruler. We have another name for this meeting when it assumed an apparently secular form: we call it *The Tragedy of Hamlet*."[92] We could also call it Johnsonian romance. In the originating anecdotal explosion that followed his death, to which I now turn, ghost and corpse are fused; to summon the former is to evoke the latter as the literary marketplace becomes both violated graveyard and haunted purgatory.[93]

## Sacred Remains

To a sophisticated audience of eighteenth-century men and women of letters, the ancestors of today's professional literary critics, the desire Johnson inspired for intimate communion through published anecdote was nothing less than a profanation of both the author's corpse and the reader's humanity. The common reader's love of Johnson, in their view, took the abstraction of critical "taste" too literally, rendering it abjectly ephemeral. In an uncanny inversion of the tradition of satiric depictions of the Christian Eucharist as cannibalism, the unprecedented demand for Johnsonian anecdote in the years following his death was denigrated as literary consumption at its most savage. As one reviewer of Thrale's *Anecdotes* put it, "An orthodox tartar may possess a sufficient degree of veneration for the *Dalai Lama,* without either worshipping or eating his excrements."[94]

The primitive idolatry that made Johnsonian worship a print phenomenon is brought closer to home in "a curious letter from a medical gentleman" appended to a shockingly scatological anonymous satiric 1787 pamphlet aptly entitled *More Last Words of Doctor Johnson.* (If Johnsonians believe in collective silence, this parody of Johnson speaks beyond the grave.) This medical Johnsonian is also a man of letters, who "while busied in the sublimest physical researches, . . . ha[s] not thought it beneath [him] to inspect the water-closets of the learned." With a relentlessly materialist vision worthy of Swift's hack in *A Tale of a Tub,* our author boasts that he is driven by "that curiosity which looks into the *bottom* of things, and which must of course be *fundamentally* learned."[95] Recognizing a kindred spirit in the "rank"-minded "*dirty fellow*" and Johnsonian biographer Sir John Hawkins, this literary acolyte plunders the spoils of the "house of office" (outhouse), "an house which has afforded me the greatest literary knowledge; not more from diving with no unhallowed hands into its sacred merdicular abyss, than from perusing the various inscriptions on its walls and windows."[96] There resides the stuff of Johnsonian anecdote, which is also, quite literally, shit. With all the scrutiny, at once scientific and devotional, given to the material remains of saints or sovereigns,[97] our author demonstrates his proficiency in a language of excrement that bears visible links to varieties of learning and literary styles:

> Thus profound and erudite authors generally emit long and sturdy ones, somewhat in the shape of ninepins; and these are either perfect or broken as their compositions are regular or unequal, and rugged or smooth according to the asperity or courtliness of their style. Johnson's, which was large, but to appearance evacuated with great labour, was surrounded with protuberances, like a cucumber or a pomegranate; so was Swift's. Pope's was extremely uniform and elegant in its structure, with some appearance, indeed, of internal roughness; but Shenstone's was as polished, as delicate, and as mournful, as a roll of the most elegant black sealing-wax.[98]

Here we have, courtesy of one of its first satirists, the ultimate Johnsonian fetishism in which the author's elusive ghost is reduced to his leavings. The "great labour" of art, disavowed by Johnsonians, as my epigraph from Raleigh epitomizes, in favor of the author himself, returns in the gift of his shit, stand-in and relic of both the authorial body and its work. Critical pleasure and instinctual repulsion, refined aestheticism and gross corporeality, meld in this excremental catalog of authorial devotion (and it should be noted here that our author bases his cloacal experiments as much on taste as on sight). In this parodic treatment of the traffic in anecdotes that John Wolcot termed "Johnsomania," the written word is transformed through contact with human filth into the very flesh it seeks to transcend. Johnsonian worship, from this skeptical perspective, inverts the sacrament, transforming the tainted text into the remains of the author "himself." Such materialist logic is reminiscent of the excremental world of Pope's *Dunciad,* while summoning up a language of resurrection, Communion, and material ingestion that is parodically and suggestively eucharistic.

Speaking of Hawkins's scandalous biography, our satirist mixes classical philosophy with a smutty dismantling of the body's wholeness at the Resurrection when he recollects "that I have seen the prediction of some philosophers verified, viz. *that all things which now owe their shape to mixture and alteration shall return to their first state;* for, in the houses-of-office of my friends and acquaintance, I generally see leaves of Sir John's book deposited on the shelves as offerings to Cloacina."[99] The letter culminates with the speaker's voyeuristic enjoyment (through a telescope) of the "extremely sublime" sight of Johnson defecating, his subsequent theft of the manuscript with which the great man has wiped his posterior, his presentation of the page to Boswell, its apprehension by the cook, who uses it to wrap a joint of meat for dinner, and the "providential discov[ery]" of the "literary *morceau*" "by a young lady's being seen to lay something like skin on the side of her plate, that she had attempted to chew in vain."

> The taste Miss ______ observed was rather strong in her mouth; but the cook persisted in saying that both the taste and colour arose from the gravy of the meat. . . . [W]e *tasted* it all round, and it felt to the palate much more bitter than gravy. Mr. Boswell was enraged, dismissed his cook at a moment's warning, and scarcely spoke a word during the remainder of the day. The company likewise were not in good spirits at their disappointment. Thus was a day's pleasure destroyed by the ignorance of a cook, whose folly deprived us of that pure gratification which we should have received from the perusal of what had perhaps never yet been printed, and what from the Doctor's posteriors might have been handed down to his posterity![100]

Johnson's aphoristic conversation, his sayings, and in this case his ephemera are literalized here as his "droppings."[101] Rather than read the great man's immortal words, his misguided fans, in their rage for private matter, ingest his excrement in futile attempts at communion with celebrity.

This aggressively excremental satire exposes the abject dimension of uncritical reading: from eighteenth-century posteriors to twentieth-century posterity, Johnson's corporeality has been both desired and disavowed. The author's ghost, resonating so powerfully in contemporary imaginations with the eucharistic "Host," is (to play on J. Hillis Miller's etymological ponderings in his classic essay "The Critic as Host") at once host and guest, stranger and friend, of matter and beyond it, at once a vehicle for the immortality of "pure" intellectual community and that immortality's fleshly, filthy remainder.[102]

In a discussion (indebted to Miller's essay) of the Johnson industry that emerged and proliferated in both serious and satiric modes after the great man's death, Donna Heiland has termed Johnson a "body god." Boswell's "anatomization" and "dissemination across England" of both Johnson's body of work and physical body are culturally analogous, she argues, to "the Dionysian ritual of *sparagmos*—in which a sacrificial body often identified with the god is torn to pieces, and then consumed, in the separate ritual of *omophagia*—as well as [to] the Christian counterpart to these two rituals, the celebration of the Eucharist."[103] We might also speculate that this stalwartly British version of eucharistic practice conflates (along the lines of the Anglican fusion of institutional and individual bonds in the administration of the sacrament) the ritual of Jewish Passover, with its affirmation of family and national ties, with the Christian annihilation of those ties in the service of individual membership in universal community.[104]

For contemporaries who saw such consumption as more profane than sacred, the resultant miracle was not Boswell's claim to have "*Johnsonised* the land," but rather a seemingly endless process of commodification that turned the host into cheap print. One reviewer listing a menu of publications that included "Hawkins' entré," Boswell's "gleanings," and Thrale's "gatherings," observed, "The Doctor's bones must be acknowledged to be the bones of a giant, or there would be poor picking, after their having furnished *Caledonian Haggis,* and a dish of *Italian Macaroni,* besides slices innumerable cut off *from the body* [by] Magazine mongers, anecdote merchants and rhyme stringers."[105] "Poor Dr. JOHNSON," another surfeited reviewer complained, "has been served up to us in every shape—We have had him boiled to a rag, *roasted, fricassed,* and now we are to have him scraped into a sermon on his wife's death."[106] In George Colman's 1786 "Posthumous work of S. Johnson," the author's ghost terrorizes Grub Street, reproaching his first biographer Thomas Tyers:

> Enough! The Spectre cried; Enough!
> No more of your fugacious stuff,
> Trite Anecdotes and Stories;
> Rude Martyrs of SAM. JOHNSON's Name,
> You rob him of his honest fame,
> And tarnish all his glories.
>
> First in the futile tribe is seen
> TOM TYERS in the Magazine,
> That teazer of Apollo!
> With goose-quill he, like desperate knife,
> Slices, as Vauxhall beef, my life,
> And calls the town to swallow.[107]

At once surgeon and priest, Tyers cuts up a hero transformed into that most British of dishes, roast beef, and feeds him to the nation.[108]

Complaining in the inscription on figure 17, Dr. Johnson, haunting his former intimate friend Hester Thrale Piozzi concludes:

When Streatham spread its plenteous Board,
I opened Learning's valued hoard,
  And as I feasted prosed.
Good things I said, good things I eat,
I gave you knowledge for your Meat,
  And thought th' Account was closed.
If Obligations still I owed,
You sold each Item to the Croud [*sic*],
  I suffer'd by the Tale:
For God's sake Madam, let me rest,
Nor longer vex your quondam Guest—
  I'll pay you for your Ale.

The debt owed by the living to the dead that animated the ghost of Hamlet's father is here reversed as the ghost attempts to pay off his tormentor for her exploitative hospitality. His immortal words, first exchanged for room and board, doom him to a debased version of purgatory when commodified as printed anecdote.

While Boswell could boast in a private letter that the *Life* "will be an Egyptian Pyramid in which there will be a compleat mummy of Johnson that Literary Monarch,"[109] Elizabeth Moody's version of Johnson's ghost (one of many such examples), in a poem of the same name, complains as if in response:

"Thy adulation now I see,
  And all its schemes unfold:
Thy av'rice, Boswell, cherish'd me,
  To turn me into gold.
"So keepers guard the beasts they show,
  And for their wants provide;
Attend their steps where'er they go,
  And travel by their side.
"O! were it not that, deep and low,
  Beyond thy reach I'm laid,
Rapacious Boswell had ere now
  JOHNSON a mummy made."[110]

Figure 17. "The Ghost of Johnson Haunting Mrs. Thrale." Originally titled "A Frontispiece for the 2d Edition of Dr. Johnson's Letters." Caricature dated 1788. Photograph from *Dr. Johnson and Fanny Burney; being the Johnsonian Passages from the Works of Mme. D'Arblay*, ed. Chauncey Brewster Tinker (New York: Moffat, Yard & Co., 1911). Note that the ghost points to a portrait of Boswell, who is pointing to an image from the *Tour.*

Figure 18. "The Bust of Johnson Frowning at Boswell, Courtenay, and Mrs. Thrale." Originally titled "The Biographers." Caricature dated 1786. Photograph from *Dr. Johnson and Fanny Burney; being the Johnsonian Passages from the Works of Mme. D'Arblay*, ed. Chauncey Brewster Tinker (New York: Moffat, Yard & Co., 1911). Note that the drawing on Boswell's manuscript depicts Johnson in familiar guise as a performing bear.

The particular stakes of this struggle over Johnson's embodiment have been erased over time as it has been successfully won. In contrast to the relatively disembodied figure of Shakespeare, the character of Johnson seems to gain its power to transcend local boundaries the more English and embodied it remains, the more vividly it can be summoned from the past to speak not through its texts as character-creating author but in characteristic style.[111] The Johnsonian monument, remembered as long as the English language endures, is built with fleshy trifles, displayed in literal and anecdotal parts, and haunted by its material remainder.

## "You shall know their fate no further"

Whether Johnsonian dissection is anecdotal or anatomical, its motive is best encapsulated by a particular anecdote about the making of anecdotes and the fragmentary "real" to which they refer. This anecdote's fascinating afterlife explicitly thematizes the vexed dynamic of Johnsonian curiosity and its object's self-conscious resistance. Boswell boasts:

> I won a small bet from Lady Diana Beauclerk, by asking [Johnson] as to one of his particularities, which her Ladyship laid I durst not do. It seems he had been frequently observed at the Club to put into his pocket the Seville oranges, after he had squeezed the juice of them into the drink which he made for himself. Beauclerk and Garrick talked of it to me, and seemed to think that he had a strange unwillingness to be discovered. We could not divine what he did with them; and this was the bold question to be put. I saw on his table the spoils of the preceding night, some fresh peels nicely scraped and cut into pieces. 'O, Sir, (said I,) I now partly see what you do with the squeezed oranges which you put into your pocket at the Club.' JOHNSON. 'I have a great love for them.' BOSWELL. 'And pray, Sir, what do you do with them? You scrape them, it seems, very neatly, and what next?' JOHNSON. 'I let them dry, Sir.' BOSWELL. 'And what next?' JOHNSON. 'Nay, Sir, you shall know their fate no further.' BOSWELL. 'Then the world must be left in the dark. It must be said (assuming a mock solemnity,) he scraped them, and let them dry, but what he did with them next, he never could be prevailed upon to tell.' JOHNSON. 'Nay, Sir, you should say it more emphatically:—he could not be prevailed upon, even by his dearest friends, to tell.'[112]

"You shall know their fate no further." In this remarkable scene, biographer and subject self-consciously ironize the inexplicable and not wholly inedible remains of orange peel as figures for their collaboration on the *Life*. Relics, metonymic fragments of consumption and of the *Life* itself (in the manner of the incorruptible corpses of saints?), the orange peels are fragrant waste turned into the stuff of mystery and posterity's communal speculation. In their resolute "thingness" and their indeterminate end, they evoke Johnson's afterlife along with his corpse. They are literature's leftovers: not the triumphant proof of Addison's open tomb that makes his texts extraneous, but rather the author's irreducible bones. At this paradigmatically self-referential moment, we are reminded that the *Life* was constructed as both monument and tomb: Boswell's text is haunted throughout by the ghost it endeavors to put to rest, by the undeniable fact and irresolvable mystery of its hero's death. "It is my design," Boswell wrote in a private letter, "in writing the Life of that Great and Good Man, to put as it were into a Mausoleum all of his precious remains that I can gather."[113]

The orange peels have in fact endured, as Boswell and Johnson intended they should. The young Samuel Beckett, deeply depressed and recovering from an unhappy love affair, took special note of the "orange peel mystery" in his research for an unfinished play on Johnson (the first he attempted) called "Human Wishes." (He also transcribed the autopsy report and many pages detailing Johnson's bodily ills.) Orange peels festoon Beryl Bainbridge's *According to Queeney,* which, we'll recall, begins with an account of the autopsy.[114] In an early poem by James Merrill, "The Flint Eye," the orange peels are the ultimate memento mori, recollected by a "matriarch with eyes like arrowheads" as she sits beneath an "orange noon":

> Ah, Dr. Johnson kept the peels, she said,
> In his coat-pocket till they withered quite.
> The rinds of noon like orange-rinds had blown
>
> Out of her lap across the bright, dazed grass,
> Lay shriveling flat upon a scorched perspective,
> As though her gaze imperial had expressed
> No wish to fix them or, since all flesh is grass,
> Fix poets, gross eccentrics who exist
> High in the shallowest stratum of the past.
>
> These learned gentlemen are frivolous soil,
> She said, that one plows up for relics—skulls
> And pottery.[115]

From the "scorched perspective" of the timeless "gaze imperial" of the poem's heroine, at once fossil herself and anti-collector, the orange peels remain un-"fixed," trivial, scattered upon the scorched grass. Shriveling in the sun, linked by syntax and the leveling truism that reduces flesh to grass in the mode of *The Vanity of Human Wishes,* they stand in for the bodies of dead poets, "gross eccentrics who exist / High in the shallowest stratum of the past." Those poets and their followers, critical and uncritical alike, indulge in archaeological digs in just such "frivolous soil," in search of "relics—skulls / And pottery," possessed by a need to defeat such truisms and "fix" the passage of time. At once sophisticated and primitive, like the "amber heads" that hang from the woman's "tribal ears," the anecdotal remnants they glean from not-so-ancient history are an attempt to defeat death, or at least to objectify it. Withered flesh, the orange peels remind us of the carapace the soul leaves behind. Dr. Johnson kept the peels just as Johnsonians keep his corpse. Their plowing for relics is an ongoing autopsy, an excavation, at once literal and metaphorical, of an interior world, a world of the spirit, otherwise closed to them and always eluding their grasp.

In their writing of "The Gospel According to Dr. Johnson" and in their summoning of his living ghost from a scrutinized corpse, Johnsonians create a secular will—their own Testament—based on a Christian paradox, that of the immortal spirit of literature dwelling in the author's mortal body. Ritually ventriloquizing his idol in "The Gospel According to Dr. Johnson" (1892), politician and man of letters Augustine Birrell shows us how Johnsonians are made:

> Death is a terrible thing to face. The man who says he is not afraid of it lies. . . . The future is dark. I should like more evidence of the immortality of the soul.
>
> There is great solace in talk. . . . Let us constitute ourselves a club, stretch out our legs and talk. . . . Sir, let us talk, not as men who mock at fate, not with coarse speech or foul tongue, but with a manly mixture of the gloom that admits the inevitable, and the merriment that observes the incongruous. Thus talking we shall learn to love one another, not sentimentally but essentially.[116]

To worship Johnson is to reverence and to conjure what a prominent Johnsonian has more recently called "some opening to life that texts do not close off."[117] If the love of art is pagan barbarity, the setting up of "jewelled idols, the work of men's hands, which are worshipped by the sacrifice of man's flesh and blood," then the love of authors is the work of the spirit. For scholarly Johnsonians, even (or perhaps especially) Johnson's own texts transcend their material status as objects, their artifice, in order to take on the human face of their author. Johnsonian morality rejects art's painful evidence of authorial labor as human sacrifice, substituting instead, through a sacramental logic, a profoundly uncritical experience of reading that produces individual revelations of communion and community. "The writing," as our modern Johnsonian puts it, "erases itself to diffuse through the reader."[118] Like the Anglican version of the Eucharist, such reading consumes symbolically, through an act of faith, one man's singular materiality; it remembers an individual life and death that cannot be repeated yet must always be imitated ("What should books teach but the art of living?").[119] And like the Anglican sacrament, such reading emphasizes the transforming power of individual *reception* of the host over the nature of its substance. What counts in both cases is the creation of a collective body of believers through individual incorporation of an embodied example. In the case of Johnson, we might call that body the profession—once imagined as personal calling and gentlemanly conversation—of English letters. Its materiality—cast off as corpse, excrement, the peel of an orange, the print on the page—remains; its spirit endures. If we are scholars, it demands our speech.

## Friendly Ghosts

> When he was about nine years old, having got the play of Hamlet in his hand, and reading it quietly in his father's kitchen, he kept on steadily enough, until coming to the Ghost scene, he suddenly hurried up the stairs to the street door that he might see people about him.
>
> HESTER LYNCH THRALE PIOZZI, *Anecdotes of the Late Samuel Johnson*

On the Market Square in Lichfield, the house where Johnson was born and spent his childhood has been renovated as the Samuel Johnson Birthplace Museum. The front of the house was restored in 1989 to its original status as a bookstore. It was here that the young Johnson, a masculine Eve in search of an apple, stumbled upon a volume of Petrarch and acquired a secret taste for romance that stayed with him all his life. The other rooms are used as galleries. Some display Johnsonian artifacts—Elizabeth Johnson's wedding ring, Samuel's shoe buckles, ivory writing tablets, a favorite china saucer Johnson nicknamed "Tetty" after his wife and used daily after her death. The museum, in what is no doubt an effort to increase attendance and raise revenue, is scattered with appeals to children—Johnson's cat Hodge introduces himself as "one of Dr. Johnson's favorite cats" and exhorts children to find the five cats hiding in the house; another flier, a page from a coloring book, challenges them to design stylish wigs for Boswell and Johnson; an elaborate electronic device festooned with portraits of Johnson bears a label reading: "Can you return the dictionary to Dr. Johnson without making a sound?" This house is haunted by a friendly ghost, the Samuel Johnson described by Macaulay as a childhood familiar, a Johnson glimpsed from the nostalgically possessive perspective of the miniature.

Unlike other tastefully empty author's houses, including the Johnson museum in London, the Lichfield museum is simply embodied. Several rooms re-create tableaux from the hero's life. In one scene, a department store mannequin dressed as Johnson's father presides over a reproduction of a bookseller's workroom. The educational fliers nearby pinpoint the display as one of general historical interest, useful for teachers taking schoolchildren on tours. Another tableau, rendering the details of an eighteenth-century kitchen, recreates a Johnsonian anecdote that inadvertently imbricates the viewer in a less-distanced form of curiosity, engaging not historical interest but literary imagination. It portrays, aptly enough for our purposes, a scene of uncritical reading (fig. 19).

A young Sam Johnson (a boy mannequin with a mop of visibly artificial dark hair) dressed in a nightshirt sits before the fire with a book. The flier describing the scene reads as follows:

> "He that peruses Shakespeare, looks around alarmed, and starts to find himself alone."
>
> The tableau depicts the famous incident, which took place in this room when Samuel Johnson was about nine years old.
>
> "Having got the play of Hamlet in his hand and reading it quietly in his father's kitchen, he kept on steadily enough, till he came to the ghost scene, he suddenly hurried upstairs to the street door that he might see people about him."
>
> Because of the lack of warmth in his family Johnson must have found a source of comfort in reading the books he discovered in his father's shop. As well as Shakespeare we know that he found a volume of Petrarch, the Renaissance poet and philosopher, and a book on Scotland, which he recalled when he made his own Scottish tour. He also became an avid reader of tales of chivalry & romance.[120]

Immortalized in awkward effigy is an exemplary scene of Johnsonian reading: a transcendence of time, enabled by literature, fixed in "this room," the very room in which the author-as-reader once sat. This tableau's paradoxical embodiment of a private moment of imagination demands the viewer's act of faith in things not seen. Driven by a lack of familial warmth to imaginary

Figure 19. Basement kitchen tableau. Samuel Johnson Birthplace Museum, Lichfield. Reproduced by kind permission of the trustees of the Samuel Johnson Birthplace.

companionship, the kitchen fire, and *Hamlet,* the impressionable young Johnson sees a ghost; in the tableau's aftermath, he rushes to the street in order to verify his place in reality and history, to erase the terror of the supernatural world of his reading that has momentarily supplanted the real world. He abandons the play and its dreadful encounter with a dead father, to "see people about him."

That ghost, the flier reminds us, still haunts the adult Johnson's criticism of Shakespeare; his individual experience of the scene from *Hamlet* comes to epitomize the universal response of the common reader of Shakespeare in general who "starts to find himself alone." In the preface, as in the epigraph from his notes to *Hamlet* with which we began, Johnson has erased fear, his youthful dash to the street, his need to "see people about him." The common reader fills that need. And so, as we gaze at the kitchen, do we: standing before the clumsy surrogate of the young Johnson who is rapt before a ghostly vision, we provide the living companionship of "people about him," becoming reassuring flesh-and-blood counterparts to the play's ghostly world.

Literature, so this tableau shows, perpetually confounds the dead with the living; by disarming us of our critical distance, it threatens to substitute one for the other. *Hamlet,* in particular, is set in motion by the obligations of the living to the restless dead, whose sin leaves their fate unconcluded. "Remember me," the ghost demands. A ghost himself, Johnson is remembered in this scene in the act of terrified encounter with his own future image.

The reading mannequin whose fear we imagine and assuage, much as Johnson imagines Hamlet's fear as he edits Shakespeare's page, reminds us of the real corpse beneath this author's ghostly figure. Like the orange peel, it at once blocks and solicits our identification. In a secular version of religious communion, Johnsonians build their monument to the author, and with it their professional and national identities, in the disfigured shape of a flawed mortal body. Obeying his injunction to remember, they reinvent their ghost as a benign father, intent not on revenge but on self-perpetuation through companionable common reading. We cannot help but recognize him—his fate is our own. Our hero of reading's romance, he allows us to know that fate no further.

CHAPTER FOUR

# THE EPHESIAN MATRON AND JOHNSON'S CORPSE

> LYCUS: Would any heart of adamant, for satisfaction of an ungrounded humour, rack a poor lady's innocency as you intend to do? It was a strange curiosity in that Emperor that ripped his mother's womb to see the place he lay in.[1]
>
> GEORGE CHAPMAN, *The Widow's Tears*, 3.1.1–4

> Possibly the thought, or talk, of the incisions of anatomy would have disturbed his imagination. But in this case, what was not prohibited was permitted. For it may be easily asked, in the words of the soldier to the Ephesian Matron in Petronius, "Id cinerem aut manes credis curare sepultos?" [Do you suppose that the shades and ashes of the dead care?]
>
> THOMAS TYERS[2]

> So excellent a king; that was, to this
> Hyperion to a satyr; so loving to my mother
> That he might not beteem the winds of heaven
> Visit her face too roughly. Heaven and earth!
> Must I remember? Why, she would hang on him,
> As if increase of appetite had grown
> By what it fed on; and yet, within a month—
> Let me not think on't—Frailty, thy name is woman!—
>
> *Hamlet*, 1.2.139–46

My first two epigraphs invoke a guilty curiosity that emerges at the intersection of two narratives: one the folktale of the Ephesian matron, which Chapman adapts for the stage as the brutally comic *The Widow's Tears;* the other, the history of Samuel Johnson's autopsy, which is the subject of this book. Both are driven by a desire to see beneath the surface, a desire that demands that surface's violation. My first epigraph comments on the curious repetition (with a difference) of a test of female chastity that reveals what the play has already proved, namely, the "truth" of feminine desire and deceit (the play's first three acts told the story of another virtuous widow's successful seduction),[3] the same truth that haunts Shakespeare's Hamlet, and that, as we will see, excludes Hester Thrale from Boswell's brand of Johnsonian devotion.

In Chapman's play, Lysander, provoked by his brother, Tharsalio, is driven to test his exemplary wife's virtue (she is aptly named Cynthia, another name for Artemis, paragon of female chastity and goddess of Ephesus) by feigning his own death and reappearing in disguise to seduce her as she mourns in his tomb. Lycus, Lysander's sensible servant, attempts to dissuade him from this project: it is heartless, baseless—"ungrounded," unmerited torture—a "racking" of his wife's innocence. While Lysander's masquerade will render him at once dead and alive, Lycus's analogy for such intrusive inquisitiveness figures it in terms of the emperor Nero's famous matricide and thus as living anatomy—a rending of the mother's dead body momentarily brought back to life by rhetoric, so that the listener might imagine the pain of her violation. Such "strange curiosity" is inherently narcissistic—Nero "ripped his mother's womb to see the place he lay in," to see the place that preceded his story, his own early grave. (It's worth noting that the play's cynical hero and the instigator of both plots, Tharsalio, is first seen holding a mirror.) Despite this warning, the test of female virtue that ensues proves itself not only through the spectacle of female inconstancy but also through the narcissistic vision of a dead male corpse.

When Johnson's first biographer, Thomas Tyers, in an addendum to his 1784 "Biographical Sketch" in the *Gentleman's Magazine,* attempts to justify the great man's autopsy (the results of which he had detailed at the close of his biography), he engages in a similar moment of hesitation and mirroring. Tyers's attempt to defend the autopsy against the public memory of a great moralist notoriously terrified of death, who had ordered his dead body to be protected from violation by a huge stone,[4] is an extraordinary moment for which I have found no equivalent. Indeed, the question of *why* Johnson's autopsy was undertaken in the first place has rarely been asked—and never answered.[5] As in Lycus's allusion to Nero's matricide, Tyers undoes the fact of death in order to tell a story, to reanimate his hero by imagining his mental distress at the thought of anatomy's violation. But if the evocation of Nero emphasizes the sadistic horror of Lysander's curiosity, Tyers's allusion to the matron story assuages the guilt that the autopsy, and his own imaginary resurrection of a Johnson terrified at the prospect, inspires. What do the dead care about the living? The allusion thus animates Johnson in order to remind us that he is dead. In this context, the Ephesian matron story evokes an uncannily complex relation of the living to the dead, an autoptic vision in which a desire to know the other (the impetus of both narratives) is supplanted by a desire to see oneself (the stuff of the Johnsonian monument)—which has kept Johnson alive to this day.

The Johnsonian romance that begins, as we saw in chapter 1, with the primal scene of the author's autopsy informs this chapter's odd marriage between a mythic female figure—at once a paragon of chaste virtue and a monster of sexual excess—and the mortal remains of a historical man of letters whose exemplary Christian virtue was shadowed by a fear of death so great as to lead some contemporaries to accuse him, posthumously and paradoxically, of the sin of suicide. Both of these paragons of virtue—the woman so renowned for fidelity that she would starve to death rather than survive her husband, the man so renowned for Christian deportment that one sarcastic newspaper writer suggested that an additional saint's day be added to the calendar after his death[6]—are balanced precariously between an exemplary embrace of death and a scandalous choice of life. The matron's shocking embrace of appetite and subsequent defilement of her husband's corpse is mirrored by Johnson's indecent fear of death and desperate deathbed self-scarification. Her substitution makes of Christ's story a lascivious joke—how did a dead man ascend the cross? His final gesture, evoking heroic Roman self-wounding but performed in the hope of postponing death, paradoxically hastened his demise. In each case, a monument rests in the balance.

## Sacred Monuments, Sullied Flesh

If we turn to the figure of the Ephesian matron—most famously known to us from Petronius's *Satyricon* and visible in her descendant, Hamlet's mother, Gertrude—we will recognize a narrative of mourning and misogyny that has endured virtually as long as literature itself.[7] The compilers of *The Types of the Folk Tale* summarize her story thus:

> No. 1510. *The Matron of Ephesus (Vidua).* A woman mourns day and night by her husband's grave. A knight guarding a hanged man is about to lose his life because the corpse has been stolen from the gallows. The matron offers him her love and substitutes her husband's corpse on the gallows so that the knight can escape.[8]

Interestingly—in light of the suppression of narrative in the service of vision with which I began—the compilers of the catalog forget narrative order and agency here in the service of gender symmetry. Their version of the story parallels the soldier's plight with the matron's response to it, when as the story goes in its unabbreviated versions, her submission to his seduction via food, wine, and sex (in that order) causes his plight. In the name of efficiency or decency or both, the catalog omits masculine agency along with the object of narrative curiosity, the "proof" readers of the story seek to confirm, the action inside the tomb: the ultimate spectacle of life juxtaposed with death—the widow who would have died of grief making love with another man on her husband's coffin.

Petronius's sources are contested: a version appears in the fables of his contemporary Phaedrus and has also been linked to an international folktale with analogues in India and China.[9] The majority of Anglo-American classicists locate the story's origins in the lost tales of Aristides of Miletus in second century B.C. One of a series of male attacks on female constancy assembled under the aegis of the Seven Sages of Rome in the Renaissance, popping up in the political philosophy of John of Salisbury, and adapted for continental novellas, the matron's tale was fitted for the stage by Chapman at the beginning of the seventeenth century and reintroduced to the fable genre by the popular fabulist John Ogilby in his 1665 *Fables of Aesop*, a book adorned with the famous motto: "Examples are the best precepts."[10] Her story becomes the stuff of religious psychology when enlisted as an example of the ill consequences of excessive grief in Jeremy Taylor's *The Rule and Exercises of Holy Dying* (1651) and of Hobbesian materialism when illustrating the rule of appetite in Walter Charleton's 1688 neo-Epicurean meditation in her name.[11] She reached the height of her popularity in Europe in the eighteenth century (at which time French and English translations of a Chinese version appeared),[12] when she became the subject of operas and burlesques as well as prose. The equally popular saga of the faithless British merchant Inkle and the devoted Native American maiden Yarico, which brought race and imperial commerce into the picture, emerged in a *Spectator* essay of Steele's in 1709 as a virtuous but frank young lady's refutation of a young man's tedious and clichéd monologue on the subject of "the Perjuries of the Fair, and the general Levity of Women," which does not cease "'till he had repeated and murdered the celebrated Story of the *Ephesian* Matron."[13]

The story varies structurally in two crucial ways, each of which foregrounds its provoking juxtapositions, indeed the potential interchangeability, of the desires of the living with the suffering of the dead. In one type (utilized by Chapman and a feature of the Chinese version), the husband feigns death and masquerades as the seducer in order to test his wife's virtue himself. In the other (particularly popular in the medieval period but recalled by Ogilby), the matron not only offers the corpse to the soldier; she mutilates it so as to make it a more convincing substitute for the stolen gibbeted or crucified body.[14] Her violation of the corpse so horrifies her lover that he responds by either abandoning her, beheading her, or (in a version extolled by John of Salisbury as fact and recommended by one of the story's less-than-amused auditors in Petronius) putting her body on the cross instead. In an intensification of the intimacy of this savaging of the corpse, in the Chinese version the lover's imminent death can only be prevented if he eats the brains of a scholar—the matron offers her husband as the main course.[15]

Transpiring in the tomb and ending on the cross, this romance in miniature ironically inverts Christ's story, replaying it in the pagan and feminine registers.[16] In its stark conjunctions of suicidal despair and sexual desire, dead bodies and murderous appetites, exemplary chastity and excessive lust (Diana of Ephesus, goddess of fertility and chastity, epitomizes both extremes),[17] materialist resignation and ironic resurrection, the story and its multiple variations have been put to a variety of uses over the centuries, throughout which the figure of woman bears a double burden—representing "humanity" in all its variability and excess, while stigmatizing femininity *as* inconstancy.

The matron story, in its multiple forms and as alluded to at the moment Johnson's autopsy is remembered, unsettles the male corpse in its fixed monumentality. Evoked at a particularly fraught moment in the building of the Johnsonian monument, the matron summons a complex and unstable economy of identification structuring the ambivalent desire to penetrate the mystery of a dead male author. The thought of Johnson contemplating his own autopsy provokes an allusion to a tale—itself most famously known in anecdotal form, and so well known as to have become the quintessential eighteenth-century anecdotal shorthand for female inconstancy—that articulates the uneasy relationship of the dead to the living at the end of the eighteenth century as a romance of anatomy.[18]

While the previous chapter looked at Johnsonian "romance" in light of the definition of the genre as "secular scripture," in this chapter I would like to situate "romance" in relation to Petronius's early and highly ironic contribution to Greek and Roman fictional prose narrative, which the matron story encapsulates as anecdote. In a series of lectures provocatively titled *Fiction as History,* the classicist G. W. Bowersock has argued that the popular genre of prose romance emerged in late antiquity as the pagan response—sometimes parodic, always imaginative—to the "miraculous narratives, both oral and written, of the early Christians."[19] The *Satyricon* thus rewrites the New Testament as a brutally literalized, cannibalistic legacy. (From this perspective, for example, the concluding fragment of the will of the poet Eumolpus, in which he demands to be eaten by his inheritors, can be read as a parody of the Last Supper and its sacrament.) The genre of romance also began, as Simon Goldhill has outlined in *Foucault's Virginity,* as an alternative forum for reflection on modes of sexual self-discipline both pagan and Christian, the narrative counterpoint to what Foucault outlined in his late work on antiquity as the "care of the self." While Foucault focused on fixed moral precepts that outlined individual techniques and disciplines of desire, the romance genre, in Goldhill's account, an inherently ironic "rhetorical strategy designed to provoke the male moral subject," put pagan and Christian precepts of chastity and self-governance, and with them a new ideology of sexual symmetry that prescribed mutual chastity in heterosexual relations, into narrative question.[20] In its cynical preoccupation with female chastity and heterosexual fidelity, along with its sophisticated repetition and demystification of the ultimate miraculous romance of Christ's death and resurrection, the matron story thus encrypts the romance genre's origins in an ironic symbiosis of pagan and Christian narrative at the limits of desire and fidelity, life and death, a symbiosis repeated at the late Enlightenment moment of Johnson's autopsy.[21]

By "romance," I also mean seventeenth-century England's "feminine" and largely female-authored descendant of the prose genre of antiquity, the early form of what William Warner has termed "the novel of amorous intrigue," distinguished by shifting identities, flat characters, and dramatically staged scenarios of desire, which preceded the masculine rise of the realist "novel" and with it the bourgeois individual that Johnson himself epitomizes so well.[22] As Helen Thompson has recently shown—in a brilliant reading of Walter Charleton's 1688 Hobbesian rendition of the matron story as a meditation on the materialist conception of the self and the infinite substitutability of the object of desire—romance is individuality's faceless antecedent and feminine ghost, monogamous teleological narrative's serial alternative. In Thompson's reading of Eliza Haywood's subversive transformation of Charleton's materialist subject, in *Fantomina* romance's love plot demands perpetual feminine masquerade in order to gratify the masculine desire for novelty. The repetitive consistency of such feminine variability makes clear that one lover is in fact (as the matron's story horrifically demonstrates from a masculine perspective on female desire) as good as another.[23] By romance, finally, I want to evoke what chapter 1 analyzed as the miraculous plot, at once individual fantasy and Christian belief, in which the anatomist wards off his identification with (and potential substitutability for) the dead body he penetrates by restoring it to imaginary life.

All of these meanings converge at the moment of Johnson's autopsy—my book's primal scene. The varied history of the matron story dissects the ways in which the unveiled secret of feminine inconstancy covers up the unspeakable story of masculine desire and death.[24] In the static narrative of the autopsy, Johnson reanimates the living men around him through anatomy's romance. But this romance preserves male identity by denying a changeability it condemns as the debased product of feminine appetite—it is unthinkable that Johnson's case could repeat the matron's story, that one man, or one corpse, could be as good as another.

The relation of the matron story to *Hamlet* elucidates its multivalent relevance to the Johnson plot. In an anonymous newspaper article, James Boswell imagines a tragicomical entertainment to celebrate Hester Lynch Thrale Piozzi's return to England from Italy with a significant allusion:

> The entertainment which is to be given by Mrs. PIOZZI on occasion of her revisiting the Brewhouse of her first husband, Mr. THRALE, will undoubtedly exceed any *fête* that this country has ever seen.
>
> It is to be truly a *Mischianza,* a medley of tragedy, comedy, music, oratory; in short, of every thing. Mr. HORNE TOOKE is to assist at these "Diversions of PURL."
>
> A grand procession is to take place, in which Mr. JOHN KEMBLE is to walk in the character of *Hamlet,* holding two miniature pictures, and repeating the well-known comparative passage,—
>
> "This WAS your husband," &c.[25]

We will return to Hester Thrale's supposed infidelity—whether to Mr. Thrale or to Johnson is never made clear—later in this chapter. For now I want to unpack a bit further the ways in which Boswell's allusion to the closet scene, in which Hamlet reproaches his mother for desecrating his father's memory, likening the juxtaposed portraits of old Hamlet and Claudius as "Hyperion to a satyr," rewrites Gertrude as a figure of the matron, forging a new link to Boswell's continuation of the *Hamlet* plot in the *Life of Johnson.* If we return to Stephen Greenblatt's reading of *Hamlet,* we can see that "the problem of the leftover" extends not just to the excremental materiality of the Host and of the printed text discussed in the previous chapter, but to the figure of Gertrude, herself a leftover consumed by and consuming another. Gertrude, like the matron, violates the sacred economy of incommensurability in which equivalences are unimaginable.[26] She is on the side of the endless repetitions of desire, a desire that Hamlet would deny along with the body's "sullied flesh."

## Opening Him Up

The more I have explored the matron story and its history, the more it seemed to provide a metaphorical framework for my work on Johnson, most obviously in its ironic highlighting of the erotic and sacramental dimensions of Johnsonian curiosity. But the story also seems to me now to allegorize my own position in relation to the Johnsonian phenomenon itself. To illustrate this, let me turn to an anecdote of my own.

As a lover of Johnson who is writing a book that rejoins the two traditions of Johnsonian investment in the great man's body as medical corpse and anecdotal ghost, I frequently risk repeating the crimes that Johnson's first biographers were accused of: disrespect, irreverence, even violation and desecration. But while (as we saw in chapters 1 and 3) eighteenth-century contemporaries savaged the shockingly intimate anecdotal curiosity of Johnson's first biographers with an equally shocking rhetoric of violent embodiment, for many twenty-first-century Johnsonians, it is offensive to talk about Johnson's body—in its particularity, its excess, its defectiveness, in short, its mortality—at all. When I first presented parts of this material at an eighteenth-century studies conference in 1996, a man in the audience demanded, "Why are you discussing Johnson's body when what really matters is the greatness of his mind?" The fact that the panel was entitled "Samuel Johnson's Body" had completely slipped *his* mind, while I suspect that *my* embodied presence as the only woman on the panel came to stand in for the master's unthinkable corporeality.

The occasional masculine outrage with which my work has been greeted is provoked, or so it seems to me now, by my disruption of the romance that has until recently largely characterized literary work on Johnson, a reverie that evokes Johnson's mortal body in order to deny his mortality; such denial, in the case of those who branded Hester Thrale "*varium et mutabile semper foemina,*" imagines a husband's afterlife through his widow's fidelity.[27] Like Steele's virtuous Arietta, I refuse to "get" the misogyny of the matron story's joke, noticing instead how the joke uses misogyny as a blind. The outraged male auditor (and cuckold) in the *Satyricon* also misses the joke when he demands that the matron pay the ultimate price by replacing her husband's body on the cross with her own.[28] William Blake, never reverent about Dr. Johnson, was more likely to get Petronius's joke, as the following dialogue between Christian and pagan philosopher makes clear:

> *Quid.* "Oho," said Dr. Johnson
> To Scipio Africanus,
> "If you don't own me a Philosopher,
> I'll kick your Roman Anus."
>
> *Suction.* "Aha," To Dr. Johnson
> Said Scipio Africanus,

> "Lift up my Roman Petticoat
> And kiss my Roman Anus."[29]

Inspired by such pagan/Christian conjunctions, I also write with the memory of Johnson's Gargantuan laughter at the insuperable fact of death, the laughter inspired by Petronius's story, an unredeemed laughter that doubts (and in Johnson's case fears) the possibility of an afterlife, that borders, in both cases, on the horrific and the tragic.

## Reanimation and Resistance

Let me return then to the Johnsonian scene of mourning—not the matron in her husband's tomb, but Johnson's friend, Bennet Langton, writing to Boswell in the presence of the corpse:

> I am now writing in the room where his venerable remains exhibit a spectacle, the interesting solemnity of which, difficult as it would be in any sort to find terms to express, so to you, my dear Sir, whose own sensations will paint it so strongly, it would be of all men the most superfluous attempt to——[30]

Alone in the room with Johnson's body, the same room in which the autopsy was subsequently performed, Langton writes to the moment in imaginary communion with his friend and fellow Johnsonian. His failure to describe the "interesting solemnity" of the "spectacle" of Johnson's corpse is matched by his imagination of Boswell's own all too successful attempt to "paint" the scene through his "own sensations." The two men join in emotional contemplation of the image of their dead friend's body at the moment when writing fails.[31]

Langton's failure would be sufficiently explained by his grief at the sight of his friend's "venerable remains," but the editors of the 1934 annotated *Life* imagine its source as an earlier scene that brings the body back to life (references included as they are in the Hill and Powell text):

> The interruption of the note was perhaps due to a discovery made by Langton. Hawkins says, 'at eleven, the evening of Johnson's death, Mr. Langton came to me, and, in an agony of mind gave me to understand, that our friend had wounded himself in several parts of the body.' Hawkins's *Life*, p. 590. To the dying man, 'on the last day of his existence on this side the grave the desire of life,' to use Murphy's words (*Life*, p. 135), 'returned with all its former vehemence.' In the hope of drawing off the dropsical water he gave himself these wounds (see *ante* iv.399). He

> lost a good deal of blood, and no doubt hastened his end. Langton must have suspected that Johnson intentionally shortened his life.[32]

The editors fill in Langton's failed description by supplementing Boswell's narrative as well. The fraught story of Johnson's death evoked here is told by a panoply of voices—the editors', Langton's, Hawkins's, Murphy's—speaking from multiply removed perspectives. The only unmediated gaze at Johnson's body, Langton's, is unable to speak. Turning to Hawkins's account of Johnson's final days, greatly compressed and largely elided by Boswell, the editors revive longstanding questions of agency that blur Johnson's desire for life with the wish for death. By speculating as to Langton's internal state at the moment he drops his pen, they give rise to even more urgent speculation as to Johnson's innermost state at the moment he seizes a scalpel to scarify himself. So ambiguous is the meaning of this final gesture that Murphy's sentence evokes its motivation as a kind of possession by "the desire of life." At that instant of self-wounding, as we have seen, Johnson became at once stoic pagan hero and Christian martyr, potential suicide and surgeon who stole the scalpel from his doctors in order to prolong life. Blurring the desire to live with the will to die, internal wish with external deed, Johnson's self-wounding eternally thwarts and perpetuates moral sympathy and Johnsonian community.[33]

Thomas Tyers testifies to the irrational underside of the motive for the autopsy, and the guilty agency of the surgeons, with this qualification, with which I began this chapter: "Possibly the thought, or talk, of the incisions of anatomy would have disturbed his imagination. But in this case, what was not prohibited was permitted. For it may be easily asked, in the words of the soldier to the Ephesian Matron in Petronius, 'Id cinerem aut manes credis curare sepultos?' [Do you suppose that the shades and ashes of the dead care?]"[34] Who, we might ask as at long last we decipher this strange analogy, is speaking to whom in place of the soldier and the matron? Johnson's "disturbed imagination" is momentarily revivified and addressed here at the moment that the quotation silences such fears by reminding the reader that Johnson is indeed, in perhaps another allusion to Johnson's own translation of Horace's Ode 4.7, "nought but Ashes and a Shade."[35] The complexly variable dialogue the allusion evokes is at once between Tyers and Johnson, and Tyers and the reader who identifies with, fears for, and thus reanimates the violated body. If we trace Tyers's quotation to Petronius's text, we realize that it is in fact a line from Virgil's *Aeneid* book 4, originally spoken to Queen Dido—smitten by Aeneas but mourning for her dead husband—by her sister Anna, recited in Petronius by a faithful female servant to another desperate widow, and transposed by Tyers to the ardent soldier in a move that uses a highly mediated series of allusions to imagine unmediated bodily desire.

While Tyers alludes to a classical scenario he constructs as a heterosexual dyad, his analogy works to construct the Johnsonian tableau as a homosexual encounter between anatomist and corpse: by the logic of his allusion, the male corpse that the soldier persuades the widow to forget becomes the object of the surgeon's sexual persuasion. The attachment of the living to the dead is transposed to Johnson's (and the sympathetic reader's) imagined fear for his own corpse, while the desire of the living is not for the mourner of the dead body, but rather for the suspension of mourning and the penetration of the dead body itself. The reader of this passage is thus suspended between two contradictory states of identification: with Johnson's corpse—momentarily reanimated by allusion, by his known fear of death, and by the imagined threat of the surgeon's knife, and with the surgeon's desire to cut into the corpse's interior, to gain complete authority over the lancet and the body. Petronius's maid's final words of persuasion bring us back to Tyers's transposed reference and to the object of Langton's gaze, reminding the reader and himself that Johnson is indeed dead: "Look at that corpse of your poor husband: doesn't it tell you more eloquently than any words that *you* should live?"[36]

In his attempt to rationalize the "incision" of Johnson's body, Tyers thus reanimates the corpse in a posthumous attempt at persuasion and ultimately successful seduction; by analogy, the surgeon's desire to know how Johnson's uncommon parts might be affected becomes the corpse's desire to submit to the knife. In all three versions of the story—Virgil's, Petronius's, Tyers's—excessive love for the dead is, through persuasion of a third party (for Tyers himself is such a third party, mediating, like the reader, between anatomist and corpse), turned toward the bodies of the living. Transposing not only death and life but also the gendered poles of desire, the quotation's insistence on the indifferent emptiness of Johnson's corpse anxiously dramatizes the ways in which for his survivors Johnson is never fully laid to rest.

This oscillation between the living and the dead, as we have seen, intensifies at the conclusion to Petronius's story. While the matron and the soldier make love in the tomb, the crucified corpse is indeed stolen (by the family for a proper burial). The penalty for the soldier's lapse is worse than death and particularly relevant to the complex web of Tyers's allusion: he must replace the lost corpse with his own crucified body, and he threatens to kill himself rather than do so. The horrific interchangeability of the living and the dead that shadows Tyers's allusion thus becomes brutally literal and is intensified by Petronius's final inversion of Christ's resurrection as another dead man ascends the cross. In the dialogue that ensues between Lysander and Cynthia in act 5 of *The Widow's Tears* as they confront the necessity of mutilating the corpse, Chapman puts this interchangeability into gendered play:

> LYS. Neither fear nor shame? You are steel to th' proof; (but[37]
> I shall iron you)—Come then, let's to work.
> Alas, poor corpse, how many martyrdoms
> Must thou endure, mangled by me a villain,
> And now exposed to foul shame of the gibbet?
> 'Fore piety, there is somewhat in me strives
> Against the deed, my very arm relents
> To strike a stroke so inhuman,
> To wound a hallowed hearse! Suppose 'twere mine,
> Would not my ghost start up and fly upon thee?
> CYN. No, I'd maul it down again with this. *She snatches up the crow.*
> LYS. How now? *He catches at her throat.*
> CYN. Nay, then, I'll assay my strength; a soldier, and afraid of a dead man?
> A soft-roed milksop! Come, I'll do't myself. (5.5.61–74)

Chapman gives his version of the matron an additional power—that of knowledge. In what the play interestingly calls a "cross caper," Cynthia sees through Lysander's act—rather than "suppose," she *knows* the corpse *is* his (while knowing full well that it isn't); it is his imaginary self-projection in the sadistic scenario he has constructed to test her virtue. Her apparent indifference to that corpse's suffering puts Lysander's deathly and self-indulgent self-regard (with its echoes of Hamlet's "Alas, poor Yorick") to the test.

The allusions that suffuse Tyers's commentary on Johnson's autopsy at once reinforce and undermine Johnson's monument.[38] Just as Johnson's self-scarification threatened subsequent narratives of his life by desperately forestalling death, a similar desperate confrontation of life with death informs the vision of the matron, who saves her lover's life, breaking the story's chain of suicidal mourning, by accepting that the dead don't care about the living, that one corpse is as good as another.[39]

When I made a joke at my own expense by announcing to an academic audience my intention to "open Johnson up," I articulated a position that at once connected me to Johnsonian curiosity and distinguished me from it. The slip revealed the ways in which this book at once dissects the idealized and immobile image of Johnson that has dominated the field for so long, on the one hand, while trying to bring him back to life, on the other. Like the matron, I am on the side of life rather than death—but I am also perpetuating a kind of violence. The network of allusion that surrounds this particular corpse also reminds us of its stubborn resistance to all such efforts—including my own—at communion.

In chapter 1, I discussed Tyers's allusion to the death of Cyrus; Jeremy Taylor's discussion of the matron story in his section "Of Visitation of the Sick" in *The Rule and Exercises of Holy Dying* reminds us of a crucial detail in that scene in Xenophon missing from Tyers's account. Recounting the story as an example of excessive emotion (both of grief and love), Taylor draws the following moral:

> When thou hast wept a while compose the body to burial; which that it be done gravely, decently, and charitably, we have the example of all nations to engage us, and of all ages of the world to warrant: so that it is against *common honesty, and publicke fame and reputation* not to do this office.
>
> It is good that the body be kept vailed [*sic*] and secret, and not exposed to curious eyes, or the dishonours wrought by the changes of death discerned and stared upon by impertinent persons. When *Cyrus* was dying he called his sons and friends to take their leave, to touch his hand, to see him the last time, and gave in charge, that when he had put his veil over his face, no man should uncover it[.][40]

In his injunction to mourners to compose themselves after weeping and then to "compose the body to burial," Taylor anticipates the excesses of autoptic vision that lovers of Johnson have committed. While the matron's infidelity serves as a foil to proper masculine devotion, Johnsonians share with her a common origin, an origin Tyers himself admits, in excess. Her excessive grief is mirrored by their excessive curiosity, an excessive desire to keep their hero alive. Taylor warns against precisely the "dishonours" and "impertinence" of such curiosity, against which Cyrus's veil protects. While he does not quote Cyrus's final prohibition, Tyers's network of allusions—as his uneasy locution "what was not prohibited was permitted" might admit—leads us to a comparable scene of inexpiable guilt and betrayal.

If we follow Tyers's evocation through Petronius of the story of Aeneas and Dido to its conclusion, we find at its heart an encounter between two men after one has died. Virgil rewrites Odysseus's meeting with a sullen and silent Ajax in the underworld of the *Odyssey* as the final encounter of Aeneas and Dido after her suicide.[41] Abandoned by Aeneas, Queen Dido falls on his sword, appropriating both the hero's weapon and the Roman hero's death; this gesture is echoed and reversed by Johnson as he wields the surgeon's scalpel against himself in a futile attempt to prolong life. When Aeneas confronts Dido's shade in the Fields of Mourning, she is marked out by her fresh wound among the dead who, pained by love for all eternity, take refuge in a hidden myrtle grove. "Was I, was I the cause?" he asks. He beseeches her to forgive him—"Do not leave my sight. / Am I someone to flee from? The last word / Destiny lets me say to you is this."—to no end. She turns away in stony silence, comforted by the husband to whom she returned in death, fleeing to the forest of shadows, where the living do not enter, where science has no purview, and literature, like Aeneas, can only follow at a distance. What endures is an endless romantic quest—the secrets of feminine desire, the secrets of the dead, remain.[42]

## "He that believes in error, never errs"

This book started as my own attempt to dissect the unique form of author love figured by the joining of literary and medical men in mutual desire and curiosity around Samuel Johnson's corpse. As we have seen, Johnson's autopsy, unexplained, unexpected, and oft reprinted, initiated two centuries of medical and anecdotal Johnsonian anatomy propelled by a commitment, on both sides, to preserving parts of the great man in order to commune with him as a whole. My decision to write such a book was fueled by several contradictory impulses: by my love for both Johnson's work and the character that Boswell created (however unsure I was about the relationship between the two); by my distance and unavoidable difference from the largely male, homosocial, and intellectually conservative[43] tradition of Johnsonian criticism (a tradition to which I remain indebted and to which, it should now be clear, this book is a kind of homage); and above all (as we saw in the introduction), by my curiosity about both the corporeal and incorporeal aspects of the Johnsonian phenomenon that supplanted the literary work with the living author, involving and eclipsing academic criticism in Johnson societies around the world. That curiosity, it seems to me now—my own belated version of the desire that motivated the autopsy—is at the heart of both the devotion and distance that gave this project life.

Two centuries of satiric and serious testimony document the ways in which Johnsonians have attempted to transcend the solitary experience of reading the literary text in order to commune with the living author. The founding event and inescapable fact of this community is Johnson's death. Faithful to their dead hero, model Johnsonian mourners over the past two centuries, like the Ephesian matron in her husband's tomb, have dwelt within his monument, a monument painstakingly constructed from anecdotal remains. At once silent and still, yet disconcertingly in motion, the figure of Johnson has stood for what Johnsonians have hoped would remain fixed and unchanging, while enlivened by their faith, about the English literary canon.

Each anecdotal fragment of Johnson, in the tradition begun by Boswell, is designed to illustrate the enduring truth of his character, to keep him alive in eternal conversation with the faithful. "Had his other friends been as diligent and ardent as I was," Boswell wrote of his efforts to procure anecdotes, "he might have been almost entirely preserved. As it is, I will venture to say that he will be seen in this work more completely than any man who has ever yet lived."[44] At the juncture of lived detail and artful fantasy, such diligent zeal transforms Boswell's text into a vision of his subject as complete and intimate as "Flemish painting," while going beyond portraiture to claim a miraculous embalming of Johnson himself.

Over the course of years of work on this project, I found myself erring from Boswellian fidelity—and not only by my attention to the materiality of the corpse that founds the Johnsonian monument. I was also straying toward an incorporeal version of Johnsonian afterlife, a living textual compliment to the medical and literary dissection and preservation of the author's dead body, a web of anecdotal allusion that connects Johnson to figures as disparate as Nathaniel Hawthorne, James Merrill, Samuel Beckett, and Vladimir Nabokov. This afterlife is also inspired by the anecdote, but it does not oscillate between the fixed poles of life and death, revivification and violation, that define the romance of Johnsonian anatomy exemplified by Boswell's monument. As manifested in multiple traces in literary texts after Boswell, the anecdote becomes something other than a textual equivalent to the relic, no longer a sort of Veronica's veil that allows Johnson himself to be "seen" or, as Walter Scott said of his style, to reproduce, prosopopoeia-like, his face upon the page.[45] Johnsonian anecdote endures not just as Boswell intended it should, in its relation to the real of Johnson himself; the very power of its immediacy also acts to compromise such a relation and such a reality. While anecdote revivifies abstract larger narratives (be they historical or biographical), it also situates all narrative at a crossroads of liminality, blurring the neat either/or distinctions of beginning and end, life and death, fiction and reality, so necessary to the Johnsonian romance of anatomy and preservation.

The anecdote thus points in two directions, which in terms of the poetics that structure this book are epitomized by two different literary plots. Boswell's melancholic devotion, as we have seen, casts him in the role of Hamlet, haunted by his father's ghost and enjoined above all to remember. Whether following in Boswell's footsteps or fighting to undo his construction of Johnson, that role has continued to structure a dominant strain of Johnsonian criticism as a kind of filial piety. My own turn away from such fidelity casts me in the role of the Ephesian matron (for whom Gertrude is a kind of double), who questions the irreplaceability of the lost father, who strays toward the future, toward change.

At this point, another personal anecdote might be in order, since my turn to a consideration of the nature of the anecdote itself transformed my view of both the book's future path and its inception. This conceptual wandering started with an essay that has haunted me from the first with its brilliance and difficulty: Joel Fineman's "History of the Anecdote: Fiction and Fiction." One could categorize (or contextualize away) this piece as a Lacanian/deconstructionist response to Stephen Greenblatt's famous essay on Shakespeare's *Twelfth Night*, "Fiction and Friction" (which in Fineman's argument stands in for the critical phenomenon of New Historicism more generally). But Fineman's tour de force makes such categorization beside the point; his uniquely literary work of history, by focusing on an occasional narrative form that has gone largely unremarked, calls conventional notions of history and literary criticism into question.

A meditation in and on anecdotal form, Fineman's prose plays with its own speaking, situated, occasional voice. (The essay masks itself as a script for a dated spoken performance, an informal preamble to an unwritten paper, hiding much of its riches in footnotes.)[46] For me, to read and reread his text is to bring a former teacher, who died tragically young, back to life and into conversation. I didn't know Fineman well, nor did I know at the time how powerful his influence on my thinking would become. But in his seminar on Shakespeare, in my first year of graduate school, I wrote an essay on Johnson's relationship to that even "greater" author. This paper, the first of many I wrote on Johnson, was also the first that followed a risky intuition, veering beyond my usual dutiful close reading of the text at hand toward the curious fascination of anecdotal detail. This fascination, in Johnson's case, seemed particularly unavoidable, and my initial account of one of the English language's most embodied writer's encounter with one of its most disembodied was the beginning of a long love affair with both Boswell and the quirky contents of the *Johnsonian Miscellanies.*

My essay raised a number of motifs with which I am still preoccupied—in particular the bodily details, delineated in a host of anecdotes, of Johnson's tics, obsessions, melancholy, fear of death, and, more broadly, issues of agency, repetition, disavowal, and the ends of literature. Typed on an IBM Selectric before the age of computers, this early effort, ironically enough, was lost. Hence this book, in which fragments of that first thinking are (however inaccurately) remembered and collected and now take on new life. Given the uncanny resurgence of Fineman's presence here, it is as if this project emerged as an attempt to remember or replace what I intuited on those pages he inspired, to fill the anecdotal void (indeed, what I remember most vividly about my lost essay is its anecdotal evidence) that loss created, a loss that echoes the void left by Fineman's tragic death, which left his own book on the anecdote, the book toward which his essay gestures, unwritten.

Fineman's history of the anecdote casts anecdote's kinship to romance, one I had previously conceived as a form of literary anatomy, in a different light:

> The anecdote is the literary form that uniquely *lets history happen* by virtue of the way it introduces an opening into the teleological, and therefore timeless, narration of beginning, middle, and end. The anecdote produces the effect of the real, the occurrence of contingency, by establishing an event as an event within and yet without the framing context of historical successivity, i.e., it does so only in so far as its narration both comprises and refracts the narration it reports.[47]

This definition of anecdote reminds us of romance's affinity with states of liminality, wandering, and erring: the two genres share an avoidance of ends. Anecdote's province, like that of romance, is the threshold: the *petit récit* inspires narrative progression yet, to use Kate Marshall's phrase, "punctures" narrative time.[48] Seen from this perspective, anecdote's romantic error, its straying from a faithful end (in our case, the preservation of Johnson himself), renders it not as anatomy's fetish but rather as something more reminiscent of the phenomenological impulse to think beyond the thing as object. This version of anecdote rewrites Johnsonian anatomy's metonymic substitution, its use of the anecdote to disavow the void of castration and death, as a productive synecdochic form of emptiness, a hole within a whole.

I use "metonymic" and "synecdochic" in order to invoke Kenneth Burke's aphoristic pronouncements on the "representative anecdote." In keeping with a move toward a phenomenological sense of the anecdotal object, Burke's definition of the ultimate anecdote blurs not only form and content, but also body and mind. Anecdote in Burke's thinking is "itself so dramatistic a conception that we might call it the dramatistic approach to dramatism"; it enlivens representation as action. In its unadulterated state as, in Burke's phrase, "'pure act' or 'pure drama,'" anecdote becomes the word made flesh; its paradigmatic form is the "Act of Creation." (We might also note that for Burke, "the basic unit of action in the human body is purposive motion," which gives Johnson's tics, in their blurring of the bounds between action and motion, a particularly indeterminate place in his anecdotal classificatory scheme.)[49] Ultimately in

Burke's formulation, anecdote encapsulates consciousness in its intersection with the material world. Burke defends his rejection of "'metonymic' anecdotes" thus: "Considering notions of mind-body parallelism, according to which a given state in consciousness has its corresponding physical state, we rejected the tactics of pure behaviorism which would treat the realm of consciousness in purely physicalist terms." Instead, "an anecdote, to be truly representative, must be synecdochic rather than metonymic; or, in other words, it must be a *part for the whole* rather than a *reduction of the mental to the physical.*" In order to define the anecdote as "synecdochic representation," rather than "metonymic reduction,"[50] Burke must conceive of the genre not as material relic of lost presence but rather as the epitome of a creative consciousness in dramatic encounter with the real.

It is just this idea of a dramatic encounter between the creating mind and the material world that Boswell evokes, curiously enough not in an anecdote of Johnson's conversation, but in the final portrait of his hero's contradictions of body and mind with which he concludes the *Life.* In an act of formal homage to Johnson's own practice of literary biography in the *Lives of the Poets*—which juxtaposed a critical evaluation of the writer's art with a survey of his "character," leaving the connections to the free play of the reader's judgment and imagination[51]—Boswell returns in the end to the beginning of his biographical project and to the earlier version of this character that began his *Journal of a Tour to the Hebrides* (the text that presaged the *Life* and that Johnson himself read while it was being composed). Subsuming singular detail into general character, shifting from present to past tense, Boswell's final summation balances a host of seemingly irresolvable opposing traits in a portrait that gains credibility by dint of the thousands of anecdotes that have preceded it. The *Life*'s end comes to life with an account of the vital reanimating power of Johnson's mind:

> But his superiority over other learned men consisted chiefly in what may be called the art of thinking, the art of using his mind; a certain continual power of seizing the useful substance of all that he knew, and exhibiting it in a clear and forcible manner; so that knowledge, which we often see to be no better than lumber in men of dull understanding, was, in him, true, evident, and actual wisdom.[52]

Johnson's transformation of knowledge—the weary, stale, flat, and unprofitable "lumber in men of dull understanding" into "true, evident, and actual wisdom"—is a form of revivification, a bringing of the dead material world to something of "useful substance." Both Boswell and Johnson emphasize the moral utility of thought—the only end of literature for both writers is to enable readers better to enjoy life or better to endure it; but underlying and haunting the translation of useless knowledge into effective wisdom—as Johnson himself acknowledges in his ambivalent criticism of Shakespeare, in his suspicion of any art that champions mimesis over morality[53]—is a power of artistic creation that goes beyond both material and moral ends. There is something eucharistic about Boswell's language here as he describes the transformation of dead lumber into living truth, a magic act of revivification staged by the author's intellect. Yet Boswell's version of Johnson's acquisitive, even predatory mind stages a model of creativity that is less Romantic than Augustan; Johnson does not give birth to new knowledge, rather, in a living version of a curiosity cabinet, he collects and displays what is of use, what (as we might recall with Johnson's declaration of his love of anecdotes in this book's introduction) we otherwise might not get. The transforming power of his mind is limited, marked—as in the case of the anecdote's allegiance to the historical real or of the unchanging face of the dead man in Petronius's joke—by the dead matter it transcends. We might also recall that the wise maxims Boswell praises here, evoked by Johnson as "aphorism"—the true wit that expresses common knowledge—are close literary relatives to the anecdotal form with which he conveys Johnson's moral representativeness. The anecdote's embodied exemplarity perfectly supplements, inverts really, the maxim's abstract universality. Indeed, both the aphorism and the anecdote—unique ways of rearticulating old wisdom—contain within themselves the seeds of self-contradiction.[54] Together both genres animate the author and his readers for centuries to come.

## Literary Reality

The anecdote lets history happen as contingency by giving history's certainty, its end, the lie. I want to link that lie, as Fineman does, to a different sort of truth, namely, that of literature.

> The anecdote, let us provisionally remark, as the narration of a singular event, is the literary form or genre that uniquely refers to the real. This is not as trivial an observation as might at first appear. It reminds us, on the one hand, that the anecdote has something literary about it, for there are, of course, other and non-literary ways to make reference to the real—through direct description, ostention, definition, etc.—that are not anecdotal. On the other hand, it reminds us also that there is something about the anecdote that exceeds its literary status, and this excess is precisely that which gives the anecdote its pointed, referential access to the real; . . . These two features, therefore, taken together . . . allow us to think of the an-

> ecdote, given its formal if not its actual brevity, as a *historeme,* i.e., as the smallest minimal unit of the historiographic fact. And the question that the anecdote thus poses is how, compact of both literature and reference, the anecdote possesses its peculiar and eventful narrative force.[55]

As Jane Gallop points out, "Fineman prizes the anecdote because it is at one and the same time literary and real." Returning to this double ontology of the anecdote at the close of her book of essays on "anecdotal theory," Gallop calls for a theoretical practice that would incorporate the present-tense and social mode of anecdotal "oral storytelling," the mode Fineman himself rehearses in his essay, into theory's factual abstraction:

> Occasional theory is about trying to bring the unpredictability and responsiveness of the flesh into writing. Abstract, disembodied theory, theory in no place or time, dreams of being the last word. Occasional, anecdotal theory, theory in the flesh of practice, speaks with the desire for a response.[56]

Gallop's desire to employ anecdote to bring writing to bodily life and to summon living speech paradoxically echoes Boswell's desire to bring Johnson to life in anecdotal conversation, thereby making his readers "not only *talk,* but *think,* Johnson." It resonates as well with Stephen Greenblatt's famous desire to speak with the dead, a wish that he identifies with the experience of literature itself, and figures as an encounter with the ghost of old Hamlet, the same ghost that haunted Boswell in the figure of Johnson.[57]

What, then, is the difference between Boswell and those anecdotal critics who, however inadvertently, echo him? It has to do with the need for the last word. Greg Clingham has argued that in his obsessive will to recollect Johnson's body, his construction of the life as mausoleum, and his imagination of his own project as a form of mummification, Boswell attempts to transcend death through an evocation—and a kind of taking possession—of the author's bodily presence (a corporeality that nevertheless, as Boswell's own morbid metaphors indicate, perpetually brings death to mind). Boswell's faith in the body is rooted, in Clingham's view, in a naive belief in the correspondence of literature and life, of author and text. Such faith empowers Boswell as a biographer but limits him as a critic, blinding him to the full complexity of Johnson's written work, motivating him to edit Johnson's writings selectively for inclusion in his own text, to find occasional fault with Johnson's style, and to give precedence to the re-creation of Johnson's living conversation (an art at which Boswell surpassed all other practitioners) over Johnson's dead letters. As

Clingham and many others have observed, Boswell's Johnson—the Johnson that has endured as a "great man, writing," rather than a "great writer," that magnificent and magical contradiction of body and mind—is more a portrait of Boswell's own psyche than an accurate record (if such a record were possible) of Johnson the author. So eager is Boswell, in other words, for the last word in his Johnsonian conversation that he would silence the aspects of Johnson "himself," and particularly of Johnson's work, that would interfere with that goal.[58] Ironically enough, however, Boswell's art of the anecdote inspired a literary conversation that took other less definitive directions.

Such divagation is the nature of the anecdotal: the impatience of many critics with the New Historicism's emphasis on anecdotal particularity (reviewed in the introduction), their need for larger more stable narratives, rehearses a movement inherent in the form of the anecdote itself. Fineman's path as a critic veers away from the positivist directness of much of such criticism. Paradoxically errant and inherently literary, Fineman attempts to write a history of anecdote's vacillations between what his essay's subtitle evokes as "fiction and fiction," a history that cannot itself be easily historicized.[59] I'd like to turn now to Fineman's analysis of the fertile and ineluctable intersection of general history and irreducible particularity at the anecdote's literary purchase on the real. Focusing on the anecdote's negotiation of the crossroads of history, literature, and science at a moment of epistemological crisis (a moment that the event of Johnson's autopsy repeats with a difference), Fineman's history of the anecdote—a genre at once rooted in the real of the flesh and inherently fictional—gives nothing the last word.

## Things in Themselves

Fineman's account of the anecdote criticizes the old historicism (which might not be very different from the new) as another version of excessive devotion (a characterization that could apply equally to the melancholy Hamlet and the paragon of feminine virtue, the Ephesian matron, who ultimately gives such devotion the lie); he critiques "pious fidelity" to a particular version of history and calls instead for an "ethics of the real" that is aware of the responsibility of choice (an ethics not unlike Johnson's call in *Rambler* 4 for the proper choice of realistic examples in fiction). In Fineman's account, the anecdote brings together two contradictory constructions of the real that meet at what Francis Bacon called "Instances of the Fingerpost," or "*instantiae crucis*," the crucial experiment that "would allow one to decide, if such an experiment could ever be devised, between competing theories." The anecdote—the etymology of which is "unpublished"—thus connects formally and historically both to the unspeakable Lacanian real and to "a specifically Renaissance translation of the real of history into the real of science."[60] Such a transfer disrupted faith in history as authentic "experience" while presaging a later crisis in the sacred truth of science as prophetic "experiment."[61] At once relic and palimpsest, the anecdote thus contains the seeds of its own undoing, its own escape from time and mortality, its abandonment of a sacred original for the infinite singularity of transformation.

The New Historicist's penchant for anecdote, divided in its loyalties between literature and other more "factual" discourses, renders him or her equally vulnerable to Fineman's critique. In the case of the Greenblatt essay under scrutiny, that other discourse is science, aptly enough evoked by the figure of the father of anatomy, Galen. Greenblatt's reliance on Galen and the discourse of Renaissance anatomy lends another historical dimension to Fineman's dissection of the aporia at the heart of the anecdote's "crucial experiment." The figure of Galen reminds us that Fineman's history of anecdotal truth can also be aligned with the rise of scientific anatomy in the Renaissance, and that the crisis of historiography he elucidates is reflected in the epistemological crisis of the mind/body split that anatomy so violently enacted (and that is at this book's heart).[62] The loss of faith in history as divine prophesy, a faith reinvested in scientific "objectivity," is paralleled by anatomy's violation of the once-sacred body. If we consider the body as the anecdote's special province, Fineman's account of the anecdote leads to a final undecidable convergence of the "things in themselves" triumphantly held aloft as the trophies of scientific anatomy finally freed from the tyranny of the medical text, and the "things in themselves" freed from the tyranny of science's empiricist gaze that both phenomenology and poetry attempt to restore to life.[63]

The late eighteenth-century rage for anecdote, as exemplified by biographers like Boswell and antiquarians such as Isaac D'Israeli, arises in part from a need to heal this rift. "Facts are anecdotes, but anecdotes are not always facts," D'Israeli wrote, long before Fineman defined anecdote as literary *historeme.* Linking history and science, D'Israeli defends his love of the small story:

> It is only the complaint of unreflecting minds, that we collect too many anecdotes. Why is human knowledge imperfect, but because life does not allow of sufficient years to enable us to follow the infinity of nature? . . . Human nature, like a vast machine, is not to be understood by looking on its superficies, but by dwelling on its minute springs and little wheels. Let us no more then be told, that anecdotes are the little objects of a little mind.[64]

Anecdote's detail gains scientific credibility here when linked to the infinite variety of nature. Echoing Johnson's own characterization of the province of biography, D'Israeli connects anecdote both to "domestick privacies" (Johnson's own phrase in *Rambler* 60) and to the province of the psychological novel. More importantly, he insists (reminding us of Boswell on the animating power of Johnson's mind) that anecdotes be viewed as objects of "utility," rather than of "idle amusement."

This utility is decidedly scientific:

> The science of human nature, like the science of physics, was never perfected till vague theory was rejected for certain experiment. An Addison and a Bruyère accompany their reflections by characters; an anecdote in their hand informs us better than a whole essay of Seneca. Opinions are fallible, but not examples.[65]

This is an inversion of Gallop's demand for a theory enfleshed by the world of situated speech. While for Gallop, theory's facts can be of use only when encased and challenged by "the unpredictability and responsiveness of the flesh," D'Israeli claims for anecdote what from our perspective (and the perspective of the postmodern critique of science evoked by Fineman) might ring as the oxymoronic *certainty* of experiment. The literary equivalent of Lockean empiricism, D'Israeli's anecdotes are the "facts" that "by slow accession" accrue into enduring human truths. Anecdotes "inform" better than the ambling aphoristic style of Seneca; while maxims—perhaps for D'Israeli the eighteenth-century equivalent of Gallop's "theory" or W. K. Wimsatt's "concrete universal"—those universal pronouncements that distinguished so much eighteenth-century (and Senecan) prose and poetry, Johnson's "true, evident, and actual wisdom" among them, are born of anecdotes: "Rochefoucault, when with such energetic conciseness, he composed his celebrated Maxims, had ever some particular circumstance, or some particular individual, before him. When he observed, that, 'It displays a great poverty of mind, to have only one kind of genius,' he drew this reflection from repeated *anecdotes* which he had collected in the persons of Boileau and Racine."[66] To paraphrase Johnson in his angry objection to Pope's theory of the best of all possible worlds in *An Essay on Man,* life must be seen, in anecdotal form, before it can be known.

In the case of scholars and men of genius, anecdotes provide "substitutes," with which "we are enabled, in no ordinary degree, to realize the society of those who are no more; and to become more real cotemporaries [*sic*] with the great men of another age, than were even their cotemporaries themselves."[67] D'Israeli's rejection of "vague theory" for anecdote's "certain experiment" can thus be read as a fetishistic return, much like Boswell's in the *Life,* to a lost embodied "real," an undoing of larger, public historical narratives in order to repair anatomy's violation of the body, a return that attempts to bring the dead, particularly the illustrious scholarly dead, to life. "Anecdotes are but squalid skeletons," D'Israeli writes, "unless they are full of the blood and flesh of reflection" and are composed by one who "must therefore possess a portion of that genius which he records."[68] These literary "facts," souvenirs of the esteemed dead incorporated into the being of those who remember them, literally link the living to the dead as we "realize their society."[69]

In just such an effort to "realize the society" of a great man, *Ravelstein,* Saul Bellow's American Jewish homage to Boswell's *Life of Johnson* (reportedly a roman à clef about Allan Bloom, that twentieth-century academic turned public moral philosopher, dubbed by the narrator a citizen of all ages), the divide between the immediacy of anecdotal vision and the objectivity of theoretical depth resurfaces in a recurrent moment of self-indictment for the narrator/Boswell figure, Chick: "I have several times mentioned that ordinary daily particulars were my specialty. Ravelstein also had several times pointed this out, not the noumena, or 'things in themselves'—I left all that kind of thing to the Kants of this world."[70] Chick admits, at times guiltily, to a postmodern poetics of surfaces reminiscent of Jonathan Swift's satiric condemnation of scientific reason's "cutting, mangling, and piercing" of the beautiful surface of life in *A Tale of a Tub*'s "Digression Concerning . . . Madness." When Chick ponders, at Ravelstein's instigation, the meaning of death, his answer is simple: "The pictures will stop." This exchange is more significant than it might first appear, since the narrator's love of the surface of things is, perhaps counter-intuitively, akin to Ravelstein's moral embrace of the preciousness of human particularity. When the two discuss a married couple who have contemplated suicide, Chick offends Ravelstein with his flippant response:

> "If you dislike existence then death is your release. You can call this nihilism, if you like."
>
> "Yes. American-style—without the abyss," said Ravelstein. "But the Jews feel that the world was created for each and every one of us, and when you destroy a human life you destroy an entire world—the world as it existed for that person."
>
> All at once Ravelstein was annoyed with me. At least he was speaking with an angry emphasis. Perhaps I was still smiling at the Battles and it might have seemed to him that I was dissociating myself from the view that you destroyed an entire world when you destroyed yourself. As if I would threaten to destroy a world—I who lived to see the phenomena, who believe that the heart of things is shown in the surface of those things. I always said—in answering Ravelstein's question

> "What do you imagine death will be like?"—"The pictures will stop." Meaning, again, that in the surface of things you saw the heart of things. (156)

Chick's quintessentially anecdotal narrative of memory that rambles, repeats, continually re-encounters the same moment, concludes without concluding: "No one can give up on the pictures—the pictures might, yes they *might* continue. I wonder if anyone believes that the grave is all there is. No one can give up on the pictures. The pictures must and will continue. If Ravelstein the atheist-materialist had implicitly told me that he would see me sooner or later, he meant that he did not accept the grave to be *the* end. Nobody can and nobody does accept this" (222–23). The novel ends, fittingly enough, with a summoning of Ravelstein to vivid anecdotal life, with the desire, above all, for a ravishing *vision* of the lost object that the anecdote fulfills: "But I would rather see Ravelstein again than to explain matters it doesn't help to explain" (231). Having summoned him once more through anecdotal conjuring, Ravelstein's designated biographer and dutiful student finally concludes on a note of incompletion: "You don't easily give up a creature like Ravelstein to death" (233).

Haunting and undermining the anecdote's magical substantiation is the fact of death as the end, as the ultimate thing in itself. As Seneca himself put it in his letter on "despising death," in which he praises Cato for a suicide that "expelled, rather than dismissed" his soul from his body, death provides the biggest challenge to the stoic, whose mandate is to "strip the mask, not only from men, but from things, and restore to each object its own aspect." Death does to us what Seneca recommends we do to it: "Death either annihilates us or strips us bare."[71] The anecdote shares with death the revelation of things in themselves, held in the balance with the possibility of their eternal loss.

Boswell's anecdotes live on, but as D'Israeli reminds us, "anecdotes are not always facts." Unlike facts, the *Life*'s anecdotes do not give their author—or should we say their collector—the last faithful word.[72] Johnson dies to live again not in the tomb of the *Life* but in literary incarnations that have a different kind of purchase on the real. The anecdotalist's awareness of death—think of Boswell and Johnson speculating about the fate of the orange peel—provides an opening into an alternative form of time. Fineman describes the "formal play of anecdotal hole and whole" as a decidedly pleasurable and productive symbiosis of opening and closure; anecdote, as distinguished from other "non-literary forms of reference," he argues, must produce an ongoing "dilation and contraction of the entrance into history," must give birth to larger closed narratives that themselves may subsequently be anecdotalized and thus reopened.[73] Of all Johnson's biographers, only Boswell accomplishes within the bounds of one text Fineman's formal narrative trajectory for the anecdote, since only Boswell enlists his own and others' anecdotes in the service of a coherent biographical whole, a whole that those anecdotes, in their literary afterlife, subsequently undo. I turn now to two counterexamples to the Johnsonian tradition that Boswell began (both as founder and grand obstacle to Johnsonian truth), gesturing toward alternative paths and alternative versions of anecdotal authorship that leave individual texts unfinished, open to other interpretations, taking us in less familiar directions.

## Hester Thrale: *Levior cortice*

By her refusal of a larger narrative arc to her anecdotal collection and by her publishing of Johnson's "domestick privacies" (to which, as we saw in the introduction, she as the mistress of Johnson's surrogate family had unique access) without Boswell's censorious regard for his hero's reputation, Hester Thrale, to give a powerful counterexample, undoes Boswell's monument.[74] Hers was not the last word, but her associational, seemingly haphazard assemblage of anecdotes has underwritten and haunted Boswell's, while continuing on a literary path of its own.

Thrale's collection is admittedly fragmentary, less a pyramid than a curiosity cabinet, less an enlistment of individual anecdotes into proof of Johnson's eternal value than a decidedly personal treasure trove of broken pieces:

> Stories of humour do not tell well in books; and what made impression on the friends who heard a jest, will seldom much delight the distant acquaintance or sullen critic who reads it. The cork model of Paris is not more despicable as a resemblance of a great city, than this book, *levior cortice,* as a specimen of Johnson's character. Yet every body naturally likes to gather little specimens of the rarities found in a great country; and could I carry home from Italy square pieces of all the curious marbles which are the just glory of this surprising part of the world, I could scarcely contrive perhaps to arrange them so meanly as not to gain some attention from the respect due to the places they once belonged to.—Such a piece of motley Mosaic work will these Anecdotes inevitably make: but let the reader remember that he was promised nothing better, and so be as contented as he can.[75]

In a deceptively modest rendition of "you had to be there," Thrale's cork model of Paris figures her version of anecdote as miniature and as souvenir, a decidedly personal memento the value of which resides in authorial recollection rather than the object itself. The souvenir links the reader to a public place enlivened only by private memory, while simultaneously, by virtue of the Horatian motto "lighter than cork," claiming a weightier literary prestige for such trivial worthlessness. Personal proof of intimacy becomes strangely confused here, by association turned to metaphor, with the public currency of classical allusion.

Susan Stewart uses a nearly identical example to articulate one aspect of the necessary incompleteness of the souvenir, explaining that the souvenir's "incompleteness works on two levels. First, the object is metonymic to the scene of its original appropriation in the sense that it is a sample." Stewart goes on to remark that such metonymy

> does not necessarily have to be a homomaterial replica. If I purchase a plastic miniature of the Eiffel Tower as a souvenir of my trip to Paris, the object is not a homomaterial one; it is a representation in another medium. But whether the souvenir is a material sample or not, it will still exist as a sample of the now-distanced experience, an experience which the object can only evoke and resonate to, and can never entirely recoup. In fact, if it *could* recoup the experience, it would erase its own partiality, that partiality which is the very source of its power.

In a summation that rehearses much that has been said about the anecdote so far, Stewart continues with the second aspect of the souvenir's incompleteness:

> The souvenir must remain impoverished and partial so that it can be supplemented by a narrative discourse, a narrative discourse which articulates the play of desire. The plastic replica of the Eiffel Tower does not define and delimit the Eiffel Tower for us in the way that an architect's model would define and delimit a building. The souvenir replica is an allusion and not a model; it comes after the fact and remains both partial to and more expansive than the fact. It will not function without the supplementary narrative discourse that both attaches it to its origins and creates a myth with regard to those origins.

Anecdotes are "partial to" facts, but after and beyond the fact. Thrale's metamorphosis of materials in her sequence of metaphors from weightless cork replica to fragments of weighty marble, along with her associational allusion, exemplifies the souvenir's transformational relationship to its original.[76]

Thrale's prose allows us to stray along with her associations—the cork model, "lighter than cork"—from diminutive versions of vast cities to notions of weightiness inverted on a lyrical scale of literary value. Her personal logic of the souvenir refers to both lived experience and dead letters that, in the case of her particular allusion, still speak. If we turn to the source of her reference, Horace's Ode 3.9, we don't find the standard Callimachean championing of lyric smallness over epic weightiness we might expect. Instead, *levior cortice* refers to an erotic faithlessness that in the case of the poem's speaker, Horace's Lydia, nevertheless inspires enduring love.

Horace's ode merits a moment of consideration, along with the historical and emotional context in which Thrale remembers it. Her reference to Italy would have reminded contemporary readers of the reason for her Italian sojourn: her "ignominious" marriage to the musician Gabriel Piozzi, a passionately chosen union (unlike her first) that caused an estrangement from family and friends, and was considered by many, as this book's introduction explored at greater length, to be a betrayal of an unspoken understanding with Johnson, whom she never saw again.[77] Consciousness of this moment of the book's composition—a moment soon after the loss of the intimacy that underwrites her authority as anecdotalist—haunts her reference to Ode 3.9's dialogue between estranged lovers, each of whom claims to be happier with another, a dialogue that exposes such claims as empty boasts by ending in reconciliation. Both speakers end on a note of hopeful desire, but the last word is the woman's:

> Quamquam sidere pulchrior
> ille est, tu levior cortice et improbo
> iracundior Hadria,
> tecum vivere amem, tecum obeam libens!
>
> [Calais is fairer than
> Any star in the sky;
> And you are lighter than
> A cork bobbing upon
> The waters of the stormy
> Adriatic Sea—
> But if you say you love me
> I'll love you truly forever.][78]

How boldly this allusion rewrites popular rumors of Thrale's infidelity to Johnson as a pledge of eternal love. This is a very different version of speaking with the dead, one that uses the levity of Horace's straying but nevertheless devoted lovers to refute charges of feminine changeability in the service of a different sort of faith. It is not surprising that Thrale—who by marrying Piozzi abandoned not only her dead husband but (to some people's minds at least) the appropriate living one—was likened to the Ephesian matron in the popular press.[79]

Thrale thus turns the masculine tradition of classical reference—and the misogyny that informs that tradition—against itself. By the logic of her allusion, *levior cortice,* the flimsy worthlessness of infidelity (her own, that of woman in the classical tradition of which Horace and Petronius both are a part, and that of the male lover in Horace's playful poem), provides a fragmentary alternative to the solemn weightiness of masculine monuments. Lydia takes her lover back despite the fact that (perhaps even because?) there have been other women after her. Thrale lovingly assembles her anecdotes despite her estrangement from their subject, anecdotes that take on additional poignancy and personal value because of her choice to love a living man rather than a dead one. Her book thus exposes by contrast the desire that underwrites Boswell's, the desire to accept no substitutes, to live with and love the "real" Johnson forever.

As we follow her metaphors, we can see how Thrale's "motley Mosaic" takes apart the weighty material of ancient monuments along with the Johnson rendered in marble in St. Paul's a few years later, while her careful assemblage, her carrying home, of each small piece proves her personal knowledge of the man, the fact of their lived intimacy. Most offensive to Boswell was her testimony to the reality of bodily care, wear, and tear that made even the composition of such a lightweight work at times a burden. Her unpublished preface to her commonplace book of Johnsoniana, part of her larger journal the *Thraliana,* begins:

> In order to accomplish that purpose, and to delight myself by committing to Paper the regard I have for Mr Johnson, I shall begin this Book by mentioning such little Anecdotes concerning his Life, his Character, and his Conversation, as I have been able to collect: All my Friends reproach me with neglecting to write down such Things as drop from him almost perpetually, and often say how much I shall some Time regret that I have not done 't with diligence ever since the commencement of our Acquaintance: They say well, but ever since that Time I have been the Mother of Children, and little do these wise Men know or feel, that the Crying of a young Child, or the Perverseness of an elder, or the Danger however trifling of any one—will soon drive out of a female Parent's head a Conversation concerning Wit, Science or Sentiment, however She may appear to be impressed with it at the moment: besides that to a *Mere de famille* doing something is more necessary & suitable than even hearing something; and if one is to listen all Eveng and write all Morning what one had heard; where will be the Time for tutoring, caressing, or what is still more useful, for having one's Children about one: I therefore charge all my Neglect to my young ones Account, and feel myself at this moment very miserable that I have at last, after being married fourteen Years and bringing eleven Children, leisure to write a *Thraliana* forsooth;—though the second Volume *does* begin with Mr Johnson.[80]

Here Thrale reminds us of the bodily labor and losses of a different sort of authorship—maternity (most of those eleven children died before she wrote those words)—which makes masculine conversation, however memorable, truly trifling.

As the published book comes to an end, speaking of one of many quarrels Johnson had provoked at her table, Thrale writes in an effort at self-justification:

> Such accidents however occurred too often, and I was forced to take advantage of my lost lawsuit, and plead inability of purse to remain longer in London or its vicinage. I had been crossed in my intentions of going abroad, and found it convenient, for every reason of health, peace, and pecuniary circumstances, to retire to Bath, where I knew Mr. Johnson would not follow me, and where I could for that reason command some little portion of time for my own use; a thing impossible while I remained at Streatham or London, as my hours, carriage, and servants had long been at his command, who would not rise in the morning till twelve o'clock perhaps, and oblige me to make breakfast for him till the bell rung for dinner, though much displeased if the toilet was neglected, and though much of the time we passed together was spent in blaming or deriding, very justly, my neglect of oeconomy, and waste of that money which might make many families happy. The original reason of our connection, his *particularly disordered health and spirits,* had been long at an end, and he had no other ailments than old age and general infirmity, which every professor of medicine was ardently zealous and generally attentive to palliate, and to contribute all in their power for the prolongation of a life so valuable. Veneration of his virtue, reverence for his talents, delight in his conversation, and habitual endurance of a yoke my husband first put upon me, and of which he contentedly bore his share for sixteen or seventeen years, made me go on so long with Mr. Johnson; but the perpetual confinement I will own to have been terrifying in the first years of our friendship, and irksome in the last; nor could I pretend to support it without help, when my coadjutor was no more. To the assistance we gave him, the shelter our house afforded to his uneasy fancies, and to the pains we took to sooth or repress them, the world perhaps is indebted for the three political pamphlets, the new edition and correction of his Dictionary, and for the Poets Lives, which he would scarce have lived, I think, and kept his faculties entire, to have written, had not incessant care been exerted at the time of his first coming to be our constant guest in the country; and several times after that, when he found himself particularly oppressed with diseases incident to the most vivid and fervent imaginations.[81]

The "accidents" Thrale describes, disruptions of domesticity caused by Johnson's incivility and conversational aggression, in Boswell's hands became the stuff of heroic anecdote. Thrale's is a different anecdotal universe where the novel also resides, the world of homely trifles infused with memory, the world of "incessant care" for a great man's distress, the world that enables the great man's writing but is omitted from his finished story. When Thrale calls attention to such omission by her escape from "the perpetual confinement" of the lived reality of Johnson's idiosyncrasies, two versions of the great man come into conflict. Thrale exposes Boswell's Johnson as a deliberately limited view, a masculine fantasy of self-sufficiency that edits out Johnson's dependency on the feminine world of household economy, and the intimacy of a friendship that would allow him to entrust her with padlocks in case he should go mad.[82] Small wonder that Thrale's version of Johnson has fascinated twentieth-century psychoanalysts, on the one hand, who have pored over the text of the French letter as if it were pornographic,[83] and writers of fictions such as Samuel Beckett and Beryl Bainbridge, on the other. Boswell's anonymous epigraph in the *Public Advertiser* is representative of Thrale's role in the public imagination as object of fantasy and punishment:

> If *Hester* had chosen to wed mighty SAM,
> Who it seems, drove full at her his BATTERING RAM*
> A wonder indeed, then, the world would have found,
> A woman who truly prefer'd SENSE to *sound.*[84]

Boswellian anecdote transforms Johnson into a monument against not only death but also the body's needs, the heart's desires. Thrale lets us know that anecdote may refer to the real, but the real itself cannot be borne, carried home, without personal cost and loss. Nor can anecdote preserve completely and eternally. Like Thrale herself, anecdote strays, errs.[85]

## Thomas Tyers: *Nugarum contemptor*

Thomas Tyers, amateur man of letters and lawyer, knew Johnson less intimately than Thrale, but he, too, provides an important foil to Boswell. If Thrale anecdotalizes the fragmentary, mortal, and taxing nature of Johnson's monument, Tyers, author of the first biography of the great man to reach print, creates in his haste something altogether different from Boswell's timeless "showing" forth of an unchanging Johnson who is as much himself at seven as he is at seventy. Tyers's focus is not on biography's monumental object but rather on himself as writing subject whose unique signature is anecdotal writing to the moment. Instead of Boswell's logic of equivalence, anecdote for Tyers is governed by an illogic of personal association and almost compulsive allusion. The conclusion to Tyers's text begins (despite over one hundred footnotes in the University of Iowa Press edition) with one of many unidentified quotations:

> "The memory of some people," says Mably very lately, "is their understanding." This may be thought, by some readers, to be the case in point. Whatever anecdotes were furnished by memory, this pen did not choose to part with to any compiler. His little bit of gold he has worked into as much gold-leaf as he could.[86]

What does it mean to equate memory with understanding? In the case of Tyers's anecdotal style, we might think of these words literally: to re-member, to collect anecdotal pieces, is to understand, stand beneath the assembled edifice those pieces construct. Anecdotal records of subjective memory, here as in the case of Thrale marked by personal possession ("this pen did not choose to part with"), come to life and coherence as proof of authorial "understanding"—comprehension, intelligence, knowledge, reason—when assembled in a text.[87] Like the cork model of Paris, what might have produced collective wealth in Boswell's hands takes on an alternative ornamental value for its unique flimsiness, virtuosic proof of the authorial labor of memory "worked into as much gold-leaf as he could."

Tyers's ramblings indeed lead to unexpected beauties. He is not only, as I've explored earlier, the only biographer to speculate as to the motive for and potential reprehensibility of Johnson's autopsy, justifying the event by allusion to the story of the Ephesian matron. He also leaves anecdotal detail to speak for itself in a way that strikes me now as strangely immediate, in a way we might call modern. Perhaps the quintessential moment of such telling randomness is in the following isolated and unexplained jewel: "The words *nugarum contemptor* fell often from him in a reverie. When asked about them, he said, he appropriated them from a preface of Dr. Hody."[88] *Nugarum contemptor*—despiser, defier of trifles, nonsense, waste. Johnson's untraceable reverie leads readers down a chain worthy of Tyers himself in its associations. The word *nugae* was used by Augustan lyric and elegiac poets as an ironically self-deprecating term for lyric poems that, in their seeming worthlessness and in their emphasis on the personal, subverted weightier public epic teleology and thereby Rome's manifest political and imperial destiny.[89]

The words have another less direct but powerful aural resonance: might Johnson be playing, however unconsciously, with the phrase Tacitus used to characterize Petronius Arbiter—author of the *Satyricon* and consul under

Nero—"arbiter elegantiae" [judge of elegance]? Arbiter of fashionable trifles rather than human affairs of life and death, Petronius is thus said to rule in the undying realm of the aesthetic rather than that of politics. The Roman satirist, whose version of the story of the Ephesian matron (itself an anecdote par excellence) haunts Tyers and structures this book, is thus involved by the historian's ironic deployment of his name in something of the same logic as the self-trivialization of Thrale and Tyers, whose intimate anecdotes also undermine official public monuments.[90]

When Johnson murmurs "*nugarum contemptor*," is he thinking of himself—who in so many places championed the trifles of everyday life as not only necessary but essential to both bearable life and edifying biography? *Rambler* 60 on biography draws a mysterious line between useful and useless anecdotal detail. "There are many invisible circumstances which," Johnson claims, "whether we intend to enlarge our science, or increase our virtue, are more important than publick occurrences." Johnson praises Sallust, "the great master of nature, [who] has not forgot, in his account of Catiline, to remark that 'his walk was now quick, and again slow,' as an indication of a mind revolving something with violent commotion." Other biographers, however, "are not always so happy as to select the most important" particularities: "I know not well what advantage posterity can receive from the only circumstance by which Tickell has distinguished Addison from the rest of mankind, 'the irregularity of his pulse.'" The anecdote's worth rests in the balance between the invisible irregularity of Addison's pulse and the visible irregularity of Catiline's walk, between intelligible and illegible bodily difference.[91]

Johnson's phrase and its multiple resonances, its balancing of the self-celebrating aesthetic *nugae*, with its rejection, *contemptor*, might have had a special meaning for the ever-judicious moralist, whose desire was to be a Christian humanist but who became instead a modern author. That these resonant words from an unidentified source by an Anglican divine should fall often from Johnson while suspended in romantic reverie renders them an appropriate epigraph not only for Tyers's biography but for this book as a whole. That they should remain untraced by Tyers or his editor to any particular text of origin sets them free.

## Fragmentary Life

Anecdote's province in the unredeemed material world—weary, stale, flat, and unprofitable—links it generically and historically to the detritus of baroque tragedy that Walter Benjamin, in a parallel consideration of the early modern crisis of faith in historical meaning that Fineman analyzes, reads as the remnants of sacred allegory. Both exemplify an unredeemed materiality, a lost wholeness that cannot be fully recovered.

But as Benjamin concluded, even, indeed especially, a melancholic contemplation of such incompleteness offers a kind of salvation.

> And this is the essence of melancholy immersion: that its ultimate objects, in which it believes it can most fully secure for itself that which is vile, turn into allegories, and that these allegories fill out and deny the void in which they are represented, just as, ultimately, the intention does not faithfully rest in the contemplation of bones, but faithlessly leaps forward to the idea of resurrection.[92]

As was the case with Thrale, faithful contemplation of the dead matter of anecdotes (in our case the matter of Johnson's embodied life) deviates toward an infidelity that contains the seeds of imaginary redemption.

Linking baroque theories of the passions and the Cartesian mind/body split to the biblical fall, Benjamin implicitly opposes science's objectivity to the "pure subjectivity" of knowledge of good and evil, the "opposite of all factual knowledge."[93] Man falls into the subject/object divide and thus into a false distinguishing of good from evil, while God saw creation and knew that it was good. The allegorical leap of faith inspired by the tragic fragments of the anecdotal real thus paradoxically come closer to a divine vision of the wholeness of things in themselves than a fallen judgment of death's remains would allow. The fragment is beautiful because it is unredeemed. "In the spirit of allegory [baroque drama] is conceived from the outset as a ruin, a fragment. Others may shine resplendently as on the first day; this form preserves the image of beauty to the very last."[94] The anecdote's dead matter contains the source of life and afterlife, and no object better epitomizes this paradox for Benjamin than the corpse: "Seen from the point of view of death, the product of the corpse is life."[95]

Lest it seem I have strayed too far from the English matters at hand, we can also note that no play more successfully accomplishes this apotheosis of life from death for Benjamin than Shakespeare's *Hamlet:* "Only in a princely life such as this is melancholy redeemed, by being confronted with itself. The rest is silence."[96] Or for our purposes, the rest is that late Enlightenment version of *Hamlet,* Boswell's *Life of Samuel Johnson.*

My alternative version of Benjamin's allegorical apotheosis is not explicitly Christian—rather it is literary. It has something in common as well with what Benjamin later, in his great unfinished ruin of a work, "The Arcades Project," will celebrate as the leftovers, the trash of everyday modern life transformed by display into a revelation of uniqueness. Anecdote can transform the divine guarantee of meaning, the fixed Platonic logic of resemblance that Hamlet mourns and that Johnsonians honor in their monument, into modern literature's romance with its own metamorphic power.[97] In the founding case of that self-styled Hamlet Boswell, as Ralph Rader and others have shown, the eternal idea of Johnson is the founding principle that guarantees the truth of a progression of anecdotes that begin and end with the great man's death. Johnsonian anecdote thus functions as both relic and icon. In its melancholy contemplation of Johnson's corpse, it makes a leap of faith.

When Boswell's anecdotes reappear in later literature, the law of the icon is broken. We are left instead with the disorienting otherness of what Gilles Deleuze, in his critique of Platonism and its enduring legacy, calls the simulacrum, an otherness that by situating the anecdote in the midst of a shifting literary context incorporates the reader's perspective, transforming anecdote into inflected allusion, deforming and transforming the original.[98] Rather than enforcing repetition as impossible resemblance, literature defies the law of the copy/icon for the simulacra. If we trace the allusive, deviating history of the orange peel anecdote (to name one important example) or, as we shall see, of Johnson's desire to protect his favorite cat, we see repetition with a difference, repetition disloyal to resemblance and erring toward singularity. Nowhere is this more evident than in the case of Vladimir Nabokov's *Pale Fire,* a Boswellian novel in search of a dead author—to which I will turn in chapter 5.

## Cutting Knowledge

Thus, ironically, unexpectedly, and I might even say humblingly, what began as an outsider's affectionate polemic against an Anglophilic and male homosocial form of author worship has ended with my re-induction into the romance of literature, poised on the brink, like the figure of Johnson himself in the multiple versions of his final hours, between life and death. This romance strays from fidelity to the author, but in so erring it revives him in new and strange forms.

Was this what William Blake had in mind when he parodied William Collins's "Ode to Evening," that paradigmatic pre-Romantic poem evoking the threshold mode of perception, that blending of light into darkness so representative of the romance mode, by evoking the great man? In a moment of levity in 1784, the year of Johnson's death, his mannerism is all that remains:

> Lo the Bat with Leathern Wing
> Winking & blinking,
> Winking & blinking,

Winking & blinking,
Like Doctor Johnson.[99]

However comically he is invoked, Blake's Johnson is not still. A half-lit spectacle of failed vision, "winking & blinking," he directs our view to a liminal world elsewhere, an afterlife other than eternal damnation. Could this alternate afterlife be what Samuel Beckett had in mind when he praised Johnson for his vision of "positive annihilation"—a repetition not dead-ending in the poles of damnation or salvation but rather undead and eternal? Was Beckett praising Johnson's heroism for the contemplation of something beyond—something perhaps even more terrifying than—a moral end?

One of this book's most representative anecdotes is therefore not the story of the autopsy but (following Burke's emphasis on action) the moment of ambiguous agency that precipitates it, a story to which I've returned throughout this book. Narrated by Johnson's servant Francis Barber and largely omitted from Boswell's *Life*, this action appears in John Hawkins's biography and is alluded to in the manuscript record of the autopsy, where it serves at once as explanation and unanswered question for the depiction of the opened body that ensues. The story that accompanies Johnson's final act both echoes and undermines Roman stoic suicide in its ambiguous exemplarity. I repeat it again in this context, this time at greater length.

> At eight in the morning of the preceding day, upon going into the bedchamber, his master, being in bed, ordered him to open a cabinet, and give him a drawer in it; that he did so, and that out of it his master took a case of lancets, and choosing one of them, would have conveyed it into the bed, which Frank, and a young man that sat up with him, seeing, they seized his hand, and entreated him not to do a rash action: he said he would not; but drawing his hand under the bedclothes, they saw his arm move. Upon this they turned down the clothes, and saw a great effusion of blood, which soon stopped—That soon after, he got at a pair of scissors that lay in a drawer by him, and plunged them deep in the calf of each leg—That immediately they sent for Mr. Cruikshank, and the apothecary, and they, or one of them, dressed the wounds—That he then fell into that dozing which carried him off.—That it was conjectured he lost eight or ten ounces of blood; and that this effusion brought on the dozing, though his pulse continued firm till three o'clock.
>
> That this act was not done to hasten his end, but to discharge the water that he conceived to be in him, I have not the least doubt. A dropsy was his disease; he looked upon himself as a bloated carcase; and, to attain the power of easy respiration, would have undergone any degree of temporary pain. He dreaded

> neither punctures nor incisions, and, indeed, defied the trochar and the lancet: he had often reproached his physicians and surgeon with cowardice; and, when Mr. Cruikshank scarified his leg, he cried out—"Deeper, deeper;—I will abide the consequence: you are afraid of your reputation, but that is nothing to me."—To those about him, he said,—"You all pretend to love me, but you do not love me so well as I myself do."[100]

The anecdote of Johnson's self-wounding is the stuff of that which Michel Foucault, following Nietzsche, calls "effective" as distinguished from "traditional" history:

> History becomes "effective" to the degree that it introduces discontinuity into our very being—as it divides our emotions, dramatizes our instincts, multiplies our body and sets it against itself. "Effective" history deprives the self of the reassuring stability of life and nature, and it will not permit itself to be transported by a voiceless obstinacy toward a millennial ending. It will uproot its traditional foundations and relentlessly disrupt its pretended continuity. This is because knowledge is not made for understanding; it is made for cutting.[101]

Motivated, like Johnson's surgeons, by disavowed and "base curiosity," Foucault's faithful historian "effaces his proper individuality so that others may enter the stage and reclaim their own speech"; he "mimic[s] death in order to enter the kingdom of the dead, to adopt a faceless anonymity."[102] The different end of unfaithful, effective, what we might call anecdotal or literary history—summoning up Johnson's angry words fraught with the cutting particularity of their moment, their refusal of the kingdom of the dead—is to disrupt such self-effacement and self-recognition in the service of error, in the service of life.

This anecdote's ambiguities culminate in the tenuousness of a present moment haunted by the awareness of impending death that motivates Johnson's final act. Such awareness opens into the space of literature. For the anecdote is not only a generic formal gesture but, as Johnson's cutting reminds us, a unique conundrum, a Möbius strip, of form and content. To trace this anecdotal afterlife, we need to take seriously Fineman's consideration of the anecdote's unique relation to the real, its status as *historeme,* the indigestible irritant at the heart of a potentially endless string of narrative pearls. Its motile multiplicity makes the anecdote at once part of a textual order but—like the couplet or the aphorism that precedes it and distinguishes Johnson's own prose in particular and early modern literary discourse more generally—independent and excerptable, a building block of meaning, a part of a whole that can never be completed. Set against teleology, the anecdote is the genre of an alternative form of romance. This romance is not with the monumental fixity of communal melancholy, with the death (and devoutly wished revival) of the singular author; instead, it is with the power—set in motion in the face of death—of a different kind of conversation, that of literature in communion with itself.

I have already discussed the scathing impact of Johnson's words upon the company of medical men surrounding him. A satirist at the last, despite a lifetime's control of satiric impulses, the dying man unmasks the doctors' sympathy with his suffering—while dismissing their concern for his life—as mere self-regard, exposing himself as fundamentally alone at the moment of death and stoically selfish in his un-stoical desire to live no matter what the physical cost. What I want to focus on once again here is the vibrating uncertainty of his final self-wounding gesture. How are we to interpret it? As an act of self-dissection that inverts the autopsy, meant to postpone death and the surgeon's knife? As stoic suicide, a possibility raised earlier? (We might remember here the classical tradition of suicide that begins with Socrates, moves through Tacitus's Seneca, Plutarch's heroes, Addison's Cato, Shakespeare's Antony, and that tradition's repeated motif of botched male wounds at the final heroic moment. These wounds, as Coppelia Kahn argues, serve to foreground the stoic control that wounds the body as performance, as the stuff of ideology.[103]) As Christian martyrdom—which shares with Johnson's self-wounding the resolution, profoundly antisocial, to endure suffering, indeed to embrace suffering as life, no matter what the cost? As an act of terror or, as Chester Chapin has argued, an act of faith in the prospect of divine intervention, a kind of self-crucifixion in hope of a reprieve from death?[104] As I've noted earlier, and as Hawkins's denial indicates, the suspicion of suicide—since loss of blood due to the wound was the technical cause of death—was the one concrete motive given for the autopsy, a suspicion its results do nothing to dispel. What remains is the act itself, at once fighting and precipitating death.

Is Johnson's self-wounding motivated by fear or by bravery? Encapsulating an ambivalence about suicide inherent in his historical moment, suspended between Christian horror and neo-Stoic respect toward the ultimate act of agency—the Romans referred to the act of taking one's own life with the phrase *sua manu* signifying "freely willed action"[105]—Johnson's cutting also exposes the anecdote's paradoxical relationship to the body's moment, summoning up the immediacy of the flesh while distancing the reader from the experience of pain. Suspended between the terror of death and a desire to live, this act—not suicide but its moral mirror image—renders his body a "bloated carcase" beyond the exigencies of physical suffering. His total command over his body, a command both rhetorical and physical in its demand for life, paradoxically anticipates, even solicits death. That paradoxical relation, that anecdotal suspension, between act and end thwarts all subsequent attempts at resolution. Hawkins tellingly recounts this anecdote, we'll recall, *after* he has narrated Johnson's death and supposed last words, those of a Roman gladiator entering the Arena: "Jam moriturus." Boswell's Johnson, by contrast dies peacefully, as a good Christian, uttering a final benediction: "GOD bless you, my dear!"[106] But in Johnson's case, there are no last words—his ghost still speaks.

CHAPTER FIVE

# CODA: ANECDOTAL ERRANCY, THREE AUTHORS

For who would bear the whips and scorns of time,
Th'oppressor's wrong, the proud man's contumely,
The pangs of dispriz'd love, the law's delay,
The insolence of office, and the spurns
That patient merit of th'unworthy takes,
When he himself might his quietus make
With a bare bodkin? Who would fardels bear,
To grunt and sweat under a weary life,
But that the dread of something after death,
The undiscover'd country, from whose bourn
No traveller returns, puzzles the will,
And makes us rather bear those ills we have
Than fly to others that we know not of?

*Hamlet* 3.1.70–82

For Johnson is immortal in a more solemn sense than that of the common laurel. He is as immortal as mortality. The world will always return to him, almost as it returns to Aristotle; because he also judged all things with a gigantic and detached good sense. One of the bravest men ever born, he was nowhere more devoid of fear than when he confessed the fear of death. There he is the mighty voice of all flesh; heroic because it is timid. In the bald catalogue of biography with which I began, I purposely omitted the deathbed in the old bachelor house in Bolt Court in 1784. That was no part of the sociable and literary Johnson, but of the solitary and immortal one. I will not say that he died alone with God, for each of us will do that; but that he did in a doubtful and changing world, what in securer civilizations the saints have done. He detached himself from time as in an ecstasy of impartiality; and saw the ages with an equal eye. He was not merely alone with God; he even shared the loneliness of God; which is love.

G. K. CHESTERTON, introduction to *Samuel Johnson*

> I trust, I shall not be accused of affectation, when I declare, that I find myself unable to express all that I felt upon the loss of such a 'Guide, Philosopher, and Friend.' I shall, therefore, not say one word of my own, but adopt those of an eminent friend, which he uttered with an abrupt felicity, superior to all studied compositions:—'He has made a chasm, which not only nothing can fill up, but which nothing has a tendency to fill up:—Johnson is dead.—Let us go to the next best:—there is nobody;—no man can be said to put you in mind of Johnson.'
>
> JAMES BOSWELL, *The Life of Johnson*

I begin this conclusion with two writers who discern Johnson is in his uniqueness at the ultimate anecdotal moment of death, taking full possession of him in his integral solitude and by virtue of his loss to history. For G. K. Chesterton at the beginning of the twentieth century, Johnson is a secular saint (and thus all the more heroic since he achieves this distinction in an "uncertain world") who in death was able to "detach himself from time," thus uniting the differing accounts of his end in startling paradox. Rather than the devout, tranquil exemplar of Christian dying that Hannah More and James Boswell imagined, Chesterfield's Johnson is most representative in his confession of the fear of death, most heroic in his timidity, most able to speak for "all flesh" in his willingness to mutilate his own flesh in life's service. Most paradoxical of all, mortal in his immortality, this Johnson unites completely with God in his utter loneliness and isolation. What is left of his death is love—the impartial love (itself an echo of the impartial judgment of Johnson's critical mind), which for Chesterton mirrors divine love, and which alleviates the solitude of those who love Johnson in return and who will continually return to him.

By contrast, Johnson's death leaves James Boswell desolate to the point of speechlessness, able only to quote the spontaneous words of another whom he deliberately leaves anonymous. The solitary isolation that for Chesterton leads to Johnson's love is for Boswell, in the wake of Johnson's death, "a chasm, which not only nothing can fill up, but which nothing has a tendency to fill up," a doubly negative unnatural void, devoid of sympathy, unique in its emptiness, that demands his labor of love. In its careful assemblage of the words of others, in its detailed anecdotal ministrations, Boswell's *Life* arises out of a desire to fill that chasm; but he must ultimately and silently acknowledge its endurance in the words of another. If "there is nobody;—no man can be said to put you in mind of Johnson," then Johnson must always be kept to mind. This chapter considers three different examples of how, in the aftermath of Boswell and in confrontation with his loss, Johnson has been kept to mind. Each demonstrates how the love of Johnson endures not despite his mortality but because of it; the image of the author is a traveler returned from that undiscovered country to share our dread and daily ills and thus alleviate them.

## Nathaniel Hawthorne
## The Uttoxeter Penance and *Our Old Home*

Over the centuries, Johnson's house in Lichfield has become a site of literary pilgrimage for amateur and professional readers alike. One of the most famous visitors, from a nation especially prone to author love "that still reads more than it writes" and from an age fascinated by the "homes and haunts" of famous authors,[1] Nathaniel Hawthorne arrived to find it "as fast bolted as the gate of Paradise."[2] Hawthorne's exclusion leads in his essay on Lichfield and Uttoxeter in *Our Old Home* (1863) to a private sojourn inspired by his lifelong obsession with a famous anecdote from the *Life*. Turned away from the bolted door of a literary paradise that is the old home of a textual father as familiar to him as "the kindly figure of my own grandfather" (122), the fallen Hawthorne wanders to the site of Johnson's private penance for filial transgression—Uttoxeter market—fifty years after the fact.[3] Hawthorne travels beyond the "Field of the Dead Bodies" (121) to an alternative site of Johnsonian memory, an anecdotal site all the more sacred because it is forgotten and thus appropriable by individual imagination. The essay that results conjoins the anecdote's historical immediacy with its afterlife of imaginative distance to convey the spirit of author love. Hamlet's father's injunction, "remember me," resounds in Hawthorne's commemoration of the Uttoxeter penance with a particularly guilty resonance. Barred from the official birthplace, Hawthorne searches for a site for which there is no original, which exists only in anecdote.

The American writer endures the solitary and "ponderous gloom of an English coffee-room" at an ancient Lichfield inn for the sake of a glimpse of the city's famous cathedral and its landmark of secular worship, "the birthplace of Dr. Johnson, with whose sturdy English character I became acquainted, at a very early period of my life, through the good offices of Mr. Boswell" (121). For Hawthorne, the cathedral and the author's house are linked through a common endeavor—the making of immortality. Johnson remains alive to Hawthorne much as he does to Thomas Macaulay in his earlier account of Boswell's hero as nursery companion. Lichfield thus becomes a site of uncanny return to a fantasy father of childhood, both Hawthorne's and America's old home.

> It is only a solitary child—left much to such wild modes of culture as he chooses for himself while yet ignorant what culture means, standing on tiptoe to pull down

> books from no very lofty shelf, and then shutting himself up, as it were, between the leaves, going astray through the volume at his own pleasure, and comprehending it rather by his sensibilities and affections than his intellect—that child is the only student that ever gets the sort of intimacy which I am now thinking of, with a literary personage. I do not remember, indeed, ever caring much about any of the stalwart Doctor's grandiloquent productions, except his two stern and masculine poems, "London," and "The Vanity of Human Wishes"; it was as a man, a talker, and a humorist, that I knew and loved him, appreciating many of his qualities perhaps more thoroughly than I do now, though never seeking to put my instinctive perception of his character into language. (122)

Through Boswell's good offices, readers will recognize that Hawthorne's portrait of his wayward childhood reading self is modeled in part on the anecdote in the *Life* of the equally solitary young Johnson, left to his own devices in his father's bookshop:

> He used to mention one curious instance of his casual reading, when but a boy. Having imagined that his brother had hid some apples behind a large folio upon an upper shelf in his father's shop, he climbed up to search for them. There were no apples; but the large folio proved to be Petrarch, whom he had seen mentioned, in some preface, as one of the restorers of learning. His curiosity having been thus excited, he sat down with avidity, and read a great part of the book.[4]

In both of these anecdotes of boyhood reading, the child's Eve-like desire for forbidden fruit paradoxically leads to legitimate masculine knowledge. Hawthorne's intimacy with the Johnson now familiar to us as a man rather than a writer, while extravagant and unsanctioned, proves just the thing for an American whose "native propensities were towards Fairy Land," not surprising given "how much yeast is generally mixed up with the mental sustenance of a New Englander" (122). While Hawthorne's childhood discovery of Johnson was illicit, it provided the "wholesome food," as quintessentially "English an article as beef-steak" (122–23), that his airy American imagination needed. Hawthorne's American literary wandering leads him, through the vivid immediacy of Johnson himself, to the moral solidity of the English past. His lively imagination fuels an ultimately chaste and chastening romance with Johnson "a man, a talker, and a humorist" rather than Johnson the writer. His childhood self—privy through ignorance to "wild modes of culture" unavailable to disciplined scholars, "going astray" within the book's solitary confines like any romance hero—embarks on an intimate adventure with a literary personage brought to life, through an "instinctive perception of his character" inspired by "sensibilities and affections" alien to intellect and to language.

In Boswell's anecdote, Johnson's potentially dangerous and romantic curiosity, his avidity, similarly results in solid food for scholarly prowess. The early encounter with Petrarch affords a happy end to an apparent idleness, a seemingly aimless wandering from book to book, that leads to an unsurpassed literary mastery:

> What he read during these two years, he told me, was not works of mere amusement, 'not voyages and travels, but all literature, Sir, all ancient writers, all manly: . . . but in this irregular manner (added he) I had looked into a great many books, which were not commonly known at the Universities, where they seldom read any books but what are put into their hands by their tutors; so that when I came to Oxford, Dr. Adams, now master of Pembroke College, told me, I was the best qualified for the University that he had ever known come there.'[5]

The shift to Johnson's own voice, and his insistence on the "manly" nature of his youthful reading, might remind us of an earlier anecdote in the *Life* relayed to Boswell by the antiquarian Bishop Percy, of Johnson's fondness for just the sort of material he denies here:

> 'When a boy he was immoderately fond of reading romances of chivalry, and he retained his fondness for them through life; so that . . . spending part of a summer at my parsonage-house in the country, he chose for his regular reading the old Spanish romance of Felixmarte of Hircania, in folio, which he read quite through. Yet I have heard him attribute to these extravagant fictions that unsettled turn of mind which prevented his ever fixing in any profession.'[6]

This counter-anecdote of immoderate literary appetite reveals a random and romantic undertone to the seeming inevitability of Johnson's march to literary success. An enduring taste for "extravagant fictions" led to an "unsettled turn of mind" and an almost feminine inability to "fix in any profession." "Dr. Johnson," Hester Thrale reminds us, "first learned to read of his mother and her old maid Catharine, in whose lap he well remembered sitting while she explained to him the story of St. George and the Dragon."[7] Like that deluded believer in his own romantic heroism Richard Savage (the hero of Johnson's first literary life), rather than being destined to rule English letters and mores, the Great Cham, "having no profession, became by necessity an author."[8] The "heavy-footed" traveler the young Hawthorne delighted to follow despite his "awful dread of death" (evidence in Hawthorne's mind of the Englishman's incapacity for "spiritual existence" [122]) walked a mazy path of error and fantasy.

Hawthorne's allusion to Johnson's melancholy fear of death leads us to two final anecdotes that delineate the dark side of reading's unreason, the same unreason that inspired the young Hawthorne to love Johnson before he understood him, and the adult Hawthorne to make his uncanny return to Lichfield. The same mother who taught Johnson romance also gave him "those early impressions of religion" that haunted him for the rest of his life. "He told me," Boswell writes, "that he remembered distinctly having had the first notice of Heaven, 'a place to which good people went,' and Hell, 'a place to which bad people went,' communicated to him by her, when a little child in bed with her; and that it might be the better fixed in his memory, she sent him to repeat it to Thomas Jackson, their man-servant; he not being in the way, this was not done; but there was no occasion for any artificial aid for its preservation."[9] Closely following this story is an account of the three-year-old Johnson's "zeal" to hear the "celebrated preacher" and Anglican reformer Dr. Sacheverell—an anecdote that Hawthorne views as an image carved on the pedestal of Johnson's monument in the Lichfield square. A final anecdote of Johnson's childhood reading, recounted by Thrale, is relegated to the *Life*'s footnotes but haunts Boswell's apparent certainty about Johnson as Christian exemplar with the shadow of superstition and doubt:

> At the age of ten years his mind was disturbed by scruples of infidelity, which preyed upon his spirits, and made him very uneasy; the more so, as he revealed his uneasiness to no one, being naturally (as he said) "of a sullen temper and reserved disposition." He searched, however, diligently but fruitlessly, for evidences of the truth of revelation; and at length recollecting a book he had once seen in his father's shop, intitled, *De Veritate Religionis, &c.* he began to think himself highly culpable for neglecting such a means of information, and took himself severely to task for this sin, adding many acts of voluntary, and to others unknown, penance. The first opportunity which offered (of course) he seized the book with avidity; but on examination, not finding himself scholar enough to peruse its contents, set his heart at rest; and, not thinking to enquire whether there were any English books written on the subject, followed his usual amusements, and considered his conscience as lightened of a crime. He redoubled his diligence to learn the language that contained the information he most wished for; but from the pain which guilt had given him, he now began to deduce the soul's immortality, which was the point that belief first stopped at; and from that moment resolving to be a Christian, became one of the most zealous and pious ones our nation ever produced.[10]

Here the "scruples of infidelity" that will plague Johnson to the point of obsession (and that will, to Joshua Reynolds's discerning eye, inform his unintelligible "antics" in polite drawing rooms years later) are alleviated by a lack of scholarly mastery and the indisputable pain of guilt. Like Hawthorne's Dimmesdale, Johnson punishes himself with solitary penance. What can be known, the pain of such self-inflicted suffering, combined with what is unknown—the inaccessible Latin that proves the soul's immortality—inspire the first step toward faith, and toward the apples that lead him to Petrarch, Oxford, and literary fame. Wayward fantasy, doubt, and ignorance thus pave the way toward scholarly authority and moral exemplarity. They provide a relief from skepticism that knowledge—the unfound apples behind the books—cannot.

While I seem to have come a long way from Hawthorne's visit to Lichfield and Uttoxeter, I have in fact returned to it through the theme of "voluntary, and to others unknown, penance." For Thrale's phrase could easily have described the later incident in Boswell's *Life* to which Hawthorne compulsively returned in his writing on Johnson. Turning from the closed door of Johnson's birthplace, Hawthorne is "a good deal consoled by the sight of Dr. Johnson himself, who happened, just at that moment, to be sitting at his ease nearly in the middle of Saint Mary's Square, with his face turned towards his father's house" (131). He refers, of course, to the statue of Johnson erected in 1838, "immensely massive, a vast ponderosity of stone, not finely spiritualized, nor, indeed, fully humanized, but rather resembling a great stone-boulder than a man" (131–32). Hawthorne aligns himself with Johnson both spatially and dramatically—both are rendered sons excluded from their father's house. Looking with "the eyes of faith and sympathy" (132), Hawthorne brings the stone to life, and he does so in anecdotal form by paying close attention to one of the three bas-reliefs on the statue's pedestal (see fig. 20).

> The third bas-relief possesses, to my mind, a great deal of pathos, to which my appreciative faculty is probably the more alive because I have always been profoundly impressed by the incident here commemorated, and long ago tried to tell it for the behoof of childish readers. It shows Johnson in the market-place of Uttoxeter, doing penance for an act of disobedience to his father, committed fifty years before. He stands bareheaded, a venerable figure, and a countenance extremely sad and woe-begone, with the wind and rain driving hard against him, and thus helping to suggest to the spectator the gloom of his inward state. Some market-people and children gaze awe-stricken into his face, and an aged man and woman, with clasped and uplifted hands, seem to be praying for him. These latter personages (whose introduction by the artist is none the less effective, because, in queer proximity, there are some commodities of market-day in the shape of living ducks and

Figure 20. Photograph (taken 2004) of Uttoxeter bas-relief panel of the Johnson statue by Richard Cockle Lucas, 1838. Reproduced by kind permission of the trustees of the Samuel Johnson Birthplace.

> dead poultry) I interpreted to represent the spirits of Johnson's father and mother, lending what aid they could to lighten his half-century's burthen of remorse. (132)

In the tableau's comic juxtaposition of commodities, "living ducks and dead poultry," Hawthorne gestures toward his essay's larger work of enlivening "creatures of mere fiction," above all "the sturdy old figure of Johnson himself," through personal reverie so that "they live, while realities have died" (130). This labor of immortality takes reading beyond the marketplace toward secular worship; the anecdote rendered tableau on the pedestal depicts a view of saintly martyrdom that leads to Johnson's literary canonization. The original anecdote in the *Life,* listed in the index under "an act of penance," is relayed by Johnson himself in the year of his death:

> He mentioned that he could not in general accuse himself of having been an undutiful son. 'Once indeed, (said he,) I was disobedient; I refused to attend my father to Uttoxeter-market. Pride was the source of that refusal, and the remembrance of it was painful. A few years ago, I desired to atone for this fault; I went to Uttoxeter in very bad weather, and stood for a considerable time bareheaded in the rain, on the spot where my father's stall used to stand. In contrition I stood, and I hope the penance was expiatory.'[11]

This brief moment in Boswell's text inspired a great deal of nineteenth-century specular curiosity, resulting not only in the sculpted scene but also in a history painting: a commemoration of the scene (exhibited in the Royal Academy circa 1869) by Eyre Crowe Sr. is on display at Johnson's House in London (fig. 21). Yet Hawthorne, in a physical embodiment of what chapter 3 discussed as a Johnsonian form of prosopopoeia, mirrors the object of his curiosity in solitude. He thus remains true to Johnson's solitary penance, itself an act of private commemoration (he searches out "the spot where my father's stall used to stand," known only to himself)—a narrative condensed by subsequent readers into a singular visual scene of sympathetic yet unintelligible interiority. Johnson's anecdote renders himself his own private monument; Hawthorne, the story's true audience, undergoes the same transformation.

In the tableau depicted on the Lichfield statue, the privacy of penance is rendered intelligible as hagiography. Johnson's "inward state" externalized both by his sad expression and the punishing storm, strikes awe in the spectators, among whom are two bystanders who "seem to be praying for him." These two know the true meaning of the scene and lend it the air of a Renaissance painting in which the knowledge of Christian revelation allows multiple

Figure 21. Eyre Crowe Sr., *Dr. Johnson Doing Penance in the Marketplace at Uttoxeter* (exhibited at the Royal Academy, ca. 1869). Reproduced by courtesy of the trustees of Dr. Johnson's House.

temporalities within the same frame.[12] Hawthorne declares these figures to "represent the spirits of Johnson's father and mother," interceding, like souls in a literary version of purgatory, "to lighten" their son's "half-century's burthen of remorse" (132). In this fantasy version of the scene, Johnson's Hamlet-like guilt is alleviated, his parental ghosts rendered benign.

The scene inspires Hawthorne to leave "Lichfield for Uttoxeter, on one of the few purely sentimental pilgrimages that I ever undertook, to see the very spot where Johnson had stood" (133). Hawthorne is very concerned to locate this scene in the appropriate spot, the spot Johnson himself sought to commemorate, and he is determined to find it where he imagines Johnson himself must have staged it, in "the middle of the market-place":

> But the picturesque arrangement and full impressiveness of the story absolutely require that Johnson shall not have done his penance in a corner, ever so little retired, but shall have been the very nucleus of the crowd—the midmost man of the market-place—a central image of Memory and Remorse, contrasting with, and overpowering the sultry materialism around him. (134)

Hawthorne seeks to realize the moral tableau on the monument's pedestal by his own reenactment. Standing on the literal site of anecdote, allegory—in the form of Johnson's embodiment of "Memory and Remorse" meets history—in the form of "the sultry materialism" of the very marketplace where he once stood. The conflation of the two in "external ceremony," given life by Johnson's "vitality and truth" (134), transforms the mundanity of the market into sacred space, informing Hawthorne's wishful thinking with the illusion of accuracy. Hawthorne's reverie becomes a form of performative and contagious identification:

> The people of Uttoxeter seemed very idle in the warm summer-day, and were scattered in little groups along the side-walks, leisurely chatting with one another, and often turning about to take a deliberate stare at my humble self; insomuch that I felt as if my genuine sympathy for the illustrious penitent, and my many reflections about him, must have imbued me with some of his own singularity of mien. If their great-grandfathers were such redoubtable starers in the Doctor's day, his penance was no light one. This curiosity indicates a paucity of visitors to the little town, except for market purposes, and I question if Uttoxeter ever saw an American before. (134–35)

Hawthorne has become a momentary ghost of the Doctor himself. This passage leaves it unclear what about the stranger is most striking to the eyes of the locals—his idle standing about in the center of the marketplace, the "singularity of mien" produced by "my genuine sympathy for the illustrious penitent," or his status as a rare American. Whatever the case, Hawthorne's affinity for Johnson and his performative commemoration of Johnson's penance paradoxically render him foreign to modern-day commercial Uttoxeter. The townspeople's curiosity convinces him both of Johnson's original suffering—his having been rendered a spectacle to an uncomprehending public—and of the erasure of his spectacular penance from historical consciousness. The artist created an awestruck audience on the Lichfield pedestal, adding the interpretive clue of Johnson's parents, interceding from beyond the grave, in order to make Johnson's penance intelligible. What renders Hawthorne most odd to actual Uttoxeter spectators is his own curious desire to commune with Johnson, to bring him beyond art and into life. By performing that communion—rather than transforming it as the anonymous Lichfield sculptor does into tableau—Hawthorne, in the midst of the marketplace, experiences it as solitude.[13] He gives up and goes to dinner—"Dr. Johnson would have forgiven me, for nobody had a heartier faith in beef and mutton than himself" (135).

The pilgrimage, Hawthorne concludes, was foolhardy; timeless anecdote is preferable to the particularity of history. "A sensible man had better not let himself be betrayed into these attempts to realize the things which he has dreamed about, and which, when they cease to be purely ideal in his mind, will have lost the truest of their truth, the loftiest and profoundest part of their power over his sympathies" (135). What Hawthorne terms "the moral sublimity of a great fact" cannot depend on "things which change and decay," if it is to be "immortal and ubiquitous." Uttoxeter's local ignorance is the proof of Johnson's universal currency. Yet Hawthorne remains unsatisfied, still wishing "that I could honestly fix on one spot rather than another, as likely to have been the holy site where Johnson stood to do his penance" (136), and cannot help but fault the locals for failing to provide a

> local memorial of this incident, as beautiful and touching a passage as can be cited out of any human life; no inscription of it, almost as sacred as a verse of Scripture, on the wall of the church; no statue of the venerable and illustrious penitent in the marketplace to throw a wholesome awe over its earthliness, its frauds and petty wrongs of which the benumbed fingers of Conscience can make no record, its selfish competition of each man with his brother or his neighbor, its traffic of soul-substance for a little worldly gain! Such a statue, if the piety of the people did not raise it, might almost have been expected to grow up out of the pavement, of its own accord, on the spot that had been watered by the rain that dript from Johnson's garments, mingled with his remorseful tears. (137)

From text—passage of life, verse of Scripture—to statue, Johnson's penance has the power to personify itself. Through Hawthorne's rapturous belief, it creates its own relic, its own monument. But such rapture must remain by necessity solitary and unintelligible. Having been told after his visit "that there were individuals in the town who could have shown me the exact, indubitable spot where Johnson performed his penance" (137), Hawthorne remains obdurately unconvinced. Such knowledge must have resulted in commemoration—to remember and not to reverence provokes both Hawthorne's contempt for others and pride in himself.

Luke Bresky situates Hawthorne's paradoxically self-ennobling disappointment in an important transatlantic context that tells us something about the history of the American version of Anglophilia of which Johnsonophilia is a subset. In this scene, he argues,

> an American pretends to have a clearer perspective on English heroism than the English people; but the American Boswell falls back on this idea in a reluctant and crestfallen frame of mind because, for one thing, it concedes the purely American importance of his biographical perspective. Having "long ago" chosen a small vic-

> tory over class-conscious arrogance [in *Biographical Stories for Children*] as the definitive moment in Johnson's life, he appears to forget that there was ever anything especially American about that choice. His fallacy . . . lies in his insistence on a hero who belongs more closely to a single place and time—not less so—than the English perceive. And having ironically forgotten the democratic spirit that moved him to canonize the penitent Johnson in the first place, he cannot even say why the common people of Uttoxeter ought to share his outlandish interest in this figure.

Hawthorne's forgetting of the populist "moral" he had drawn from the anecdote in his earlier didactic text, for Bresky, is also a question of proper scale and perspective worthy of Swift's Gulliver: "Dwarfed by a 'colossal' statue of Johnson in Lichfield, the former patriotic biographer prefers to focus on the bas-relief built into the pedestal, and hastens away to Uttoxeter, where he paces off the little town's market-place. . . . [T]he close-up perspective ties Johnson punitively to a patch of English soil, but fails to link him with the people who inhabit it."[14] This failure, I would argue, is the result of a particular sort of autoptic anecdotal vision that renders Hawthorne, not merely viewing but replicating an anecdotal moment he despairs of locating historically, at this moment a Johnsonian. Would he have been pleased or disappointed to learn that the Lichfield society, in its annual celebration of Johnson's birth, publicly commemorates Johnson's penance? Like the singing of the Johnson hymn composed by their hero in the days before his death, such a ritual repeats an inscrutable interiority that at once solicits and repels sympathy. The moment of Johnson's penance—an intensely private period of atonement for an untold and, to many readers of Boswell, negligible sin, a ritual that does not seem proportionate to its own historical origins—encapsulates both in content and in its effect on generations of readers the paradoxes of solitude and community that the figure of Johnson inspires.

As in the anecdotes of early uncritical reading with which I began, ignorance reaps unexpected rewards. Uttoxeter's absurd and materialist amnesia perpetuates rather than commemorates Johnson's penance—forever will he remain a curious figure of "Memory and Remorse," amidst the workaday world of the market, discernible only to those who love him from a distance. This distance, in Hawthorne's case both historical and geographical, enables sympathy, creating immortality through necessarily solitary communion. Transforming commemoration to consecration, it leaves history behind.

## Vladimir Nabokov
## Hodge and *Pale Fire*

> The imaginary is not formed in opposition to reality as its denial or compensation; it grows among signs, from book to book, in the interstice of repetitions and commentaries; it is born and takes shape in the intervals between books. It is a phenomenon of the library.
>
> MICHEL FOUCAULT, "Fantasia of the Library"

Think of Samuel Johnson in his London garret, copying out passages for his definitions for the *Dictionary,* a self-styled humble drudge giving up, in his magisterial and mournful preface, his ambition to embalm the language, facing instead its perpetual mutability and his own personal mortality. Or think of him addressing Falstaff as a living being in his commentary on Shakespeare's text, closing his eyes to the end of *King Lear,* or remember him running into the street when he encounters for the first time, through solitary reading, the ghost in *Hamlet.* Recollecting that moment in conversation with Hester Thrale, he warns her to "make your boy tell you his dreams."[15] Think of Boswell in his study, depressed and in debt, assembling the materials for his great ghost story, the *Life,* copying, sometimes literally pasting, the great man's words into its pages, assembling his anecdotal monument in "infinitesimal fragments." "The Renaissance explores the universe," writes Walter Benjamin, that great modern archivist, "the baroque explores libraries."[16] Suspended between the baroque rage for encyclopedic order and the modern fantasia of the archive, the Enlightenment, the age of vigilant doubt, saw ghosts everywhere, particularly in and through the impossible worlds of print.[17] The uncanny animating power, what Foucault calls the "visionary experience," of print informs the figures of Johnson and Boswell, as they rehearse *Hamlet* in a late Enlightenment key.

Vladimir Nabokov's *Pale Fire,* that most self-conscious and allusive of fictions, seems to contain whole libraries within its frame of reference, bound together in infinite combination like the index cards on which the novel's poet and chief object of desire, John Shade, composes his poem, bundled by a rubber band that when it falls to the floor, always lands in the shape of infinity. Diligent critics have hastened to the archive in Nabokov's wake, tracing sources from Anglo-Saxon etymology to Nabokov's own four-volume scholarly edition of Pushkin; to Shakespeare (and Russian editions and translations thereof); to Viking, Russian, and English history; to Alexander Pope (who is the subject of John Shade's monograph, *Supremely Blest,* and whose

*Essay on Man* and *Rape of the Lock* form a delicate counterpoint to Shade's own poem); to Wordsworth, Matthew Arnold, and Goethe; to modern science (including Nabokov's private code of lepidoptery); as well as to the mirrored scholarly pairs of Pope and Swift, Wordsworth and Coleridge, and Johnson and Boswell, who haunt the novel's central pairing of writer John Shade and commentator Charles Kinbote. That central doubling of poet and critic/devotee/parasite has been the subject of vehement critical debate in which opposing sides have argued among other things that one member of the pair in fact imagined the other. *Pale Fire*'s linking of textual scholarship with delusion is more than just the parody of academe some critics have been content to take it for; Nabokov's "centaur-work, half poem, half prose"[18] provides us with a quintessential example of the madness of the library, as a crazed Kinbote absconds with the manuscript of his hero's poem, and through devoted (mis)reading informed by desire, paranoia, nostalgia, and sheer creative brilliance, transforms the genre of the scholarly edition into "the monstrous semblance of a novel,"[19] a novel in which he, as the exiled king of Zembla (the land of mirror reflections), a distorted mirroring of the dethroned English Charles, is the hero.

Kinbote and Shade's symbiotic authorial endeavors are haunted, in an endlessly variable counterpoint that is Nabokov's own authorial signature, by the allusive, associative, multiplicitous play of language across time and space. We need only to think about the various resonances and reflections of each character's name to enter into this dimension of the novel. Kinbote has his own mirror image and possible alter ego, the real-life Russian Shakespeare scholar (also an émigré professor of Russian in the novel) Botkin. "Kinbote" is also the Anglo-Saxon word for "compensation for the murder of a relative," while "botkin" reminds us of Hamlet's bare bodkin or as Kinbote's index puts it, "Danish stiletto," as well as king-bot, or king killer, defined in the index, "king-bot, maggot of extinct fly that once bred in mammoths and is thought to have hastened their phylogenetic end," or as the poet's wife, Sybil Shade, describes Kinbote, "an elephantine tick; a king-sized botfly; a macaco worm; the monstrous parasite of a genius."[20] With "stiletto" we are brought back to John Shade's initial vision of "svelte/Stilettos of a frozen stillicide" (34–35), Nabokov's nod to Coleridge's "Frost at Midnight" and the creative, preservative, and deathly powers of the imagination. As for Shade himself, his name resonates both with shadowy reflection, and—more on this later—shade or ghost; he is, as it turns out, like Boswell's Johnson, neither dead nor alive.

The connection to *Hamlet* signals a final pun on Botkin, namely "bodykins"—an Elizabethan term for the Host. But the magic of *Pale Fire,* as we shall see, is more anecdotal than transubstantiative. It is more in sync with the line that follows Hamlet's response to Polonius's resolve to treat the players according to their deserts: "God's bodykins, man, much better. Use every man after his desert, and who should scape whipping?" (2.2.533–34). While in Stephen Greenblatt's reading Hamlet's words take on theological resonance: "It is the commemoration, the anamnesis, of [Christ's] sacrifice, that lies at the heart of the Mass, and it is this anamnesis that has evidently called forth Hamlet's exclamation: 'God's bodykins,'"[21] Nabokov's act of anamnesis rejects divine reparation for human sin and limitation, exchanging the Mass for communion with the printed page. His view of the pun thus comes closer—though Nabokov would be horrified to think so—to Adam Phillips's interpretation of Freud's quotation of the passage in a letter on the futility of biography. As is often the case in Freud, Hamlet is "brought on . . . to tell us the truth about truth":

> The truth is—whatever else may also be true—that people are fundamentally punishable; that they are guilty because they are both criminals and imposters. . . . Hamlet's suicidal browsing and brooding is, among other things, about how difficult it is to die; or even to die in one's own fashion, according to one's own wish.[22]

For Freud, the biographer's desire, and this applies to the even more fervent author love of which Kinbote is a practitioner, is always ambivalent, precisely because the biographer wishes not "to depose his hero, but . . . to bring him nearer to us. That means, however, reducing the distance that separates him from us: it still tends, in effect, towards degradation."[23] The self-aggrandizing delusion of the desire to bring the author near, to have him die according to our wishes rather than his own, initiates what Phillips calls an "infinite fictive regression" (a regression echoed in the original referent of Hamlet's words—players who may or may not be legitimate). This desire fuels Boswell's *Life,* and it is *Pale Fire*'s subject; this desire is the source of art. Despite the famous anecdote about to be discussed, Hodge will indeed be shot, but somehow he will live on in the epigraph, in the fantasy of print, in the romance of the reader's imagination.

. . .

In its hallucinatory confusion of art and life in the interstices between books, *Pale Fire*'s landscape of novelistic "reality" becomes a semblance of the printed page. In his poem's opening, John Shade gazes out the window at his own reflection, only to puzzle over the text of the snowy ground:

. . . Whose spurred feet have crossed
From left to right the blank page of the road?
Reading from left to right in winter's code:
A dot, an arrow pointing back; repeat:
Dot, arrow pointing back . . . A pheasant's feet! (20–25)

Allied with Shade's vision of world turned print here, Nabokov's punning wit throughout makes us constantly aware of every dimension of the text as seething with language. As his poem's quietly domestic consideration of love, loss, and the possibility of an afterlife concludes, Shade makes a note to himself that wittily marks the poem itself as at once perfectly ordered and incomplete:

*Man's life as commentary to abstruse*
*Unfinished poem.* Note for future use. (939–40)

He concludes with a quiet affirmation of literary design, and a particular insistence on the historical moment:

I feel I understand
Existence, or at least a minute part
Of my existence, only through my art,
In terms of combinational delight;
And if my private universe scans right,
So does the verse of galaxies divine
Which I suspect is an iambic line.
I'm reasonably sure that we survive
And that my darling somewhere is alive,
As I am reasonably sure that I
Shall wake at six tomorrow, on July
The twenty-second, nineteen fifty-nine,
And that the day will probably be fine (970–82)

Shade does wake "tomorrow" after writing those lines, but he does not live through the day. *Pale Fire,* for all its art (and I have barely scratched the surface of its brilliance here), confronts chaos in the form of a random bullet that races, beyond Shade's marking of the date, through space and time—fired by an escaped lunatic prisoner who is imagined by Kinbote to be a political assassin, Gradus, whose movement toward his final goal (in Kinbote's imagination he seeks to kill him, the exiled king of Zembla, but murders Shade instead) Kinbote hears in the meter of Shade's iambic lines,[24] the same bullet that in 1922, also aimed at another, killed Vladimir Nabokov's father, exiled from Communist Russia to Berlin.[25] That bullet is the random yet overly determined anecdotal catalyst, Nabokov's little piece of the traumatic real, the detail that makes the difference between, in Michael Wood's terms, an author's theoretical death and his actual murder.[26] A novel that cycles between imagined reparation and repeated loss, the art of *Pale Fire,* which pits Shade's Popian faith in orderly iambs against Kinbote's wildly imaginative Shakespearean prose, can never come to an end.[27] Its infinite artfulness spirals around that "opening to life that texts do not close off,"[28] an opening that is also an opening to death, and that is the source of the Johnsonian phenomenon as well.

In its endless play with mirrors and in its tragicomic exploration of author love, *Pale Fire* has been acknowledged by critics for a host of creative transformations of literary sources, but it has never been read as perhaps the ultimate homage to Boswell's autoptic vision and proprietary devotion to his dead hero in the *Life.* We might also think here of art's mirroring in *Pale Fire,* reading's solitary assuagement of solitude, in relation to Johnson's relief in chapter 3 in finding a "mirrour in every mind" in Gray's "Elegy," forgetting the art he already knows, forgetting death. I turn now to Boswell and Johnson because my brief consideration of the novel here is meant to restore this particular strand of Nabokov's complicated pattern of allusion and imitation to light.

· · ·

Just before James Boswell self-consciously (and incompletely) departs the text of his *Life of Johnson,* "reliev[ing] the readers of this Work from any farther personal notice of its author," he recounts a final and unfinished epistolary exchange between himself and his hero. He confesses to having been

> so much indisposed during a considerable part of the year [the last year of Johnson's life], that it was not, or at least I thought it was not, in my power to write to my illustrious friend as formerly, or without expressing such complaints as offended him. Having conjured him not to do me the injustice of charging me with affectation, I was with much regret long silent. His last letter to me then came, and affected me very tenderly:
>
> 'TO JAMES BOSWELL, ESQ.
>
> 'DEAR SIR,
>
> 'I HAVE this summer sometimes amended, and sometimes relapsed, but, upon the whole, have lost ground very much. My legs are extremely weak, and my breath very short, and the water is now encreasing upon me. In this uncomfortable

> state your letters used to relieve; what is the reason that I have them no longer? Are you sick, or are you sullen? Whatever be the reason, if it be less than necessity, drive it away; and of the short life that we have, make the best use for yourself and for your friends. . . .'

Johnson's letter speaks with the urgency of an impending end at once metaphorical and corporeal (remember, dropsy was his disease): "the water is now encreasing upon me." What can mere melancholy mean to a drowning man in need of epistolary relief? After an ellipsis, Boswell resumes:

> Yet it was not a little painful to me to find, that in a paragraph of this letter, which I have omitted, he still persevered in arraigning me as before, which was strange in him who had so much experience of what I suffered.[29]

Johnson's chastising paragraph has never been recovered, though its contents would have been familiar to readers of the *Life*. The man whom Boswell characterized as both fellow sufferer and stoic gladiator against melancholy had often reproached his friend for affecting or indulging a malaise that ought to be resisted, the cure for which was society. "*Be not solitary*" was Johnson's motto from the *Anatomy of Melancholy*, "*be not idle*." In his Register of Letters for this period, Boswell notes that he himself was "under a cloud of inactivity," enduring "a long affliction of bad spirits."[30] Melancholy, so Boswell's silence during this period self-consciously acknowledges, imprisons the self, reducing social companionship to solitary complaint. Rather than offend his friend with such solipsism, Boswell is mute. Yet his plaintive recording of this frustrated exchange points to the biographer's impossible desire in the *Life* for the paradox of melancholy communion. Such desire can only be satisfied after Johnson himself—in desperate need of living companionship to save him from contemplation of the end—has died.

The original manuscript of the *Life* records and then deletes a mirroring of the two men through the metaphor of melancholy (see fig. 22). Omitted from the paragraph beginning "I unfortunately was so much indisposed" after "offended him" is the following: "I felt the force of Shakspeare's [*sic*] words 'How weary stale flat and unprofitable seem to me all the uses of this world.'" This passage from *Hamlet* is scored out, but later, after the phrase "so much experience," Boswell considers using the phrase to describe Johnson, appending "of that inexplicable state in which he felt the force of Shakspeare's 'how weary stale flat and unprofitable seem to me all the uses of this world.'" Hamlet and *Hamlet* once again become a literary web binding the psyches of biographer and subject—at first meeting glimpsed through the bookstore window

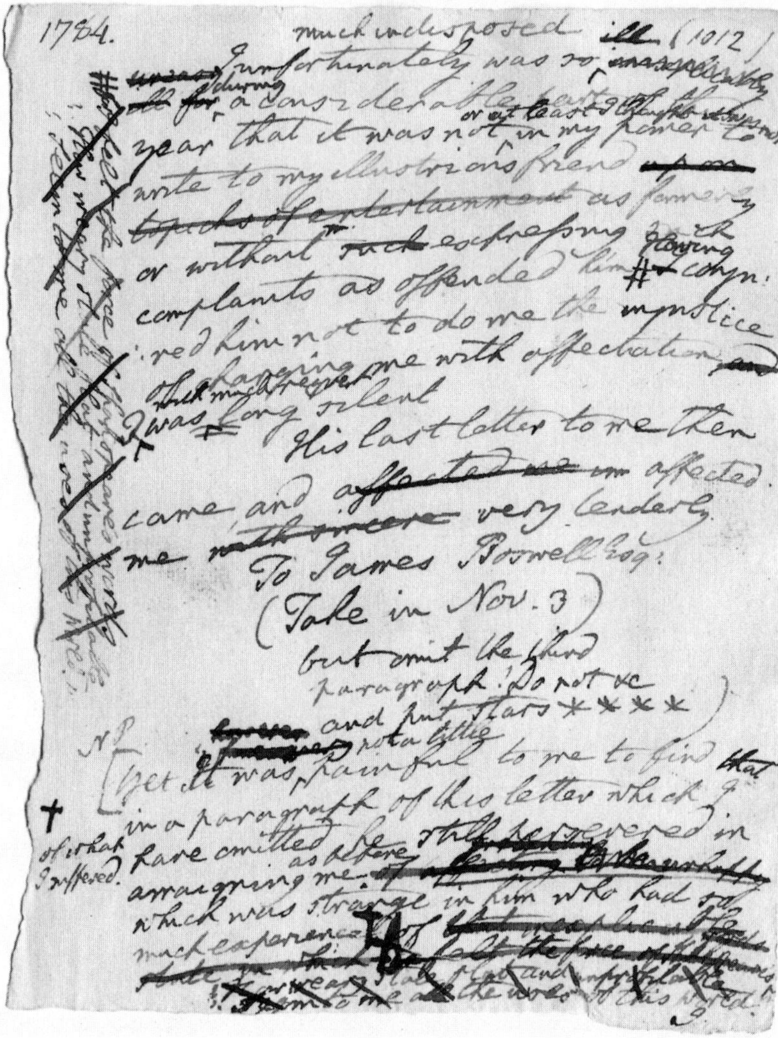

Figure 22. Page 1012 of the manuscript of James Boswell's *Life of Johnson.* Beinecke Rare Book and Manuscript Library, Yale University.

as haunted son and paternal ghost, at last epistolary meeting seen through a melancholy glass, darkly, as thwarted doubles, each of whom feels the force of the soliloquy that joins them in solitary misery. Disavowal rendered visible, the crossed-out passages blacken the margins of Boswell's manuscript like mourning crepe covering a mirror, evoking both death and its denial. This is the allusive equivalent of the autoptic vision that looks for and at itself in death.[31]

The most effective version of Johnson's life is the one that allows the reader to "still see" him in touching detail and to translate that vision into the mundane details of his own life. Hamlet's litany—weary, stale, flat, and unprofitable—paradoxically revives the detritus of the everyday through the oxymoron of melancholy communion.

Melancholy vision may empty out the uses of this world, but through the death of the author it illuminates another world of art. Nabokov's *Pale Fire* and Boswell's *Life* are both blocked mirrors of readerly desire, a desire that creates art out of the death of the author. The novel's mirroring theme begins with the title, taken from Shakespeare's *Timon of Athens:*

> I'll example you with thievery:
> The sun's a thief, and with his great attraction
> Robs the vast sea; the moon's an arrant thief,
> And her pale fire she snatches from the sun;
> The sea's a thief, whose liquid surge resolves
> The moon into salt tears. (4.3.435–40)

This motif of resemblance as both theft and transformation (staged by Nabokov/Shade's self-conscious borrowing of this passage in the poem with which the novel begins) is rendered deathly by the first lines of the poem "Pale Fire":

> I was the shadow of the waxwing slain
> By the false azure in the windowpane.

To repeat an image, to borrow its "pale fire," is not only to steal but to deceive, with potentially deathly consequences. But while art's resemblances can kill, they also can transform death.

As an alternate subtext to Shade's title indicates, the pale fire of representation also has something to do with reparation. One of the passages from Shakespeare's *Hamlet* that the young Nabokov translated into Russian is the following:

> The glow-worm shows the matin to be near
> And gins to pale his uneffectual fire.
> Adieu, adieu, adieu. Remember me. (1.5.89–91)

This admonition from a paternal ghost clearly haunts the novel, as does the following line, also treated by Nabokov:

When he himself might his quietus make
With a *bare bodkin*? (3.1.75–76) (emphasis mine)

In reading a novel that explicitly treats the question of the afterlife and the morality of suicide (commenting on the central event of the poem, the suicide of Shade's daughter, Hazel, Kinbote philosophizes most eloquently upon the question), while playing throughout with motifs of translation and distortion (see, for example, Kinbote's references to Zemblan [mis]translations of Shakespearean texts, including the title passage from *Timon of Athens*),[32] and in relation to which Michael Wood has powerfully argued that the greatest loss Nabokov mourns is that of the Russian language itself, it is worth noting that in his version of these lines from *Hamlet,* Nabokov specifically restores the "bodkin" to its proper place in the Russian text, a place to which Kinbote's multiple names remind us to return.[33]

Through the multiple motifs in multiple languages raised by the "botkin": death and afterlife, reparation for a father's death, and, veering now in a new direction, parasitism—*Hamlet* thus links Nabokov not only to Shakespeare but to Boswell. But while Nabokov's debt to Shakespeare is well known, his debt to Boswell, while evident from the beginning in Nabokov's (or is it Shade's? or is it Kinbote's?) choice of epigraph, has been often overlooked.[34] Yet between the table of contents and the foreword, we encounter the following passage alone on the page:

> This reminds me of the ludicrous account he gave Mr. Langton, of the despicable state of a young gentleman of good family. 'Sir, when I heard of him last, he was running about town shooting cats.' And then in a sort of kindly reverie, he bethought himself of his own favorite cat, and said, 'But Hodge shan't be shot: no, no, Hodge shall not be shot.'
>
> JAMES BOSWELL, the *Life of Samuel Johnson*

What are we to make of this anecdote? Whatever the critical debate about who "authored" the epigraph—the novel's Boswell figure Kinbote, Shade himself, or Nabokov have all been argued—the passage's position asks that it first be interpreted in isolation. We might start by noticing its juxtaposition of Johnson's "ludicrous account" of tragicomic randomness (a young gentleman of good family reduced to a despicable state, not unlike the derangement of the social outcast Kinbote himself) with his apotropaic "kindly reverie." The anecdote shifts from sociable narrative to solitary fantasy, and the impetus of such fantasy is the impending possibility of random death, the very possibility with which the novel *Pale Fire* ends and that enables its beginning. (The arbitrary and seemingly unmotivated death of the poem's author John Shade precipitates the theft of the poem by Kinbote, who writes the crazed commentary with which the novel begins.)

We might also notice the way in which the passage shifts from an indirect summation of Johnson's own gossipy narration, culminating in direct quotation, to Boswell's voyeuristic gaze at his private reverie, a reverie rendered intelligible proleptically ("he bethought himself of his own favorite cat") by Johnson's near-incantatory utterance. As the tense shifts from past to future, from perfect to conditional, we enter the realm of fond imagination, of magical thinking in the face of death. We enter too, however comically, the realm of human attachment and inevitable loss.

And it is the spectacle of an author's mind in confrontation with loss that both Boswell and Nabokov are concerned with. In Nabokov's text, however, it becomes clear that the most brilliant author, the one with the richest imagination, might in fact be the critic/reader. Kinbote, having absconded with Shade's manuscript, closes a foreword that inches from scholarly decorum to crazed paranoia with a moment of anecdotal preservation and adoration that is profoundly Boswellian:

> We never discussed, John Shade and I, any of my personal misfortunes. Our close friendship was on that higher, exclusively intellectual level where one can rest from emotional troubles, not share them. My admiration for him was for me a sort of alpine cure. I experienced a grand sense of wonder whenever I looked at him, especially in the presence of other people, inferior people. This wonder was enhanced by my awareness of their not feeling what I felt, of their not seeing what I saw, of their taking Shade for granted, instead of drenching every nerve, so to speak, in the romance of his presence. Here he is, I would say to myself, that is his head, concerning a brain of a different brand than that of the synthetic jellies preserved in the skulls around him. He is looking from the terrace (of Prof. C.'s house on that March evening) at the distant lake. I am looking at him. I am witnessing a unique physiological phenomenon: John Shade perceiving and transforming the world, taking it in and taking it apart, re-combining its elements in the very process of storing them up so as to produce at some unspecified date an organic miracle, a fusion of image and music, a line of verse. And I experienced the same thrill as when in my early boyhood I once watched across the tea table in my uncle's castle a conjuror who had just given a fantastic performance and was now quietly consuming a vanilla ice. (17–18)

This passage rehearses the form of the anecdote as visual and temporal close-up beginning with Kinbote's general description of his privileged "wonder" at "the romance of [Shade's] presence." This romance takes the temporal form of multiple past repetitions ("here he is, I would say to myself") and quickly narrows down to an imaginative reinhabiting of a revived and specifically marked ("of Prof. C.'s house on that March evening") moment. Kinbote's resurrection of an immediate vision of authorial creation—"I am looking at him. I am witnessing a unique physiological phenomenon"—lapses into the past tense, marking by contrast those precious seconds of recollection during which he raptly beholds, nearly inhabits, the "organic miracle" of the author's creating mind. While Shade in the anecdote's eternal present tense stores the present for the future production of poetry, the reader Kinbote ultimately must return to his own inner resources, the world of his past, of childhood magic, lost to the "personal misfortunes" that the romance of Shade's presence had helped him momentarily to transcend. This fragile nostalgic presentness links the anecdote—which has a strong visual component in many of its manifestations—to the photograph.

As Roland Barthes articulates it in *Camera Lucida,* the temporality of the photograph—like that of the Johnsonian anecdote—transpires between two deaths. Barthes's singular science enables us to see how the anecdote and the photograph share a complex temporality that wounds those who view them. Whether viewing a photograph of his now-dead mother as a child or of a nineteenth-century assassin awaiting hanging, Barthes comes to see time, and with it mortality, as the ultimate distinguishing characteristic or *punctum* of any photograph:

> The *punctum* is: *he is going to die.* I read at the same time: *This will be* and *this has been;* I observe with horror an anterior future of which death is the stake. By giving me the absolute past of the pose (aorist), the photograph tells me death in the future. What *pricks* me is the discovery of this equivalence. In front of the photograph of my mother as a child, I tell myself: she is going to die: I shudder, like Winnicott's psychotic patient, *over a catastrophe which has already occurred.* Whether or not the subject is already dead, every photograph is this catastrophe.[35]

He is dead and he is going to die—both the photograph and the anecdote inhabit that space between the past (in its definitive aorist tense) and the "anterior future of which death is the stake." Both reproduce, by suspending, an impending awareness of "a catastrophe which has already occurred." In the case of Hawkins's narrative of Johnson's self-wounding, that awareness is also Johnson's, and his gesture fights off the death that his audience of doctors at once fears and desires, the death his biographer and readers know has already happened. In the case of the Hodge anecdote that begins *Pale Fire,* that awareness takes the form of a reverie, an imaginary warding off of death that is a kind of magical thinking: "no, no, Hodge shall not be shot." In the case of Kinbote's lingering gaze at the mortal man, John Shade, from whom issues the miracle of a poem that outlasts its author, that awareness itself produces its own form of devotional art, a wound that transpires as mutual inspiration and animation.

For Johnson, who warned the ambitious scholar to "pause a while from letters to be wise" and who often wrote about the particularly tragic vanity of an author's human wishes, the "transition from an author's books to his conversation" was usually a disappointment, the sad revelation of a little man behind literature's magic curtain.[36] In *Rambler* 14, such a shift from dead letter to living author is similarly articulated as a kind of visual close-up:

> A transition from an author's books to his conversation, is too often like an entrance into a large city, after a distant prospect. Remotely, we see nothing but spires of temples, and turrets of palaces, and imagine it the residence of splendor, grandeur, and magnificence; but, when we have passed the gates, we find it perplexed with narrow passages, disgraced with despicable cottages, embarrassed with obstructions, and clouded with smoke.[37]

Fantasies of authorial royalty—"splendor, grandeur, and magnificence"—much like Kinbote's own, are dispelled with the perplexity, disgrace, and embarrassment of the poverty and squalor of his conversational presence; just as a "distant prospect" seen from up close is "clouded" by a smoke that has a similar effect to the deluding "mist" that obscures human vision in *The Vanity of Human Wishes.*

If we consider what Nabokov described, while toiling as translator of and commentator on Pushkin's *Eugene Onegin,* as the "dead ends" of great literary texts—ends that eclipse the critic's desire to revive the living author—we can begin to understand how *Pale Fire* reflects the *Life of Johnson* in an inverted mirror.[38] While Boswell's insistence on preserving the author in death is fueled by his longing for the author's presence, Nabokov meditates on a different sort of death of the author, namely, his disappearance into the enduring presence, what Shade playfully calls the "texture," the magical web of the text. The dead ends of art, Brian Boyd has recently and boldly argued, make of John Shade quite literally a shade, a benevolent ghost in the novel's machine who reflects the possibility of some life beyond death, some semblance of divine order in a doubt-ridden universe.[39]

*Pale Fire* thus reveals an additional facet to the mirror of reflection—the moon's pale fire in *Timon* and the glowworm's fire in *Hamlet* are fraught as well with the self-consuming pale fire of the earlier drafts of Shade's poem that fuel its ultimate completion.[40] For Nabokov, the divinity that shapes our ends is finished art, yet his novel testifies, through the commentary of Kinbote, to the longing for an opening, for communion, that possesses every solitary reader. The consummate "link-and-bobolink," the "correlated pattern in the game" (812–13) that reassures John Shade about life's overarching design, connects the ends of art to the endless refractions of art within individual imaginations. Art becomes an encounter fraught with desire: whether it be Kinbote's fantasy that he has "impregnated Shade" with the true inspiration for his poem or Boswell's claim that he alone is uniquely "impregnated with Johnsonian aether," art remains unfinished; it does not end—rather it wounds, and in wounding it comes to life.[41]

This link between ends and endlessness, the hole within artistic wholes created by the reader's longing for connection, is what distinguishes Shade's poem and Nabokov's novel from Pope's *Essay on Man.* It is what causes Johnson to live on beyond the opening of his own "conclusion, in which nothing is concluded," to inspire and to undo the Boswellian monument in books like *Pale Fire.* While it's difficult to think of two writers more different in style than Nabokov and Johnson, it's interesting to contemplate how coincident Lawrence Lipking's claim that "we wake from a book like *Rasselas* to discover that we are in it" is with Brian Boyd's description of *Pale Fire* as a book from which we wake to call it life.[42]

While Shade calls his monograph on Pope *Supremely Blest,* his appraisal of *An Essay on Man* is critical to the point of dismissiveness:

> *"See the blind beggar dance, the cripple sing"*
> Has unmistakably the vulgar ring
> Of its preposterous age. (419–21)

Kinbote's note, undoing art's dead end, reveals a rejected variant:

> . . ."Such verses as
>
> 'See the blind beggar dance, the cripple sing,
> The sot a hero, lunatic a king'
> Smack of their heartless age.". . .
>
> This is, of course, from Pope's *Essay on Man.* One knows not what to wonder at more: Pope's not finding a monosyllable to replace "hero" (for example, "man") so as to accommodate the definite article before the next word, or Shade's re-

> placing an admirable passage by the much flabbier final text. Or was he afraid of offending an authentic king? In pondering the near past I have never been able to ascertain retrospectively if he really had "guessed my secret," as he once observed (see note to line 991). (145)

Kinbote's aesthetic disapproval of Shade's final draft mirrors Shade's moral disapproval of Pope and his "heartless" age. In the service of meter, Kinbote would render Pope's lines even more heartless—making the ironic juxtaposition that between "hero" and "man" rather than "hero" and "sot." In his (mis)recognition of himself in the omitted second line of the passage—is Kinbote a king, a lunatic, or both?—we are reminded of the next couplet in Pope's passage, omitted by both poet and critic:

> The starving chemist in his golden views
> Supremely blest, the poet in his muse. (*Essay on Man,* 2.269–70)

If we remember Shakespeare's equation of lunatic, lover, and poet, we can begin to discern how Shade's reference to Pope levels multiple fantasies of power and imagination into common delusion.[43] As Robert Alter puts it:

> The conjunction of lunatic king and poet is of course the conjunction in the novel of those antithetical doubles, Kinbote and Shade. There is, moreover, a thematic rather than a structural revelation about the novel in Pope's tartly satirical yoking, shrewdly reinforced by a rare enjambement, of the starving chemist (that is, alchemist) and the poet. The alchemist deludes himself by thinking he can effect a magical transformation of lead to gold (in the novel, Kinbote . . . refers to writing as "blue magic"). The poet, for Nabokov's purposes at any rate, deludes himself by imagining he can transform the death-sodden mire of existence into a pellucid artifice of eternity. But perhaps, we are made to feel through both "Pale Fire" and its Commentary, the delusion is strictly necessary to make life livable; perhaps in some way it is not altogether a delusion.[44]

John Shade and Vladimir Nabokov were not the only authors to search for a weak spot in the armor of the confident order of Pope's *Essay on Man,* a theodicy that famously asserts:

> All Nature is but Art, unknown to thee;
> All Chance, Direction, which thou canst not see;
> All Discord, Harmony, not understood;
> All partial Evil, universal Good:

> And, spite of Pride, in erring Reason's spite,
> One truth is clear, "Whatever IS, is RIGHT." (I.289–94)

Samuel Johnson's passionately satiric response to these sentiments (in his review of Soame Jenyns's *A Free Enquiry into the Nature and Origin of Evil*) is reminiscent in its destructive energy of Kinbote's transgressive reopening of Shade's neatly finished couplets. The vanity of such couplet art in Johnson's account expands by analogy, emptying out the divinely ordered world. The couplet in this view becomes an empty means with which to confront and control emptiness.

So important is Johnson's desire to "see life" in all its painfully embodied reality before "knowing" it, that he is willing to dismantle the Great Chain of Being, the chain that, topped by divine omniscience, still orders Pope's *Essay* and justifies his command to the reader to "Submit" without trying to know beyond his or her station.

> Between the lowest positive existence and nothing, wherever we suppose positive existence to cease, is another chasm infinitely deep; where there is room again for endless orders of subordinate nature, continued for ever and for ever, and yet infinitely superior to nonexistence.
>
> To these meditations humanity is unequal. But yet we may ask, not of our maker, but of each other, since on the one side creation, wherever it stops, must stop infinitely below infinity, and on the other infinitely above nothing, what necessity there is that it should proceed so far either way that beings so high or so low should ever have existed? We may ask; but I believe no created wisdom can give an adequate answer.
>
> Nor is this all. In the scale, wherever it begins or ends, are infinite vacuities. At whatever distance we suppose the next order of beings to be above man, there is room for an intermediate order of beings between them; and if for one order then for infinite orders; since every thing that admits of more or less, and consequently all the parts of that which admits them, may be infinitely divided. So that, as far as we can judge, there may be room in the vacuity between any two steps of the scale or between any two points of the cone of being, for infinite exertion of infinite power.[45]

Johnson's voice in this passage stops and starts again in a rhythmic mimicking of the compulsion to speculate that he analyzes. Humanity is unequal to such questioning, but we may ask. We may ask, but only God can answer. This vision of infinite levels of being punctuated by infinite vacuity "for ever and for ever" is an intense condensation of Johnson's lifelong rhetorical and imaginative habit of contemplating and filling vacuity, a habit he saw as the imperative of the human condition. Life, he was fond of saying, "must be filled up." Hester Thrale termed infinite vacuity "[Johnson's] favourite hypothesis, and the general tenor of his reasonings," and the prospect of infinite gradation punctuating a leap from a precipice into nothingness appears throughout his work.[46]

From this perspective, Pope's hierarchy in *An Essay on Man* thus falls apart because of the gaps between its levels, the very gaps that make it possible. The ideal procession from man to God with which Pope and Jenyns vindicate the ways of a newly distant deity to their readers is for Johnson an endless procession of sufferers and spectators not unlike the *Vanity* itself. By putting his trust in analogy and the ordering power of art, Jenyns attempts to prove that within this closed system, as Johnson puts it, "the evils suffered on this globe may by some inconceivable means contribute to the felicity of the inhabitants of the remotest planet":

> He imagines that as we have not only animals for food, but choose some for our diversion, the same privilege may be allowed to some beings above us, *who may deceive, torment, or destroy us for the ends only of their own pleasure or utility.* This he again finds impossible to be conceived, *but that impossibility lessens not the probability of the conjecture, which by analogy is so strongly confirmed.*[47]

The same "scruples of infidelity" that caused the eight-year-old Johnson to search "diligently but fruitlessly, for evidences of the truth of revelation" and that ultimately convinced him by their very pain of the soul's immortality,[48] cause him to apply his rigorous logic to that which the sentimental Jenyns finds impossible to be conceived. At the heart of analogy is infinite, uncontrollable repetition experienced as causeless suffering. By turning that repetition into divine art, Pope reveals the theatricality, the infinite fictional regress, of the human condition and the accompanying possibility of a merciless, unmoved audience. The divinity that shapes our ends might do so with only an aesthetic effect in mind.

> Many a merry bout have these frolic beings at the vicissitudes of an ague, and good sport it is to see a man tumble with an epilepsy, and revive and tumble again, and all this he knows not why. As they are wiser and more powerful than we, they have more exquisite diversions; for we have no way of procuring any sport so brisk and so lasting as the paroxysms of the gout and stone, which undoubtedly must make high mirth, especially if the play be a little diversified with the blunders and puzzles of the blind and deaf.[49]

The prospect of human suffering as aesthetic spectacle leaves Johnson in a double bind. To disprove Jenyns, he must take degree and its latent dangers away; he must eliminate the ungovernable licentious aspect of analogy and in so doing eliminate the language of comparison and identity altogether. But this rejection forces him to leap off the precipice of degree into infinite vacuity, into the hole within the whole posited by the lived experience of suffering. When Johnson descends from this vision of infinity back to the human world, he returns as well to the anecdotal realm of the body's pain. It is this world that must be seen to be known; it is this world that leaves the largest unanswered question, the largest hole, in Pope's theory.

Having thus done away with the possibility of an ending, Johnson asserts another sort of passionately desired end:

> The only end of writing is to enable the readers better to enjoy life, or better to endure it; and how will either of those be put more in our power by him who tells us that we are puppets, of which some creature not much wiser than ourselves manages the wires. (536)

Johnson's insistence on a worthwhile end or purpose to a progression of analogy that takes human relationship to a higher order than that of "flies to wanton boys" is consistent with his decision to accept Nahum Tate's happy ending to *King Lear* at the acknowledged expense of the play's unendurably limitless suffering and indeterminacy. "I cannot easily be persuaded," he writes in his own defense, "[that] the audience will not always rise better pleased from the final triumph of persecuted virtue."[50] The key word here, I would suggest, is "final." In responding to both Jenyns's inquiry and Shakespeare's play, Johnson turns away from a vista of meaningless repetition to impose a moral closure that, however arbitrary, allows him to distinguish between what at the end of *Lear* remains indeterminate, the promised end or image of that horror. In the service of seeing and knowing human things in themselves, Johnson turns to poetic justice in conscious disavowal of the play's unending truth, thus providing writing with the "only end" possible.

Despite Shade's dismissal of Pope, the poet to whom he owes a great stylistic debt, as "heartless," Nabokov reminds us through allusion that his *Essay on Man* is not as morally certain as Johnson's response might have it. When Pope attempts to map human morality, trying to locate "th' Extreme of Vice," he is faced with relativity and reflection:

> Ask where's the North? At York, 'tis on the Tweed;
> In Scotland, at the Orcades; and there,

At Greenland, Zembla, or the Lord knows where:
No creature owns it in the first degree,
But thinks his neighbour farther gone than he. (2.222–26)

These multiple puns on geography and morality, culminating in the universal attribution of "farther gone" to another, demonstrates how Pope's diamond-sharp maxims vibrate with uncertainty, radiating multiple meanings. Yet as Michael Wood writes in his reading of this passage, Nabokov goes Pope one better. For the eighteenth-century satirist, "Zembla . . . is a figure of plausible illusion. Vice is not really like the north, a matter of relative location. It is a sort of ruined or muffled absolute, our arguments are evasions, its proclaimed relativity merely the work of our hypocrisy." By contrast, Nabokov's "great trick" is to render "vice itself both judge and jury, [to have] the lamentable, self-aggrandizing, self-accusing Kinbote, master of displacements, speak to us from the heart of the deluded moral universe Pope so ruthlessly identifies."[51] He asks of us not judgment but compassion. God's bodykins, man, treat one of us according to our deserts and who would scape whipping?

In its Möbius strip of poem and commentary, *Pale Fire* rehearses this alternation between a potentially heartless art that shapes our ends and an endlessness born of individual suffering and desire. But Nabokov is a writer of his own moment, and poetic justice of the sort that allows Cordelia to live on happily as Edgar's bride is not his province. Rather than Johnson's moral end, Nabokov chooses another sort of endlessness: instead of the "heartless" certainties of Pope's theodicy, we are given the heartfelt question of a novel that ends where it began, with the senseless and violent death of one of its authors. While the *Vanity* concludes with Johnson's public prayer to a disembodied "Celestial Wisdom," *Pale Fire* leaves us with the representative talisman of its epigraph, an affectionate anecdotal revivification of a beloved author's private reverie, suspended at romance's threshold between life and death. These semiconscious words are not declaimed to a divinity but overheard by an audience of readers, who, beginning with Boswell, despite all they know of the story's end, take heart: "Hodge shan't be shot; no, no, Hodge shall not be shot."

## Samuel Beckett
## The Autopsy and "Human Wishes"

Anecdote brings things closer to us in space, allows them to enter into our lives. Anecdote represents the extreme opposite of history—which demands an "empathy" that renders everything abstract. Empathy amounts to the same thing as read-

> ing newspapers. The true method of making things present is: to imagine them in our own space (and not to imagine ourselves in their space). Only anecdote can move us in this direction.
>
> WALTER BENJAMIN, "Benjamin the Scrivener"[52]

When I first saw Samuel Beckett's transcription, in a frenzied hand, of Samuel Johnson's autopsy record into one of the three 1937 notebooks in which he compiled material for his play "Human Wishes," the effect was not unlike what Benjamin attributes to the anecdote.[53] Like Benjamin himself transcribing quotations under the artificial sky of the Bibliothèque nationale, imagining—as he planned "The Arcades Project" that would end (or never end) as a magnificent ruin—a book that would consist of nothing but lists of things in themselves, Beckett transcribed the autopsy record without comment or context, without the "empathy" that would locate it in a narrative. Absorbed as he writes, he lets the report speak in all its bodily particularity, while repeating it with a difference. He imagines it in his own space.

Reading those notebooks (or at least their Xeroxed copies) in the tiny room behind the stacks at the Reading University Library, puzzling over Beckett's nearly illegible scrawl (which had apparently reduced readers before me to tears), was to imagine an alien mind who had similarly puzzled over Johnson's printed remains in the form of both anecdote and autopsy, in my own space. It was to feel an uncanny and, yes, uniquely Johnsonian connection. I was no longer the only literary scholar I knew who had become fascinated with Johnson's monstrous and diseased body, or who had pored over the autopsy record. The version of Johnson that Beckett assembles in the notebooks was uncannily close to my own.

The notebooks, studded with inky crosses marking the death of each person researched, are the ideal terminus for this book because they bring the Johnsoniana of the eighteenth and nineteenth centuries to a different sort of life. Beckett, who had made his own pilgrimage to Lichfield in 1935, was at the time an unemployed academic and a very thorough researcher; he transcribed anecdotes from a wide variety of sources, periodically speculating about their fit into an alternate narrative context (the years between the death of Henry Thrale and Johnson's own death—the time, in other words, of heightened speculation about Johnson's own human wishes), underlining important details in colored crayon and entering crucial notes on the backs of pages. On the pages of the manuscript of the play itself, he executed an elaborate doodle of a secular Golgotha that exists in uneasy relationship to what precedes it (fig. 23). This disappointed love story was never written, the play never finished, but the version of Johnson that emerges from these pages—plagued by hope

Figure 23. Samuel Beckett's doodle on the unfinished manuscript of "Human Wishes." The Archive of the Beckett International Foundation at the University of Reading, MS 3458. Johnson's "ghastly smile," perhaps depicted on the crucified thief on the right whose hands point downward toward damnation, with which Beckett intended to end the play, haunts this image of a secular Golgotha, with its two thieves, one saved and one damned. All quotations and reproduction from the works of Samuel Beckett by permission of the Samuel Beckett Estate.

and fear, at once desiring and postponing an end to both—never left Beckett. The well-versed Johnsonian will recognize such moments as the following without needing Boswell:

> Much as he dreaded the next world he dreaded annihilation still more. "Mere existence" he said on one occasion "is so much better than nothing, that one would rather exist even in pain than not exist." He went on to say, in answer to an objection that was raised, "The lady confounds annihilation, which is nothing, with the apprehension of it, which is dreadful. It is in the apprehension of it that the horror of annihilation consists."[54]

But many Beckett scholars have puzzled over the following pronouncement:

> Still there is a mass of material that would be useful, e.g. in the Annals, his recollection of the first time that the heaven-hell dichotomy was brought to his mind

> when he was in bed with his mother after 18 months. Heaven she described as the happy place where some people went, hell was the sad place where the rest went. . . . All this would come in quite naturally in the last act, i.e. his fearing his death, when he was being reproached by his clerical friend, Taylor, for holding the opinion that an eternity of torment was preferable to annihilation. He must have had the vision of *positive* [Beckett's emphasis] annihilation. Of how many can as much be said.[55]

Christopher Ricks—who eloquently explicates in *Beckett's Dying Words* a prose that solicits, indeed "incarnates," death—characterizing Beckett's English language as "not 'abstracted to death'" but rather "an abstract of death,"[56] notes that the "He" of Beckett's final sentence "floats free rather, but it might fit Taylor better than Johnson (and Boswell better than Taylor?)." Ricks quotes the full exchange in Boswell that Beckett quotes above and includes Boswell's remark: "If annihilation be nothing, then existing in pain is not a comparative state, but is a positive evil, which I cannot think we should choose," noting that "Boswell's comment . . . that 'existing in pain . . . is a positive evil' may have colored Beckett's phrase 'the vision of *positive* annihilation.'" "But," he continues, "Beckett may also have remembered the italicized *positive,* with the adjacent paradox of 'nihil,' from T. S. Eliot: 'Mr. Lewis proposes a Shakespeare who is a *positive* nihilist.'"[57]

The concept of "positive annihilation" as death without an end to consciousness is at the heart of Stephen Dilks's vision of Johnson, whose stylistic and philosophical example (balanced between Christian faith and stoical skepticism, between "the strength of his voice" and a mind "in constant vacillation," between the thief on Golgotha who was saved and the one who was damned), taught Beckett to write sentences that in their "equivocating balance" have "the shape of doubt." While Dilks acknowledges that "Johnson's horror at the prospect of an active, everlasting, and conscious process of decomposition after death has a theological weight that is as absent from Beckett's work as God," he suggests "that Beckett transformed Johnson's theological vision into his own aesthetic vision," a vision that culminates in the secular crucifixion he doodles in the manuscript of "Human Wishes," a map, in the view of Dilks and others, of Beckett's future work.[58]

In her discussion of the sketch as an early prototype of Beckett's fascination with and deployment of the figure of crucifixion, Mary Bryden describes this version of Christ as one "of the *Ecce Homo:* a human being to be looked upon and pitied, caught in a death-trap with no escape route. . . . [T]his Christ is discerned in his passivity, in what Beckett in an early essay referred to as the 'choseté' (thingness) of the brutalised Christs depicted by Georges

Rouault."[59] We might recall from the previous chapter that Beckett's goal in this essay on "Endgame Painters" was to define art as "an unveiling that approaches the undisclosable, the nothing, the thing itself," a "thingness" at the heart of the anecdotal hole within the whole, a thingness that can never be fully incorporated into an end. Such refusal of closure helps to account for what Bryden calls "the ambiguity of the Christ-motif within Beckett's writing," a refusal to allow the motif to function outside the "zone of desolation prompting the Christ-cry 'My God, why have you deserted me?'"[60] Bryden continues:

> While there is profound empathy with the psychological dereliction to which that outburst testifies, an underlying resentment is often to be detected against the same Christ who, in the Johannine account only, utters the words: "It is accomplished" (*John* 19.30), as if rubber-stamping a debt paid in full. Indeed, for Murphy, "pondering Christ's parthian shaft: 'It is finished,'" those final words are seen not so much as a consummation but as a parting shot to all those whose sufferings are still in progress.[61]

Instead of debts paid in full, we have Johnson's despair over the impossible debt enjoined by the parable of the talents in Matthew (as well as his envy of Dr. Levet's "single talent well-employed")[62] and (as we shall see shortly) Johnson's housekeeper, the blind poetess Mrs. Williams's refusal to forgive the dead Goldsmith's unpaid debt in "Human Wishes" itself. What Beckett most admires about Johnson, and what distinguishes Johnson from Mrs. Thrale, is the capacity for suffering:

> . . . there can hardly have been many so completely at sea in their solitude as he was or so horribly aware of it. Read the Prayers and Meditations if you don't believe me. . . . [S]he [Mrs. Thrale] had none of that need to suffer or necessity of suffering that he had. . . . [H]e, in a sense was spiritually self-conscious, was a tragic figure, i.e., worth putting down as a part of the whole of which oneself is part.[63]

Especially evocative here is Beckett's recognition of the ambiguously volitional nature of Johnson's suffering, "that need to suffer or necessity of suffering," which, as is the case with his later evocations of Christ, and as we have seen to be the case with Johnson's final self-wounding, is at once inexplicably arbitrary yet self-inflicted.

If Beckett's vision of Christ functions, in Bryden's words, as "an enduring image of enduring," an unfinished and unembellished representation of suffering that, in its resistance of "crucifixion cliché, retains the capacity to be shocked by both the gibbet and by the syndrome which transforms it into the gold-plated yet anodyne stock-in-trade of the jeweler," then Samuel Johnson's theological vision of "positive annihilation," the ultimate thing in itself, suffering without end, is at that vision's heart.[64] This art, like religion, addresses suffering with self-conscious inadequacy, but unlike religion "shrink[s] from attempting to valorise that distress, or incorporate it into a divine economy."[65]

· · ·

Beckett—recently returned from a trip to Germany during which he was recovering from a newly ended affair while nostalgic for an old one, knowing return would be impossible, severely depressed and possessed by what he described as a "coenaesthesic of mind, a fullness of mental self-aesthesia that is entirely useless,"[66] plagued by his own host of "idiosyncratic illnesses,"[67] unemployed, broke, and living with his mother in a contentious death grip of mutual dependence, anticipating a major breakdown—nevertheless managed to glean a great deal of "once-private" Johnsoniana as he sat beneath the glass roof of the reading room of the National Library of Ireland.[68] What we read in those notebooks is Beckett's record of his own autoptic gaze at a historical figure who differs profoundly from himself and who thus functions all the more effectively as a mirror. What Beckett wrote in his early study of Proust is equally true of his version of and relationship to Johnson: "We cannot know and we cannot be known."[69]

Beckett intended his Johnson to be an anecdotal creation, partly because he drew upon largely anecdotal sources in the notebooks—relying heavily, for example, upon biographies by John Hawkins and Leslie Stephen, anecdotal compilations by George Birkbeck Hill and others, as well as Johnson's *Prayers and Meditations*—but more importantly because his Johnson reminds us of the original sense of "anecdote" as "unpublished" and, in Beckett's twist upon the ana genre, previously unspeakable:

> There won't be anything snappy or wisecracky about the Johnson play if it is ever written. It isn't Boswell's wit and wisdom machine that means anything to me, but the miseries that he never talked of, being unwilling or unable to do so. The horror of annihilation, the horror of madness, the horrified love of Mrs Thrale, the whole mental monster ridden swamp that after hours of silence could only give some ghastly bubble like "Lord have mercy upon us." The background of the *Prayers and Meditations*. The opium eating, dreading-to-go-to-bed, praying-for-the-dead, past living, terrified of dying, terrified of deadness, panting on to 75 bag of water, with a hydracele on his right testis. How jolly.[70]

Suspended between life and death and death's possible aftermath or lack of an aftermath—"past living, terrified of dying, terrified of deadness"—reduced to a "bag of water, with a hydracele on his right testis," rendering him "a Platonic gigolo or house friend, with not a testicle, auricle or ventricle to stand on when the bluff is called,"[71] Beckett's Johnson is indeed a "bloated carcase," not the master of the printed word but devoid of all speech but for prayer.

Yet that "ghastly bubble . . . Lord have mercy upon us" was all Beckett in 1937 thought that a work of art could say. Earlier that year he had written, "The art . . . that is a prayer sets up prayer, releases prayer in onlooker, i.e. *Priest:* Lord have mercy upon us. *People:* Christ have mercy upon us." This idea of art as prayer, while not often associated with Beckett, was, as James Knowlson has shown, "essential to his view of art at the time, whether this was the art of the writer, painter, or musician."[72] In its rewriting of artistic authority as an abdication of authorial will, Beckett's secularized devotion is reminiscent of the end of Johnson's *Vanity of Human Wishes* (the text that of course most haunts Beckett's "Johnson fantasy"), the poem that, were it not for its turn to prayer as last resort in the search for proper objects for hope and fear, would never have come to an end.

Beckett had intended to write what would have been his first play about Johnson's unrequited love for Hester Thrale, a love that, to his mind, was as much an object of fear for Johnson as of hope. He began with questions about the Johnson-Thrale relationship:

> what would have happened to the Thrale ménage without the Johnsonian buffer? Did not the whole domestic edifice at Streatham rest on a base of Johnsonian platitude?

> what emotional relations between Mrs. T. and J at this period (immediately after brewery crisis, i.e. 1773–6)? What kind of emotions did each feel for the other? . . . a definitely emotional quality in situation, at least on side of doctor.[73]

What follows is a representative example of his psychological speculation about both parties:

> Johnson morbidly melancholy, dependent on goodness of Mme, terrified, equally afraid of going or staying . . . almost frantic appeal to mercy . . . as though he apprehended the breaking up of front? Of his own abandonment and the terrible loss of "Mme." Johnson sounds almost like suppliant lover, Mrs. T. like petulant mistress. . . . But, Johnson in love (whether he knew it or not) with Hester Thrale.

> His morality the typical bulwark of neurosis. Could not admit a situation (i.e. love for Mrs. T.) that would have exiled him from Streatham.
>
> Rationalizes his dependence on Mrs. T.
>
> Mrs. T. in love with no one. The "tethered nag" of maternity. Baretti accused her of cruelty to her children, of stupidity and criminal ignorance. Her daughter hated her. She was jealous of Queeney even as a child.
>
> Johnson devoted to Queeney.[74]

In a purely conjectural but particularly suggestive speculation, Beckett presumed that Johnson was impotent:[75]

> His impotence was mollified by Mrs Thrale so long as Thrale was there, then suddenly exasperated when the licensed mendula was in the connubial position for the first time in years, thanks to *rigor mortis.*[76]

This is a brutally literal version compared to Macaulay's desexed childhood familiar and Boswell's straitlaced moralist who went to his grave tormented by private guilt.[77] Beckett's abject version of Johnson exposes on all levels—physical as well as spiritual—the lack against which the love of Johnson was meant to defend, a lack brought into focus by the unmediated proximity of Thrale. Deriding Johnson as "the Harmless Dandy" and "en-Thraled,"[78] Beckett puts his view more tragically in the following letter:

> I have been working . . . on the Johnson thing to find my position . . . more strikingly confirmed than I had dared hope. It seems now quite certain that he was rather absurdly in love with her all the fifteen years he was at Streatham, though there is no text for the impotence. It becomes more interesting, the false rage to cover his retreat from her, than the real rage when he realizes that no retreat was necessary, and beneath all, the despair of the lover with nothing to love with, and much more difficult.[79]

This view of Johnson emerges at Thrale's expense. By contrast with Beckett's scathingly and excruciatingly particular anecdotal Johnson, the specific object of Johnson's affection is gradually effaced in the notebooks by an all-too-familiar abstraction.

Upon noting the letter from Johnson to Thrale on her marriage to Piozzi,

Beckett observes: "In this transaction, J. seemed to have forgotten the story of the Ephesian Matron related by Petronius."[80] The notebooks show detailed sympathy for the physical suffering and emotional deprivation of Hester Thrale's life: Beckett counts her pregnancies, observing that "taking into account a few miscarriages, Mrs. T. almost continuously pregnant during period of her friendship with Johnson"; notes her isolation at Streatham and Henry Thrale's domination of everyone in the household including Johnson, who called Hester Thrale a "kept mistress"; puzzles over Johnson's insistence that she call Thrale "master";[81] and, most strikingly, seems to be consistently looking for evidence of Johnson's own selfishness in relationship to her (e.g., exclamation marks punctuate the margins of his transcription of Johnson's July 1784 letter on Thrale's remarriage in response to the line "What you have done, however may I lament it, I have no pretence to resent, as it has not been injurious to me."[82] Yet Beckett ultimately loses interest in Thrale as a character, concluding that she was "in love with no one."[83] He abandons Thrale altogether, evoking the aspect of Petronius's joke that is as old as misogyny itself.

But in the wake of the failed love story, the other aspect of Petronius's joke, the pagan joke on Christian resurrection and the finality of death, comes to dominate the play; misogyny yields once again to solitary community in death. Beckett's letters at the time recount what Knowlson calls "his first subject transforming itself into his second": "His horror at loving her I take it was a mode or paradigm of his horror at ultimate annihilation, to which he declared in the fear of his death that he would prefer an eternity of torment."[84] Johnson's supposed physical impotence becomes both symptom and cover of his existential vision of "positive annihilation," a vision that is mirrored by Thrale's annihilating femininity, a type of the ultimate nothingness.

Beckett's erasure of Thrale as particular object of desire demonstrates his affinity for what Stephen Dilks calls "Johnson's desire for something to remove desire," which he links to "Johnson's vision of positive annihilation." Beckett's recording of Johnson's sentiment that "Life is a progress from want to want"[85] underscores the fact for Dilks that in 1937 Beckett was "developing an aesthetic of desire and the annihilation of desire."[86] The desire that remains for Beckett we will recognize as distinctly Johnsonian.

In *Beckett's Eighteenth Century,* Frederik N. Smith has recently argued that Johnson shared with Beckett's characters what Boswell praised as "philosophic heroism." Drawing on Bertrand H. Bronson's essay "The Double Tradition of Dr. Johnson" (discussed in chapter 3), Smith suggestively claims that Beckett's version of Johnson, reappearing throughout his career in various guises and various texts (particularly in the Trilogy), took

> the popular image and combine[d] it with the learned image, thus giving us the typical Beckettian character: a slovenly, rebellious, uneven conversationalist, who nonetheless is a powerful—though confused—intellect and moral force. Beckett casts his figure as one tormented by hopes and fears, by a mind which will not shut down (the learned Johnson) and a body that is unkempt, very much in decline (the popular Johnson). We might say that, in effect, the eighteenth-century Johnson is compelled by Beckett to confront those very horrors which stalked him—loneliness, powerlessness, mortality—without the solace of religious belief.[87]

Referring to the "contemporaneity" of Johnson, and citing Macaulay's evocation in his Encyclopaedia Britannica article on Johnson of the popular image of "the old philosopher" "blinking, puffing, rolling his head," which Beckett mentions in the notebooks and which ends: "No human being who has been more than seventy years in the grave is so well known to us," Smith concludes:

> Although the sentimentality of these statements would have been off-putting to Beckett (he refers in the notebooks to Macaulay's "romantic bilge"), the immediacy of Johnson in the late nineteenth and early twentieth centuries must have had an impact on him. Johnson's presence had been sensed by others. But few perhaps sensed Johnson's presence quite so deeply and personally as Beckett.[88]

Almost thirty years after his Johnson fantasy, Beckett wrote in a letter: "They can put me wherever they want, but it's Johnson, always Johnson, who is with me. And if I follow any tradition, it is his."[89] This is author love as communion in solitude, author love as inspiration, author love as mutual powerlessness and despair. But above all it is love.

While Beckett's sensitivity to Johnson's presence, his "contemporaneity," was almost preternatural—"it's Johnson, always Johnson, who is with me"—for Smith, the universality of Beckett's characters, the fact that they "are human beings like us," persists "in spite of their origin—at least in part—in the historical personage of Johnson."[90] That locution gives more away than Smith intended and brings us back to the epigraph from Benjamin with which this section began. While Smith implies that literature's roots in the "historical" are precisely what compromise its "humanity," rendering it potentially alien and untranslatable, for Beckett (as was the case in a very different way and at a different moment for Macaulay) the same particularities that link Johnson to his historical moment and to the torments of the body are what render him most human, most familiar. Macaulay's puffing and rolling Johnson is the stuff of what Benjamin condemned as the empathy of historical abstraction, in this case of a Whiggish idealization of an "old England" more beloved because more eccentric. This is a safe Johnson, completely known, from a past successfully redeemed for the nursery. Beckett's Johnson, by contrast, uniting suffering body with tormented mind, is unknown to himself and to his heirs, confronting twentieth-century readers with an image unsettlingly familiar in its strangeness.

This Johnson demands not Macaulay's knowing empathy but rather disturbing recognition of a historical other present in our space. The notebooks practice this anecdotal history of detail, devoid of narrative purpose, devoid of causes, devoid of the confident positioning of foreground and background that characterize Macaulay's timeless portrait. What is left is the record of pure incoherence of "times and men and places," that body, that man.[91]

. . .

When asked years later why he had abandoned the play, Beckett claimed that he was unable to integrate Johnson's English as rendered by Boswell into the living speech of the Irish actors who would have spoken his words. Johnson remained alien for him: suspended between life and death in the "proper English" of Boswell's printed page, he could not come alive on the stage.[92]

The one scene of "Human Wishes" Beckett did write (perhaps as late as 1940) takes place in Johnson's absence, as the three contentious women of his household, whom Beckett termed Johnson's "seraglio"—the blind poetess Mrs. Williams (who meditates), Johnson's Lichfield relative Mrs. Desmoulins (who knits like one of the fates, the "second of a pair" of mittens, the first of which is absent), and Polly Carmichael, a prostitute Johnson took in from the streets (who reads)—ruminate about death. (The cat Hodge, it should also be mentioned, "sleeping—if possible," becomes a member of the cast.)[93] Even the knitting—or knotting as it has also been called—has a referent in Beckett's Johnsonian research:

> "Next to mere idleness, I think knotting is to be reckoned in the scale of insignificance; though I once attempted to learn knotting; Dempster's sister endeavoured to teach me, but I made no progress." Strange that he never took to smoking. "I cannot account why a thing which requires so little exertion and yet prevents the mind from total vacuity, shd. have gone out. Every man has something by which he calms himself; beating with his feet or so."[94]

The dialogue, itself haunted by print, vacillates between quotation of print and living speech, between the text consumed by the eye and what the blind Mrs. Williams can hear. Mrs. Williams asks Miss Carmichael what she is reading to no avail; when she asks her to write down an extemporaneous poem, which she recites and then repeats, Miss Carmichael responds that she has taken it down, "In what will not dry black and what was never white," handing her a blank page.

> MRS. W. (*fingering the sheet tenderly*). I did not hear the scratch of the quill.
> MISS C. I write very quiet.
> MRS. W. I do not feel the trace of the ink.
> MISS C. I write very fine. Very quiet, I write, and very fine.
> *Silence.* (157–58)

Even the silence of the characters has a double life here—both as the blankness of the page that Mrs. Williams fingers so tenderly and in its resonance as the word "silence" in Beckett's hand, written down just as are the words of Mrs. Williams's poem, lost to the stage audience but present on the page.

When the drunken Dr. Levet (in regard to whom Beckett in his notes interrupts the sentimental piety of Leslie Stephen's opining that "Levett admired J. because others admired him, & Johnson in pity loved Levett, because few others could find anything in him to love" with the exclamation in verso "Balls. He was a symptom of J's anxiety"[95]) enters the scene, hiccups with "such force that he is almost thrown off his feet," and staggers off, the women's response reminds us that the play itself is a text:

> MRS. W. Words fail us.
> MRS. D. Now this is where a writer for the stage would have us speak no doubt.
> MRS. W. He would have us explain Levett.
> MRS. D. To the public.
> MRS. W. The ignorant public.
> MRS. D. To the gallery.
> MRS. W. To the pit.
> MISS C. To the boxes. (160–61)

The general public who would encounter the play on the printed page (the primary way the play has been encountered, as it turns out) only slowly turn into the social demographic of the theater. Throughout one is aware of the variety of printed texts that haunt the scene. Mrs. Desmoulins quotes *Hamlet* in reference to the death of Goldsmith (who died owing Mrs. Williams money), pardoning "the frailties of a life long since transported to that undiscovered country from whose—" only to be refuted: "None of your Shakespeare to me, Madam. The fellow may be in Abraham's bosom for aught I know or care, I still say he ought to be in Newgate" (162). Mrs. Williams, in a private reverie, echoes a passage from Johnson's *Prayers and Meditations*:[96]

> For years, for how many years every day, dead, whose name I had known, whose face I had seen, whose voice I had heard, whose hand I had held, whose—but it is idle to continue. Yesterday, in the flower of her age, Mrs Winterbotham, the greengrocer of the Garden; Monday, Mr Pott of the Fleet; Sunday, in great pain, in his home in Islington, after a lingering illness, surrounded by his family, the Very Reverend William Walter Okey, Litt.D., LL.D; Saturday, at Bath, suddenly, Miss Tout; Friday,—but it is idle to continue. I know— (164)

Here we see the roots of the most famous Beckettian trope—"I can't go on, I'll go on"—in the Johnsonian paradox of death as the end without end.

The scene concludes with Miss Carmichael reading, with frequent interruptions by Mrs. Williams, the following sentence:

> Death meets us every where, and is procured by every instrument, and in all chances, and enters in at many doors: by violence, and secret influence, by the aspect of a star, and the stink of a mist, by the emissions of a cloude, and the meeting of a vapor, by the fall of a chariot, and the stumbling at a stone, by a full meal, or an empty stomach, by watching at the wine, or by watching at prayers, by the Sun or the Moon, by a heat or a cold, by sleeplesse nights, or sleeping dayes, by water frozen into the hardnesse, and sharpnesse of a dagger, or water thawed into the floods of a river; by a hair, or a raisin, by violent motion, by sitting still, by severity, or dissolution, by Gods mercy, or Gods anger, by every thing in providence, and every thing in manners, by every thing in nature and every thing in chance.[97]

The sentence, encapsulating in its magisterially balanced oppositional rhythms the same impulse of closure and repetition that Mrs. Williams rehearsed in her litany of obituary clichés, is followed by silence, which of course begins again. In the ensuing dialogue, the women continue to puzzle over the relation between reading, hearing, comprehension, and memory; this exchange ends with the proper textual attribution, returning us to print:

> MRS. W. Brown for a guinea.
> *Miss Carmichael rises.*
> MRS. W. I say: Brown for a guinea.
> MISS C. I hear you, Madam.

MRS. W. Then answer me. Is it Brown or is it not Brown?

MISS C. Brown or Black, Madam, it is all one to me.

MRS. W. Is it possible she reads and does not know what she reads.

MISS C. I read so little, Madam, it is all one to me.

MRS. W. Turn to the title page, my child, and tell me is it Brown.

MISS C. (*turning to the title page*). Taylor. (166)

As in the riddle of the blank page that will not dry black and was never white, Beckett's puns—staging the gap between the blind woman of letters and the reformed prostitute to whom letters are "all one"—remind us of the materiality, the "thingness" of words. If we turn to the passage Miss Carmichael reads from Jeremy Taylor's *Rule and Exercises of Holy Dying,* the sentence that immediately follows this one will bring us up short:

> *Eripitur persona, manet res,* we take pains to heap up things useful to our life, and get our death in the purchase; and the person is snatched away, and the goods remain.[98]

Taylor is quoting Lucretius, "the mask is torn off, and the truth remains behind."[99] Yet he draws a different moral, one more suited to his didactic intent: material things are useless in the face of death. If we isolate this aphorism, removed from its context as easily as Beckett's quotation of Taylor, or as any anecdote from Boswell, and apply it to our own purposes, the moral we can draw is closer to that of Taylor's source: death strips away illusions, leaving the thing—the primary meaning of *res*—itself. Or as close as we are able to come—the record of the autopsy, the feel of the ink on the page, the sound of a sentence read and repeated.

When Johnson, "a poet doomed at last to wake a lexicographer," disillusioned in his *Preface to A Dictionary* at his unsuccessful attempt to order the chaos of a living tongue, observes, "*words are the daughters of earth,*" while "*things are the sons of heaven,*" this claim might at first seem paradoxical, even counter-intuitive.[100] The materiality of things, according to the memento mori tradition that Johnson knew so well and that Taylor exemplifies, would seem to connect them to earth and to mortality. Words, the abstract realm of language, by contrast, one would think would endure. Immersed in an endless struggle to put words in their proper place, to assign them to the material world of things whose existence, we'll recall, he refused to doubt, gave Johnson a different perspective, one much closer to Beckett's in its awareness of the thingness of words, their paradoxically solid mutability. Like death itself, like the conundrum of "positive annihilation," words insist on continuing when it is idle to continue.

"Taylor" was the last word of "Human Wishes" that Beckett wrote, but how did he want the play to end? The following cryptic and near-final notes, written with special emphasis on the page facing numerous transcriptions of accounts of Johnson's death, along with Beckett's comments quoted earlier about the play's "final act," will give us an idea:

> "panted on to 90," spoken toward end of last act, when, with J. panting in silence after "sent to hell; Sir etc.," curtain falls.
>
> Conditional Sálvation motif . . .
>
> Johnson's "ghastly smile"[101]

Johnson's "ghastly smile," Dilks observes, is repeated in the smile of the crucified thief who points "towards the earth, towards hell," in the sketch that interrupts the manuscript.[102] What I want to focus on here is a different conflation of corporeal and verbal, namely, the words "panted on to 90," underlined and repeated on the facing page, originally written by Johnson in a letter in reference to the asthma specialist Sir John Floyer, who "panted on to 90 as was supposed."[103] What Beckett envisioned, it seems, was that Johnson's penultimate words would be the impassioned and fearful cry to the polite Dr. Adams: "sent to Hell, Sir, and punished everlastingly." The last words we would hear would echo in a different material form the repetitive background of Johnson's labored breath, clinging desperately to life while confronting death without an end, as unredeemed as the crucified men in bowler hats that smile from the manuscript page. This breath will resound as well through Beckett's short play *Breath,* first performed in 1969, in which no words are spoken and all that punctuates the sound of respiration are cries (of pain?).[104] This breath informs our gaze at Johnson's lung—lost but reproduced in print (see fig. 2). One could almost say that this imagined breath, inspiring Beckett's quotation of Johnson's words, quite literally animates the imagined lung. As long as there are readers, this breath will not end.

NOTES

## Introduction

I allude in this introduction's title to the title of the final chapter of Johnson's apologue *Rasselas* (1759), an Abyssinian tale of a prince's escape from Paradise in search of "the choice of life," in which process is all, melancholy is prevalent, and closure is impossible.

1. This epigraph is one of several "Appreciations and Testimonia" included in Alice Meynell and G. K. Chesterton, *Samuel Johnson* (London: Herbert & Daniel, 1911), 261–62.
2. Jack Lynch, ed., *Samuel Johnson's Dictionary: Selections from the 1755 Work that Defined the English Language* (New York: Levenger Press, 2002).
3. Mildred C. Stuble, *A Johnson Handbook* (New York: F. S. Crofts & Co., 1933); Jack Lynch, ed., *Samuel Johnson's Insults: A Compendium of Snubs, Sneers, Slights, and Effronteries from the Eighteenth-Century Literary Master* (New York: Levenger Press, 2004).
4. Samuel Johnson, *Rambler* 60, in *The Rambler,* ed. W. J. Bate and Albrecht B. Strauss, *The Yale Edition of the Works of Samuel Johnson,* 16 vols. (New Haven: Yale University Press, 1969), 3:321. For a recent anthology of essays on this theme, see David Wheeler, ed., *Domestick Privacies: Samuel Johnson and the Art of Biography* (Lexington: University of Kentucky Press, 1987). For Johnson's career-long interest in "everyday life," a rubric very close to that domestic privacy and similarly on the border of art and life, and for the transformation of this category in Boswell's *Life,* see Kevin Hart, *Samuel Johnson and the Culture of Property* (Cambridge: Cambridge University Press, 1999), chap. 6.
5. Johnson's first literary biography, *An Account of the Life of Mr. Richard Savage* (1744), published anonymously early in his career, provides a founding example of a complex attempt at once to judge and sympathize with a writer who was "not so much a good man, but the friend of goodness." Samuel Johnson, *An Account of the Life of Mr. Richard Savage,* in *Selected Poetry and Prose,* ed. Frank Brady and W. K. Wimsatt (Berkeley: University of California Press, 1977), 594. See Helen Deutsch, "'The Name of an Author': Moral Economics in Johnson's *Life of Savage,*" *Modern Philology* 92, no. 3 (February 1995): 328–45. For good introductions to the balancing of art and life in Johnson's *Lives,* see Greg Clingham, "Life and Literature in Johnson's *Lives of the Poets,*" in *The Cambridge Companion to Samuel Johnson,* ed. Greg Clingham (Cambridge: Cambridge University Press, 1997), 161–91; and Robert Folkenflik, *Samuel Johnson, Biographer* (Ithaca, NY: Cornell University Press, 1978).
6. Hester Thrale, entry for May 1, 1779, *Thraliana; the Diary of Mrs. Hester Lynch Thrale (Later Mrs. Piozzi), 1776–1809,* vol. 1, *1776–1784,* ed. Katherine C. Balderston (Oxford: Clarendon Press, 1942), 384–85. For astute and comprehensive analyses of the Johnson/Thrale relationship, see John C. Riely, "Johnson's Last Years with Mrs.

Thrale: Facts and Problems," *Bulletin of the John Rylands University Library of Manchester* 57 (Autumn 1974): 196–212; and Riely, "Johnson and Mrs. Thrale: The Beginning and the End," in *Johnson and His Age,* ed. James Engell, Harvard English Studies (Cambridge, MA: Harvard University Press, 1984), 12:55–81. A mysterious letter, written in French, from Johnson to Thrale on the subject of confinement has also been the subject of much critical speculation. I think Riely is right to conclude (following Walter Jackson Bate) that the "secret" and the substance of all these private details was not Johnson's masochistic desire for Thrale but rather his past experience and present fear of madness. Riely, "Johnson and Mrs. Thrale," 64–70; for the masochism argument, see Katherine C. Balderston, "Johnson's Vile Melancholy," in *The Age of Johnson: Essays Presented to Chauncey Brewster Tinker,* ed. Frederick Whiley Hilles (New Haven: Yale University Press, 1949), 3–14; and for its explicit refutation, see Walter Jackson Bate, *Samuel Johnson* (New York: Harcourt Brace Jovanovich, 1977), 384–89, 439–41.

7. Hester Lynch Piozzi, *Anecdotes of the Late Samuel Johnson, LL.D., during the Last Twenty Years of His Life,* ed. Arthur Sherbo (London: Oxford University Press, 1974), 102.
8. The first of many such detractors was Thrale's rival, James Boswell, whose "animadversions" against her have generated much discussion. For the most comprehensive discussion of this relationship, see Mary Hyde, *The Impossible Friendship: Boswell and Mrs. Thrale* (Cambridge, MA: Harvard University Press, 1972).
9. Samuel Johnson to Hester Thrale, Friday, July 2, 1784, *The Letters of Samuel Johnson,* vol. 4, *1782–1784,* ed. Bruce Redford (Princeton: Princeton University Press, 1994), 338.
10. Samuel Johnson to Hester Thrale, Thursday, July 8, 1784, ibid., 343–44.
11. Frances Burney, in *Dr. Johnson and Fanny Burney; being the Johnsonian Passages from the Works of Mme. D'Arblay,* ed. Chauncey Brewster Tinker (New York: Moffat, Yard & Company, 1911), 184.
12. Alice Meynell gives us a different narrative of Thrale's reputation, in which public condemnation culminates with Thomas Babington Macaulay, and then wanes into acceptance. Her concern, as her essay's title shows, is with another remarrying widow, "Mrs. Johnson," Elizabeth Porter, who is similarly criticized by Macaulay for expressing her own desire when desire should die with a first marriage, and to whom succeeding generations have not been as kind. This theme will resurface with the discussion of the Ephesian matron story, applied by Macaulay to Thrale's case, which follows. Alice Meynell, "Mrs. Johnson," in *Essays,* 6th impression (London: Burnes Oates & Washbourne Ltd., 1925), 213–14.
13. For classic examples of a divided Johnson, see W. B. C. Watkins, *Perilous Balance: The Tragic Genius of Swift, Johnson, and Sterne* (Princeton: Princeton University Press, 1939), chaps. 2–4 (of the three authors treated, Johnson is the one to whom Watkins gives the title's laurel); Bertrand H. Bronson, *Johnson Agonistes and Other Essays* (Berkeley: University of California Press, 1965), chap. 1; Leopold Damrosch Jr., *Samuel Johnson and the Tragic Sense* (Princeton: Princeton University Press, 1972).
14. George Birkbeck Hill, *Wit and Wisdom of Samuel Johnson* (Oxford: Clarendon Press, 1888), xxii–xxiii.
15. Perhaps this explains the quotation from Johnson that Damien Guerrero, on trial for murdering his former girlfriend, chose for his entry in the Redlands East Valley High

2002 yearbook: "He who makes a beast of himself loses the pain of being a man." *LA Weekly,* January 9–15, 2004, 25.

16. Anna Seward to Mrs. Mary Knowles, April 20, 1788, *Letters of Anna Seward: Written Between the Years 1784 and 1807,* 6 vols. (Edinburgh: George Ramsay & Co., 1811), 2:103. Seward comments that "cupboard-love" and "Platonic love" "are certainly heterogeneous; but Johnson, in religion and politics, in love and in hatred, in truth and falsehood, was composed of such opposite and contradictory materials, as never before met in the human mind" (2:103–4)—hence his enduring popularity as object of public speculation.
17. Frith based this image on a moment in Siddons's memoirs in which she describes Johnson as "extremely, though formally polite" in her visits to him shortly before his death. A hugely popular and prolific history painter, Frith was fascinated with depicting the strange loves of authors, also depicting Lady Mary Wortley Montagu rejecting Pope, and (more to our purposes) *Dr. Johnson's Tardy Gallantry* (1886), a scene based on a moment in Boswell's *Life* in which Johnson, "eager to show himself a man of gallantry," rushes out of his house to assist Madame de Boufflers, "a foreign lady of quality," to her carriage, dressed "in a rusty brown morning suit, a pair of old shoes by way of slippers, a little shriveled wig sticking on the top of his head, and the sleeves of his shirt and the knees of his breeches hanging loose. A considerable crowd of people gathered round and were not a little struck by this singular appearance." Both paintings are cataloged along with their anecdotal origins in Siddons and Boswell (as quoted in the Royal Academy catalogs) in David Layne Montgomery, *William Powell Frith (1819–1909): A Reevaluation of His Artistic Career* (Ph.D. diss., University of Missouri, Columbia, 1997; Ann Arbor: University of Michigan Microfilms, 1997), 701. Frith characterized his "subject pictures" as imaginary moments from both fictional and historical narrative distilled into (almost Boswellian) "scenes from real life." His painting of a moment from Boswell's *Life* depicting Johnson and Garrick in the midst of many from Johnson's famous Club sold for the "largest price that had been paid for the work of a living artist at that time." His discussions of these paintings nearly always involve sustained attention to the realism of his mimesis, what models he chose (human or artificial), and how closely those models resembled the actual characters. W. P. Frith, *My Autobiography and Reminiscences,* 3 vols. (London: Richard Bentley, 1887), 1:81, 219–21, 385–89; 2:274.
18. Thrale responds to Johnson's admonishment to sympathize with the poor enjoying the (to her distasteful) smells of Porridge Island by commenting: "These Notions—just as they doubtless are;—seem to me the faeculancies of his low Birth, which I believe has never failed to leave its *Stigma* indelible in every human Creature; however exalted by Rank or polished by Learning." Hester Thrale, entry for December 1777, *Thraliana,* 1:186. My characterization of Johnson's "pure manners," his art without nature, is indebted both to Claudia L. Johnson, "The Divine Miss Jane: Jane Austen, Janeites, and the Discipline of Novel Studies," *Boundary 2* 23, no. 3 (Fall 1996): 143–63, in whose account the masculinization of Austen and novel studies purges the profession of literature of "the homosexual's Jane Austen, the decadent's Jane Austen, damning all persons for whom manners bear no relation to nature—which, here, is shorthand for bourgeois morality and heterosexual desire" (161); and to D. A. Miller, *Jane Austen, or The Secret of Style* (Princeton: Princeton University Press, 2003), who positions Austen's style in a complex tension with nature: "'All style and

no substance': the formula helps us recognize not that style is *different,* or even *opposite,* to substance . . . , but that the one is incompatible with, and even corrosive of the other. Style can only emerge at the expense of substance, as though it sucked up the latter into the vacuum swollen only with the 'airs' it gives itself" (17).

19. "*Ultimus Romanorum*" from "The Hero as Man of Letters. Johnson, Rousseau, Burns" (1841), in *On Heroes, Hero-Worship, and the Heroic in History,* ed. H. D. Traill, vol. 5, *The Works of Thomas Carlyle* (New York: Scribner's, 1903), 184; second quotation from *Essays,* included in Meynell and Chesterton, "Appreciations and Testimonia," 261.
20. Marianna Torgovnick, *Gone Primitive: Savage Intellects, Modern Lives* (Chicago: University of Chicago Press, 1990), 235.
21. Jane Gallop, *Anecdotal Theory* (Durham, NC: Duke University Press, 2002), 7, 8.
22. In *The Academic Postmodern and the Rule of Literature* (Chicago: University of Chicago Press, 1995), 41–71, David Simpson links Boswell's preservation of Johnsonian anecdote and seemingly infinite conversation to the modern academic preference for the personal voice—both, he argues, are the result of the alienation that characterizes modernity and the subsequent disavowal of larger wholes—of public discourse, knowledge, and history. I owe much to Simpson's argument, but I deploy it in a much less critical vein.
23. Peter Pindar [John Wolcot], *A Poetical and Congratulatory Epistle to James Boswell, Esq. on His Journal of a Tour to the Hebrides, with the celebrated Dr. Johnson,* 8th ed. (London, 1788), 9–10. On the complicated politics of Wolcot's hugely popular pseudonymous satire—at once conservative, populist, anti-French, and derisive of the personal foibles of the monarch—as symptomatic of the complicated literary milieu of the 1790s, see Gary Dyer, *British Satire and the Politics of Style, 1789–1832* (Cambridge: Cambridge University Press, 1997), 31–37.
24. See Janet Sorensen, *The Grammar of Empire in Eighteenth-Century British Writing* (Cambridge: Cambridge University Press, 2000), chap. 2, for an interesting analysis of metaphors of food, consumption, and regurgitation in Scottish critics of Johnson's *Dictionary* in particular and cultural authority in general.
25. Carolyn Steedman puts this well when she describes how Freud, in *Jokes and Their Relation to the Unconscious,* "had made plain the very great pleasures made available by *seeing* the absurdity of the world's dislocations: the charming unexpectedness of knowing oneself part of the effect produced, when time and event miss a beat, stumble; and then the restoration of their timing." Carolyn Steedman, *Dust: The Archive and Cultural History* (New Brunswick, NJ: Rutgers University Press, 2002), 11.
26. Letter to Susan Phillips, December 1784, quoted in O. M. Brack Jr. and Robert E. Kelley, eds., *The Early Biographies of Samuel Johnson* (Iowa City: Iowa University Press, 1974), 6–7.
27. *A Poetical Epistle from the Ghost of Dr. Johnson, to His Four Friends: The Rev. Mr. Strahan, James Boswell, Esq., Mrs. Piozzi, J. Courtenay, Esq. M.P. from the Original Copy in the Possession of the Editor. With Notes Critical, Biographical, Historical, and Explanatory. "My Little Fame May Heav'n Defend, / From Ev'ry Feign'd or Foolish Friend!" Editor's MS.* (London, 1786).
28. Brack and Kelley, *Early Biographies,* 62.
29. Sigmund Freud, *Jokes and Their Relation to the Unconscious,* trans. James Strachey, vol. 6, The Pelican Freud Library, ed. Angela Richards (1976; repr., Harmondsworth: Penguin, 1983), 46.

30. Ralph Waldo Emerson, "Uses of Great Men," in *Representative Men: Seven Essays*, ed. Douglas Emory Wilson et al. (Cambridge, MA: Harvard University Press, 1987), 17.
31. James Boswell, *The Life of Samuel Johnson, LL.D.*, ed. George Birkbeck Hill, rev. L. F. Powell, 6 vols. (Oxford: Clarendon Press, 1934–50), 1:421.
32. Boswell, *Life*, 1:5.
33. I will have more to say later on the relationship between icons, which follow a logic of resemblance in the service of identity, and simulacra, which in the service of singularity repeat with a difference. Here I rely on Gilles Deleuze's analysis of Plato's foundational logic of representation in "The Simulacrum and Ancient Philosophy," in *The Logic of Sense*, trans. Mark Lester with Charles Stivale, ed. Constantin V. Boundas (New York: Columbia University Press, 1990), 253–66. I am grateful to Page duBois for this reference.
34. From *The Works of Petronius Arbiter, in Prose and Verse. Translated from the Original Latin, by Mr. Addison. To which are prefix'd the Life of Petronius, Done from the Latin: and a Character of his Writings by Monsieur St. Evremont* (London, 1736), 227.
35. Few would disagree that despite important exceptions such as (to name two of the most influential examples) Mary Hyde—to whose collection of Johnsoniana and work on Hester Thrale this book and many others are indebted—and Isobel Grundy, "female Johnsonian" continues to be largely an oxymoron in the profession of eighteenth-century studies. Two essays by Grundy that inform my own thinking on Johnson and Boswell are "Samuel Johnson: Man of Maxims," in *Samuel Johnson: New Critical Essays*, ed. Isobel Grundy (London: Vision, 1984), 13–30; and "'Over Him We Hang Vibrating': Uncertainty in the *Life of Johnson*," in *Boswell: Citizen of the World, Man of Letters*, ed. Irma S. Lustig (Lexington: University of Kentucky Press, 1995), 184–202. The recent work of Norma Clarke and, especially, Felicity A. Nussbaum, in its resituation of Johnson in relation to his female literary contemporaries, seems to me radically and usefully un-Johnsonian in its dismantling of the solitary masculinity of Johnson's monument.
36. I am indebted to Margaret Anne Doody's reading of the episode in *The True Story of the Novel* (New Brunswick, NJ: Rutgers University Press, 1996), 110–13, for help in forming these oppositions.
37. Boswell, *Life*, 3:209.
38. Thus, James Boswell, in his periodical paper *The Hypochondriack*, rewrites Juvenal's dictum of "sound mind in a sound body," embraced by John Locke in *Some Thoughts concerning Education*, by characterizing his own and Johnson's common malady, melancholy's eighteenth-century heir hypochondria, as an unhealthy mind in a healthy body, proud indiscernible mark of the excessive sensibility of the man of feeling. James Boswell, *The Hypochondriack* (1777), ed. Margery Bailey, 2 vols. (Stanford: Stanford University Press, 1928), 2:236–38.
39. The identification of men of letters with Hamlet as the quintessential melancholic hero is part of a history of the glorification of male melancholy that begins with Aristotle and culminates in Freud's evocation of Hamlet as harbinger of human truth in "Mourning and Melancholia." See Juliana Schiesari, *The Gendering of Melancholia* (Ithaca, NY: Cornell University Press, 1992), who argues that such ennobling of the masculine "lack of a lack" depends upon the silencing of women. My work focuses on a particular chapter of this association, during which professional authorship in

eighteenth-century Britain, free from patronage and bound to the marketplace, financially flourishes and physically ails. The association between authorship and disease, and the communal self-reflection of authors and doctors during the eighteenth century, is in fact so overdetermined during this period that it is difficult to think of a single canonical male figure who was not remarkably ill and/or a student of illness. Such masculine affliction can be read contra Schiesari as a response to the unprecedented volubility of women in print during this period. See Helen Deutsch, "Symptomatic Correspondences: Engendering the Author in Eighteenth-Century England," *Cultural Critique* 42 (Spring 1999): 35–80.

40. For Beckett's explicit preference for the Johnson of the *Prayers and Meditations,* see the letter to Mary Manning quoted in James Knowlson, *Damned to Fame: The Life of Samuel Beckett* (New York: Simon & Schuster, 1996), 250; for his claim that it is Johnson "who is always with me," see Deirdre Bair, *Samuel Beckett: A Biography* (1978; repr., New York: Simon & Schuster, 1990), 257.
41. Julian Barnes, *England, England* (New York: Vintage, 2000) 212–18, quotations from 218.
42. For one early and still influential example, see William C. Dowling, *The Boswellian Hero* (Athens: University of Georgia Press, 1979). Johnson was in fact made an honorary doctor of laws by Trinity College, Dublin, in 1765 and a doctor of civil law by Oxford in 1775; Boswell's "Dr. Johnson," however, is a timeless construction that epitomizes an entire lifetime.
43. For this detail, see Hart, *Johnson and the Culture of Property,* 68. Hart brilliantly tells the story of the organizational power of Johnson's monument—rendered public property by Boswell in an "economy of death" (178) in which the man, rather than the writer, lives beyond the grave—in shaping the institutional rise of the "Age of Johnson" in British culture within and beyond the academy.
44. John Bailey, *Dr. Johnson and His Circle,* 2nd ed. with assistance of L. F. Powell, vol. 64 of the Home University Library of Modern Knowledge (1913; repr., London: Oxford University Press, 1945). Narrating an encounter with a London cabman "quoting" Johnson, Bailey concludes, "You would not find a cabman ascribing to Milton or Pope a shrewd saying that he had heard and liked" (10–11).
45. *The R.B. Adam Library Relating to Dr. Samuel Johnson & his Era. Printed for the author at Buffalo, New York,* 3 vols. (London: Oxford University Press, 1929), 1:i. Adam notes that this edition is "an elaboration, and a continuation of a 'Catalogue of the Johnsonian Collection of R.B. Adam' (2nd), privately printed (fifty copies) in the year 1921." No page numbers.
46. One might also consider Augustine Birrell's 1899 pronouncement on the transmission of Johnson's personality through the repeating of his sayings as a "confirmed habit of the British race" (in George Whale and John Sargeaunt, eds., *Johnson Club Papers by Various Hands* [London: T. Fisher Unwin, 1899], 3) in relation to the American Thomas Marc Parrott's 1901 essay "On the Transmission of Dr. Johnson's Personality," which argues that "Johnson's dictatorship was due to his personality rather than to his productions, to his spoken rather than to his written words" (134). Parrott also credits Johnson with holding England back, by sheer force of personality, from an inevitable "revolution in . . . literature, in politics, and in religion," a fighter in a losing battle against what he implies is American modernity (171). While Parrott clearly views Johnson as an artifact of an outmoded British age and

culture, like Birrell he evokes the overwhelming "charm" of a personality that merges spoken with written word. Thomas Marc Parrott, *The Personality of Dr. Johnson,* reprinted from *Studies of a Booklover* (1901) in pamphlet form (New York: James Pott & Co., 1906).

47. For the fascinating story—an epic of riches lost and found—of the discovery and acquisition of the Boswell papers, see David Buchanan, *The Treasure of Auchinleck: The Story of the Boswell Papers* (New York: McGraw-Hill, 1974); particularly germane to my interest in the anecdote and the relation of American collector to English academic is the story of Isham's returning to Oxford Professor Emeritus David Nichol Smith, upon their meeting in his New York apartment to dine and view his collection, a scrap of paper (believed lost) bearing Johnson's emendation (at Boswell's suggestion) to a couplet from *The Vanity of Human Wishes*; when Smith tells Isham of Boswell's original intention to deposit the scrap in the Bodleian Library, Isham "sat quiet for a moment, and then he said firmly, 'take it back with you.'" Smith carried it with him on the plane to England, wondering "what the observations of Johnson and Boswell would have been had they known of its flight over the Atlantic. My first duty when I got back to Oxford was to do what Boswell had intended. I deposited the slip in the Bodleian" (ibid., 259–61). In this utopian anecdote, lost objects are actually found and origins are restored.

48. William H. Epstein, "Counter-Intelligence: Cold-War Criticism and Eighteenth-Century Studies," *ELH* 57 (1990): 63–99; and Epstein, "Assumed Identities: Gray's Correspondence and the 'Intelligence Communities' of Eighteenth-Century Studies," *The Eighteenth Century: Theory and Interpretation* 32, no. 3 (Fall 1991): 274–88. Epstein links (both metaphorically and literally) the disembodied, ahistorical, and seemingly "objective" nature of the American New Criticism of the mid-twentieth century to the seemingly non-ideological politics and surveillance practices of the early Cold War. The author that most interests Epstein in this regard is Thomas Gray, whose dense network of literary allusions and evasive strategies of authorial self-representation certainly lend themselves to Epstein's construction of twentieth-century American critical reading as a process of counterintelligence and code cracking, while also providing a symbiotic counterexample to Johnson's own brand of literary accessibility that I will discuss later in this introduction. The story of the organizational power of Johnson's name in the institutional rise of literary criticism in general and eighteenth-century studies in particular is incisively told by Hart in *Johnson and the Culture of Property.*

49. Johnson himself was accused of indiscreet overindulgence in anecdotal detail in his *Lives of the Poets* and was certainly an innovator in his cautious reliance on anecdotal evidence, the closest he could come to the ephemeral circumstances of an individual life. For the classic essay on the topic, which sees Johnson wrestling with the relation between particular facts and general examples "at the fork of the roads leading separately to science and art," a conjunction at which, as we shall see, the genre of the anecdote has long been located, see Clarence R. Tracy, "Johnson and the Art of Anecdote," *University of Toronto Quarterly* 45 (1945–46): 86–93, quotation from 93. For a respectful refutation of Tracy that sees Johnson's use of detail as more aesthetically integrated, see Robert Folkenflik, "Johnson's Art of Anecdote," *Studies in Eighteenth-Century Culture* 3 (1975): 171–81; quotations from Johnson's *Dictionary* taken from Folkenflik, 181n4.

50. For Boswell's narrative innovations, see Adam Sisman, *Boswell's Presumptuous Task* (2000; repr., New York: Farrar, Straus, and Giroux, 2001); and John A. Vance, ed., *Boswell's* Life of Johnson: *New Questions, New Answers* (Athens: University of Georgia Press, 1985).
51. Boswell, preface to the second edition, *Life,* 1:17.
52. This emphasis on the empirical as source and corrective of art is one side of Johnson's impulse as critic and moralist; the other is moral utility. For an excellent introduction to Johnson's critical principles, see Jean H. Hagstrum, *Samuel Johnson's Literary Criticism* (Minneapolis: University of Minnesota Press, 1952). I quote here from Johnson's 1757 review of Soame Jenyns's *A Free Enquiry into the Nature and Origin of Evil,* in *The Oxford Authors: Samuel Johnson,* ed. Donald Greene (Oxford: Oxford University Press, 1984), 527. By refuting Jenyns's theodicy, Johnson also refuses Pope's moral certainty in *An Essay on Man,* a topic to which I will return in chapter 5.
53. Claudia L. Johnson, "Austen Cults and Cultures," in *The Cambridge Companion to Jane Austen,* ed. Edward Copeland and Juliet McMaster (Cambridge: Cambridge University Press, 1997), 212, 216.
54. Consider, for example, the exclusionary judgment of Q. D. Leavis: "I started by saying that Miss Austen is and has long been regarded with affection as well as admiration and respect, and it is worth asking why, since, if one thinks of it, it is *only* Jane Austen (no longer Charles Lamb to the same extent or even Dickens, and we don't *love* Dr. Johnson though his life is incessantly being investigated)—only Jane Austen who is loved and esteemed by her readers as a *person.*" Q. D. Leavis, "Jane Austen: Novelist of a Changing Society," in *Collected Essays,* ed. G. Singh, vol. 1, *The Englishness of the English Novel* (Cambridge: Cambridge University Press, 1983), 27–28. Leavis's "we" here is quite Austenian in its assumption of a sympathetic audience.
55. On the exclusivity of Austen love, see Deidre Shauna Lynch, introduction to *Janeites: Austen's Disciples and Devotees* (Princeton: Princeton University Press, 2000), 7–12; and Mary Anne O'Farrell, "Jane Austen's Friendship," in ibid., 45–62.
56. For Austen's forgotten queerness, see C. L. Johnson, "The Divine Miss Jane."
57. Johnson, "Life of Gray," in *Selected Poetry and Prose,* 642.
58. John Guillory, *Cultural Capital: The Problem of Literary Canon Formation* (Chicago: University of Chicago Press, 1993), 99.
59. Ibid., 92. I will return to Johnson's forgetting of Gray's sources in the service of his imagination of the common reader in chapter 3 in the context of Neil Hertz's essay "Dr. Johnson's Forgetfulness, Descartes' Piece of Wax," *Eighteenth-Century Life* 16 (November 1992): 167–81.
60. D. A. Miller describes a similar effect—the abstraction of style into an apparently universal "nature"—when he gives us "the first secret of Austen Style: its author hates style, or at any rate, must always say she does; she must always profess the values, and uphold the norms, of 'nature,' even as she practices the most extraordinarily formal art the novel had yet known." Miller, *Jane Austen,* 26–27. In Austen's novels, Miller argues, this has the effect of the author writing herself out of the text entirely since its "nature" has no place for a single woman who writes. By contrast, Johnson the masculine moralist, rather than disappear from his work, comes to supplant it. Chapter 2 will discuss at greater length Johnson's own embodied conflation of nature and art, eternal substance and singular style (a style whose knowing impersonality influ-

enced Austen's own), while his own struggle as a critic with the ethical implications of an art for its own sake that threatens an unnatural and amoral endlessness is the subject of the Johnsonian divagations of chapter 5.

61. Eric O. Clarke, "Shelley's Heart: Sexual Politics and Cultural Value," *Yale Journal of Criticism* 8 (1995): 199.
62. In *Johnson and the Culture of Property,* Hart calls the capaciously noted 1887 G. B. Hill edition of Boswell's *Life,* revised definitively by L. F. Powell in 1934, "annotated scripture for eighteenth-century literary scholars," noting that Hill's *Life* is a "sacred text not because its hero can offer a secular salvation by teaching us to avoid cant (as others will claim in effect) but because the age viewed there has a special value . . . ; it is a golden age, a world with which [Hill] has an imaginary relationship" (88). Hill's version of Boswell's Johnson is therefore "exemplary, and therefore no longer singular" (86). Boswell's construction of Johnson, noteworthy in Hart's account for being the first to focus "most self-consciously on 'minute particularities'" in order to represent "individuality as a prime value" (76) is, by contrast I would argue, exemplary *because* singular. It's worth noting as well that the Oxford scholar Hill, who also edited the *Johnsonian Miscellanies,* sees himself following in Johnson's footsteps (both attended Pembroke College) and models his preface to the third edition of the *Life* after Johnson's preface to the *Dictionary.*
63. For the Romantic equation of anecdote and experimental science, see Isaac D'Israeli, *A Dissertation on Anecdotes* (1793; repr., New York: Garland, 1972), 27. For the seminal essay on the anecdote's genesis at the Renaissance crossroads of history and medical diagnosis, see Joel Fineman, "History of the Anecdote: Fiction and Fiction," in *The New Historicism,* ed. H. Aram Veeser (New York: Routledge, 1989), 56. For an unequivocal endorsement of Fineman's reading of the anecdote's unique purchase on the real, see Stephen Greenblatt, *Learning to Curse* (New York: Routledge, 1990), 5–11. For two powerful critiques of the New Historicism's anecdotal evasions of history, see Alan Liu, "Local Transcendence: Cultural Criticism, Postmodernism, and the Romanticism of Detail," *Representations* 32 (Fall 1990): 75–113; and Simpson, *Academic Postmodern,* esp. 55–58. For a well-balanced consideration of the differences between the literary New Historicism and the practice of history that turns upon the use of the anecdote (the former emphasizes irregularity and wonder, the latter coherence to a larger framework), and that finds much common ground, see Sara Maza, "Stephen Greenblatt, New Historicism, and Cultural History, *Or,* What We Talk about When We Talk about Interdisciplinarity," *Modern Intellectual History* 1, no. 2 (2004): 249–65.
64. This was in part because the first edition "through hurry or inattention" had allowed "some obscene jests" to be credited to Johnson. Boswell, *Life,* 2:432. Donald Greene has traced a number of Johnson's "smart sayings" in the *Life* to popular jest books predating Boswell's text. See, for example, "The World's Worst Biography," *American Scholar* 62, no. 3 (Summer 1993): 376–77.
65. Boswell, *Life,* 3:325.
66. Ibid., 3:326.
67. Edmund Burke, *Reflections on the Revolution in France,* ed. Conor Cruise O'Brien (Harmondsworth: Penguin, 1969), 172.
68. Boswell, *Life,* 1:17.
69. *Gentleman's Magazine* 54 (December 1784): 883.

70. J. C. D. Clarke, *Samuel Johnson: Literature, Religion and English Cultural Politics from the Restoration to Romanticism* (Cambridge: Cambridge University Press, 1994), quotation taken from prefatory blurb. For a sensible and measured outline of this controversy, see Robert Folkenflik, "Johnson's Politics," in *The Cambridge Companion to Samuel Johnson,* ed. Greg Clingham (Cambridge: Cambridge University Press, 1997), 102–12.
71. John Buchan, *Midwinter* (1923; repr., Boston: Houghton Mifflin, 1929).
72. For a discussion that illuminates this debate's connection to the controversy over the ability of Johnson's *Dictionary* to speak for the entire British nation, see Sorensen, *Grammar of Empire,* chap. 2.
73. Lawrence Lipking, *Samuel Johnson: The Life of an Author* (Cambridge, MA: Harvard University Press, 1998), 10.
74. George Macaulay Trevelyan, "Clio, A Muse," in *Clio, A Muse: and Other Essays Literary and Pedestrian* (London: Longman, Green, and Co., 1913), 46, quoted in Hart, *Johnson and the Culture of Property,* 63; Trevelyan, in A. S. Turberville, *Johnson's England,* 2 vols. (Oxford: Clarendon Press, 1933), 1:6.
75. I should note here that I am interested specifically in the literary anecdote, the plethora of biographical "ana" that flooded the presses at the end of the eighteenth century, of which Boswell's *Life* is the enduring canonical remainder. It is important for my purposes, as it is in a different way for the New Historical critic, that such literary anecdotes have a certain amount of truth value. If we turn to philosophy, say the work of J. L. Austin or Wittgenstein, anecdote, however absurd its content, is useful for what it can demonstrate, not for its relationship to the "real." Toril Moi, talk at MLA Convention, New York, December 2002.
76. Roland Barthes, *Camera Lucida: Reflections on Photography,* trans. Richard Howard (New York: Hill and Wang, 1981), 8.
77. Ibid., 8–9.
78. Quoted in Hart, *Johnson and the Culture of Property,* 91.
79. For an extensive though incomplete list taken at the time of his death, see Boswell, *Life,* 4:421.
80. This literature is reviewed in chapter 1.
81. Philip Neve, *A Narrative of the Disinterment of Milton's Coffin, in the Parish-Church of St. Giles, Cripplegate, on Wednesday, 4th of August, 1790; and of the Treatment of the Corpse, during that, and the following day. The second edition, with additions* (London, 1790). For Richardson's more recent disinterment, see J. L. Scheuer and J. E. Bowman, "The Health of the Novelist and Printer Samuel Richardson (1689–1761): A Correlation of Documentary and Skeletal Evidence," *Journal of the Royal Society of Medicine* 87 (June 1994): 352–55. For Swift's postmortem and nineteenth-century disinterment, see the aptly named literary neurologist Russell Brain, *Some Reflections on Genius* (London: Pitman Medical Publishing, 1960), 25–27. For the fate of Ben Jonson's skull, see Frank Buckland, *Curiosities of Natural History,* 3rd ser., 2:181–89, quoted in A. P. Stanley, *Historical Memorials of Westminster Abbey* (1867), 6th ed., 3 vols. (Philadelphia: George W. Jacobs, 1882), 2:21.
82. Michael Paterniti, *Driving Mr. Albert: A Trip Across America with Einstein's Brain* (New York: Dial Press, 2000), 25.
83. Lydia Davis, *Samuel Johnson Is Indignant* (Brooklyn: McSweeney's Books, 2001), 44.

84. *Philadelphia Public Ledger,* February 20, 1916, 2.
85. For the classic essay on the phenomenon (which began in his own day) of Johnson as text-transcending personality, see Bertrand H. Bronson, "The Double Tradition of Dr. Johnson," in *Johnson Agonistes,* 156–76.
86. This fetish has its own history. See, for example, the sarcastic "Elegy on the Loss of Dr. Johnson's Oak-stick," *Gentleman's Magazine* 55, no. 2 (December 1785): 978.
87. Freud puzzles over illogical anecdotes founded on impossibilities of absence, such as "Lichtenberg's knife without a blade which has no handle," in a suggestive footnote to *Jokes and Their Relation to the Unconscious,* 98. Such anecdotes baffle Freud and make him doubt the very taxonomy of the "joke" that he is trying to define. That these absences bring Freud to the limits of linguistic self-consciousness make them especially telling for our critical purposes.
88. Davis, *Johnson Is Indignant,* 50.
89. Richard Hunter and Ida Macalpine, eds., *Three Hundred Years of Psychiatry, 1535–1860: A History Presented in Selected English Texts* (Hartsdale, NY: Carlisle Publishing, 1982), 417. Hunter and Macalpine make explicit the connection that for Johnson's contemporaries was implicit in his authority as "man of letters," namely, the author's identity as both subject and object of the clinical case (a genre that was just emerging in the eighteenth century), both exhibitor and skillful interpreter of the external signs of internal suffering. From the medical perspective as well, then, Johnson becomes a living and complex configuration of the eighteenth-century mind/body problem, at once an individual and a type. The anecdote, as we shall see in chapter 4, is similarly positioned at the crossroads of literature and science, of subjective and objective experience.
90. Boswell, *Life,* 4:374.
91. Ibid., 4:400; William Windham, "Narrative of the Last Week of Johnson's Life," in *Diary of the Right Hon. William Windham* (1866), extracted in George Birkbeck Hill, ed., *Johnsonian Miscellanies,* 2 vols. (Oxford: Clarendon Press, 1897), 2:386. The "&c." refers to Johnson's scrotum, an act of near self-castration to which we will return.
92. Sir John Hawkins, *The Life of Samuel Johnson, LL.D.* (1787), ed. Bertram H. Davis (New York: Macmillan, 1961), 275.
93. Boswell, *Life,* 4:380.
94. William Cooke, *The Life of Samuel Johnson, LL.D.* (1785), in O. M. Brack Jr. and Robert E. Kelley, eds., *The Early Biographies of Samuel Johnson* (Iowa City: University of Iowa Press, 1974), 122 (emphasis Cooke's).
95. Referring to the Homeric adjective "'*pukinai*' (a word used of things close together or of close texture, e.g. a thicket, the twigs and branches of a tree or the stones of a wall), which fits admirably the multitude of branching passages and veins within each lung and the intricate tracery, the polygonal lobules of the outside," Richard Broxton Onians writes, "Moving about in the passages of the lungs and conditioned by them was the *thumos,* the vital principle that thinks and feels and prompts to action. Here also perhaps is the explanation of the epithet *lasios,* 'bushy, shaggy,' which could also be applied to a sheep, a tree or a thicket and which Homer applies to the heart, between which and the *phrenes* the consciousness seems to be shared." Richard Broxton Onians, *The Origins of European Thought: About the Body, the Mind, the Soul, the World, Time, and Fate* (Cambridge: Cambridge University Press, 1954), 28. This

passage is a good example of the ways in which Onians accords to the language of ancient texts about the body a material purchase on the corporeal reality they construct.

96. Ibid., 33, 32.
97. Ibid., 33–34.
98. Ibid., 35, 40.

## Chapter One

1. Underlying this play on holes and wholes is Joel Fineman's characterization of the anecdote as a "hole within a whole," in "History of the Anecdote: Fiction and Fiction," in *The New Historicism,* ed. H. Aram Veeser (New York: Routledge, 1989), 49–76, to which I will return at greater length in chapter 4.
2. Thomas Babington, Lord Macaulay, "Samuel Johnson," review of *The Life of Samuel Johnson, LL.D., by James Boswell, Esq.,* ed. John Wilson Croker (1831), in *Critical and Historical Essays,* ed. Hugh Trevor-Roper (New York: McGraw-Hill, 1965), 115.
3. Luke Wilson, "William Harvey's *Prelectiones:* The Performance of the Body in the Renaissance Theater of Anatomy," *Representations* 17 (Winter 1987): 62–95, 89.
4. Ibid., 89.
5. For the shift in the cultural definition of dissection from "public spectacle to private event," see Katharine Young, *Presence in the Flesh* (Cambridge, MA: Harvard University Press, 1997), 119; the quotation is Young's from Francis Barker, *The Tremulous Private Body* (London: Methuen, 1984), 13. Wilson, "William Harvey's *Prelectiones,*" discusses the seventeenth-century British anatomy theater as a site in transition from one definition to the other. For details on public anatomy in England, see Fiona Haslam, *From Hogarth to Rowlandson: Medicine and Art in Eighteenth-Century Britain* (Liverpool: Liverpool University Press, 1996), 256–57. For the connection between public anatomy and the carnival during the time of Vesalius, see Giovanna Ferrari, "Public Anatomy Lessons and the Carnival: The Anatomy Theater of Bologna," *Past and Present* 117 (1987): 50–106; and Andrea Carlino, *Books of the Body: Anatomical Ritual and Renaissance Learning,* trans. John Tedeschi and Anne Tedeschi (Chicago: University of Chicago Press, 1999). The historian of science Anita Guerrini is restoring the culture of eighteenth-century public anatomy to scholarly view, a culture that the current controversial *Body World* exhibit is trying to revive. Her work demonstrates just how difficult it is to draw the line between public and private anatomy during this period: professional medical lecturers such as Alexander Munro attracted curious civilian auditors, while anatomy beyond the lecture hall was considered popular entertainment in London. People viewed wax works such as one by the surgeon Abraham Chovet depicting a woman chained to a table and opened alive, and also witnessed animal experimentation and vivisection; both were spectacles designed to provoke moral sentiment while providing pleasurable frissons, often making the former difficult to distinguish from the latter. See Anita Guerrini, "Alexander Munro *primus* and the Moral Theater of Anatomy," *The Eighteenth Century: Theory and Interpretation,* forthcoming, 13–15; and Guerrini, "Anatomists and Entrepreneurs in Early Eighteenth-Century London," *Journal of the History of Medicine and Allied Sciences* 59 (2004): 219–39.

6. *Morning Chronicle and London Advertiser,* December 17, 1784, no. 4863, p. 3, col. 3, lines 67–81. I am much indebted to Kimberly Garmoe for tracking down this reference. When the Company of Surgeons separated from the Barber-Surgeons in 1745, private dissections, which previously could not be performed outside the Barber-Surgeons Hall, were given free rein and largely supplanted public instruction. While William Hunter and Percival Pott were made Masters of Anatomy to the Company of Surgeons in 1748, no formal courses of instruction were ever arranged. The Company of Surgeons dissolved in 1796 after much dissension, and the Royal College of Surgeons was at last founded in 1800. Haslam, *From Hogarth to Rowlandson,* 18–19. For a detailed discussion of the private British anatomy school's "*de facto* unification of the different branches of the medical arts," paving the way for the clinical medicine of nineteenth-century France, see Othmar Keel, "The Politics of Health and the Institutionalization of Clinical Practices in Europe in the Second Half of the Eighteenth Century," in *William Hunter and the Eighteenth-Century Medical World,* ed. W. F. Bynum and Roy Porter (Cambridge: Cambridge University Press, 1985), 228–30.
7. *Gentleman's Magazine* 54 (December 1784): 957. The practice of publishing autopsy results was not uncommon in the later eighteenth century. George II's dissected heart is imaged in the *Gentleman's Magazine* 32 (November 1762) in color, along with a detailed account of the circumstances of both autopsy and death. See fig. 5. The actor John Henderson's autopsy, also performed by James Wilson, was published in the *Morning Herald* for December 5, 1785. Other such autopsies took place in a private dissection room at William Hunter's school of anatomy at Great Windmill Street (founded a year after the formation of the Company of Surgeons in 1745, and henceforth the primary London site of anatomical training).
8. Guerrini notes that the anonymity of the corpse could be suspended in the service of anatomy's "moral message": "The skeletons in the Leiden theatre were identified according to the crimes of their possessors, and those in the London College of Surgeons were named." Guerrini, "Alexander Munro *primus,*" 13–14; see also Guerrini, "Anatomists and Entrepreneurs," 232. What distinguishes Johnson's case is the singling out of a corpse that cannot be so safely distanced from the doctors themselves.
9. I take my cue here from Eve Sedgwick's call for a more nuanced, descriptive, textually and historically sensitive mode of queer reading that would deploy but also unsettle psychoanalytic categorizations of sexual identity. See her "Is the Rectum Straight?: Identification and Identity in *The Wings of the Dove,*" in *Tendencies* (Durham, NC: Duke University Press, 1993), 73–103. For a thoughtful consideration of the limitations of "paranoid" queer reading and a "hermeneutic of suspicion," see her introduction to *Novel Gazing* (Durham, NC: Duke University Press, 1997), 1–37. Such sensitivity to ambiguity is particularly necessary for scholars of the British eighteenth century, an era when middle-class masculine identity was in flux and forming itself against the newly emergent biological/sexual others of woman and sodomite. See, for example, Michael McKeon, "Historicizing Patriarchy: The Emergence of Gender Difference in England, 1660–1760," *Eighteenth-Century Studies* 28 (1995): 295–322; and Randolph Trumbach, "Sodomy Transformed: Aristocratic Libertinage, Public Reputation and the Gender Revolution of the 18th Century," *Journal of Homosexuality* 19, no. 2 (1990): 105–24. This is not to deny, as Hamlet's bunghole indicates, the decidedly erotic undertones of the autopsy's violation and penetration of

its object. For a fascinating analysis of the decidedly ambiguous sexuality of the actor as deployed in the publication of his autopsy results in relation to the construction of professional medical authority during this period, see Cheryl Wanko, "Dissecting the Actor's Authority: Barton Booth's Final Act," in *Roles of Authority: Thespian Biography and Celebrity in Eighteenth-Century Britain* (Lubbock: Texas Tech University Press, 2003), 90–109.

10. For a reading of the "queering" of John Gay that locates the origins of Gay's exclusion from the canon partly in Johnson's rejection of the surface ironies and social ambiguities of Gay's Tory satire for exemplary depth of individual masculine character in the *Lives of the Poets,* see Dianne Dugaw, "Dangerous Sissy: Gendered 'Lives,' John Gay and the Literary Canon," *Philological Quarterly* 75 (1995): 339–60. Dugaw locates the kind of ghettoizing queer reading that I am trying to avoid as a historical development of the late eighteenth century. Similarly, in his *Romantic Genius: The Prehistory of a Homosexual Role* (New York: Columbia University Press, 1999), Andrew Elfenbein reads the ambiguous gendering of the modern definition of original genius as a creation of the late eighteenth century but uses Johnson as an unimpeachably masculine counterexample to genius's queerness. My version of Johnson and Johnsonian desire attempts to uncover the queerness of this version of masculine community. On the late eighteenth-century author as a representative figure of self-made bourgeois subjectivity, see also Michael McKeon, "Writer as Hero: Novelistic Prefigurations and the Emergence of Literary Biography," in *Contesting the Subject,* ed. William Epstein (West Lafayette, IN: Purdue University Press, 1991), 17–41.

11. For a classic statement of the recurrent trope of Johnson as exemplary Briton, see Lawrence Lipking's *Samuel Johnson: The Life of an Author* (Cambridge, MA: Harvard University Press, 1998), and the qualified but still enduring claim for the Age of Johnson in the history of authorship that Johnson represents "by a sort of popular vote, a large community of writers and readers, if not print culture or the nation itself" (302). On the fine line between individuality and monstrosity in the case of the author in general and Johnson in particular, see Helen Deutsch, "Exemplary Aberration: Samuel Johnson and the English Canon," in *Disability Studies: Enabling the Humanities,* ed. Sharon L. Snyder, Brenda Jo Brueggemann, and Rosemarie Garland-Thomson (New York: MLA Press, 2002), 197–210, as well as chapter 2 of this book. On Boswell's construction of Johnson as a representative masculine subject in the *Life,* see Felicity A. Nussbaum, *The Autobiographical Subject* (Baltimore, MD: Johns Hopkins University Press, 1989), 103–26. For a meditation on Johnson's confrontation with a modernity that renders exemplarity impossible, see Thomas Reinert, *Regulating Confusion: Samuel Johnson and the Crowd* (Durham, NC: Duke University Press, 1996). For one of many discussions of the "contradictions, inequities and atrocities" that the imperial project brought into the English national consciousness, see Kathleen Wilson, "Citizenship, Empire, and Modernity in the English Provinces, c. 1720–1790," *Eighteenth-Century Studies* 29, no. 1 (1995): 69–96; for a more sanguine view of the successful mobilization of such differences into common patriotism, see Linda Colley, *Britons: Forging the Nation, 1707–1837* (New Haven: Yale University Press, 1992). Eve Sedgwick's meditation on Oscar Wilde's literal embodiment of "the consciousness of foundational and/or incipient national *difference* already internal to national *definition*" is also germane to the consideration of Johnson's body at a period when both national and sexual iden-

tities were in even greater flux. Sedgwick, *Tendencies,* 151. My thinking on Johnson as representative of the nation is most indebted to Joseph Roach's discussion of surrogacy in *Cities of the Dead: Circum-Atlantic Performance* (New York: Columbia University Press, 1995), and is enhanced by the meditation on the uncannily embodied nature that characterizes rituals of love for the dead in Roach's essay "History, Memory, Necrophilia," in *The Ends of Performance,* ed. Peggy Phelan and Jill Lane (New York: New York University Press, 1998), 23–30.

12. Elisabeth Bronfen, *Over Her Dead Body: Death, Femininity and the Aesthetic* (New York: Routledge, 1992), 11.
13. Samuel Richardson, *Clarissa; or, The History of a Young Lady,* ed. Angus Ross (Harmondsworth: Penguin, 1985), 1383–84.
14. Bronfen, *Over Her Dead Body,* 95–97, 102–3.
15. Abigail Solomon-Godeau, *Male Trouble: A Crisis in Representation* (London: Thames and Hudson, 1997). For a related discussion of the dangers of effeminacy involved in viewing the male nude for eighteenth-century gentlemen on the Grand Tour, see Chloe Chard, "Effeminacy, Pleasure and the Classical Body," in *Femininity and Masculinity in Eighteenth-Century Art and Culture,* ed. Gil Perry and Michael Rossington (Manchester: Manchester University Press, 1994), 142–61.
16. Freud uses these cases to rule out the theory of "scotomization," roughly synonymous with complete visual erasure "so that the result is the same as when a visual impression falls on the blind spot on the retina," in describing fetishism's psychic structure: "The two young men had no more 'scotomized' the death of their fathers than a fetishist scotomizes the castration of women. It was only one current of their mental processes that had not acknowledged the father's death; there was another which was fully aware of the fact; the one which was consistent with reality stood alongside the one which accorded with a wish." This insightful digression remains a detour only—Freud resumes, "To return to my description of fetishism." Sigmund Freud, "Fetishism" (1927), in *Collected Papers,* authorized translations under the supervision of Joan Riviere, 5 vols. (London: Hogarth Press and the Institute of Psychoanalysis, 1950), 5:198, 202–3. My use of the term "disavowal," and my sense of the larger ideological structures it implies, is dependent upon Kaja Silverman's analysis in *Male Subjectivity at the Margins* (New York: Routledge, 1992), 42–48. For Silverman, "Conventional masculinity can best be understood as the denial of castration, and hence as a refusal to acknowledge the defining limits of subjectivity. The category of 'femininity' is to a very large degree the result"; Freud's essay "implicitly shows [fetishism] to be a defense against what is in the final analysis *male* lack" (46). From this perspective, the viewers of Johnson's corpse align themselves with "the 'ideal' female subject" who "refuses to recognize male lack." If "the female subject's recourse to disavowal and fetishism can be characterized as 'pathological,' that pathology must be attributed to the dominant fiction" (47), namely, the ideology of phallic wholeness and female lack.
17. For Bronfen, the alternative to this iconic fetishization of the feminine is the repetition and temporality of the text; the alternations between them rehearse the fort-da game of Freud's *Beyond the Pleasure Principle.* Bronfen, *Over Her Dead Body,* 108.
18. James Boswell, *The Life of Samuel Johnson, LL.D.,* ed. George Birkbeck Hill, rev. L. F. Powell, 6 vols. (Oxford: Clarendon Press, 1934–50), 1:471.
19. Elaine Scarry, *Dreaming by the Book* (Princeton: Princeton University Press, 1999), 7, 12–13.

20. Scarry's comment on the image of the ghost as one "whose own properties are second nature to the imagination" illuminates Boswell's paradoxical achievement: "It is not hard to imagine a ghost successfully. What is hard is successfully to imagine an object, any object, that does *not* look like a ghost." Ibid., 23, 24. Johnsonians crave a vivacity and solidity to their hero that is haunted by an awareness of the evanescent nature of the object of their imagination.
21. Edgar Allan Poe, "Letter to B__," July 1836, in *Essays and Reviews,* ed. G. R. Thompson (New York: Library of America, 1984), 11.
22. *Transactions of the Johnson Society,* ed. Alethea Reazon Acasio Malley (Lichfield, 1999), 13.
23. Beryl Bainbridge, *According to Queeney* (New York: Carroll & Graf, 2001), 1–3, 216. Bainbridge locates the scene of the autopsy in Hunter's school on Great Windmill Street, either in unintentional error or in order to begin with the moment of Mrs. Desmoulins bidding the corpse farewell.
24. *Transactions,* iii, 61.
25. Bainbridge, *According to Queeney,* 3. The novelist casts the sculptor James Hoskins as the taker of the impression for the death mask, while others (see Korshin below) claim that Joshua Reynolds did this himself, and still others claim that the surgeon William Cruikshank supervised Hoskins's work in making the impression. See *Samuel Johnson: 1709–1784. A Bicentenary Exhibition,* introduction and catalog by Kai Kin Yung (London: The Arts Council of Great Britain and the Herbert Press Ltd., 1984), 137. In either case, Hoskins created the classical bust based on the mask, now in the National Portrait Gallery in London and the subject of a BBC documentary hosted by Bainbridge.
26. *Transactions,* biographical endnote for Spinks, 61.
27. Paul J. Korshin notes that John Hoole makes this observation in a private letter but does not repeat it in his narrative of Johnson's death. Paul J. Korshin, "Johnson's Last Days: Some Facts and Problems," in *Johnson After Two Hundred Years,* ed. Paul J. Korshin (Philadelphia: University of Pennsylvania Press, 1986), 60.
28. John Caspar Lavater, *Essays on Physiognomy, Designed to Promote the Knowledge and Love of Mankind,* trans. Henry Hunter (London: Printer for John Murray, H. Hunter, and T. Holloway, 1789), fragment XVIII, 194. The well-known portrait is most likely a 1773 image by Ozias Humphrey, but it also resembles a 1769 Reynolds portrait of Johnson that renders him in mid-gesticulation as classical orator, which I discuss in chapter 3. Both these portraits are remarkable for their realistic recording of Johnson's squint. For a discussion of Lavater's use of Johnson's example, see Kevin J. H. Berland, "'The Air of a Porter': Lavater and the Details of Johnson's Physiognomy," unpublished ms. presented at American Society for Eighteenth-Century Studies, Nashville, TN, 1997.
29. Lawrence C. McHenry Jr., "Art and Medicine: Dr. Johnson's Dropsy," *JAMA* 206, no. 11 (December 9, 1968): 2507.
30. Ibid.. According to one authority, Percy, who specialized in individual busts, did not start making group full-length portraits until the 1790s. G. Bernard Hughes, "Wax Portraits by Samuel Percy," *Country Life* 117 (May 12, 1955): 1257. An authority at the London Museum where this piece resides concluded, based on the age of the glassware used in the piece, that it was most likely made in the nineteenth century.
31. Hughes, "Wax Portraits," 1256.

32. Neil Hertz focuses on the elegiac fleshiness of wax as he recounts a series of sublime textual moments when art collides with an awareness of its relation to the "actual" and to death in "Dr. Johnson's Forgetfulness, Descartes' Piece of Wax," *Eighteenth-Century Life* 16 (November 1992): 167–81. My analysis of the imaginary dynamics of the miniature and the collection are greatly indebted to Susan Stewart's *On Longing: Narratives of the Miniature, the Gigantic, the Souvenir, the Collection* (Baltimore, MD: Johns Hopkins University Press, 1984).
33. James Wilson, *Dissections,* MS 655, p. 12, Historical Collections of the Royal College of Physicians of London. This autopsy record, with its shift from narrative to impersonal display, enacts the desire for "epistemological sovereignty" over the body characteristic of the "humanitarian narrative" that Thomas Laqueur posits for the autopsy and for institutional expression of medical compassion from the late eighteenth to the nineteenth centuries. See Thomas W. Laqueur, "Bodies, Details, and the Humanitarian Narrative," in *The New Cultural History,* ed. Lynn Hunt (Berkeley: University of California Press, 1989), 176–204.
34. "It was thought advisable to obtain evidence that death was due to natural causes, because, as a result of his action in incising his legs on the morning of the day he died, rumours got about that he had committed suicide." Sir Humphrey Rolleston, "Samuel Johnson's Medical Experiences," *Annals of Medical History,* n.s. 1 (1929): 543. "That this act was not done to hasten his end, but to discharge the water that he conceived to be in him, I have not the least doubt." Sir John Hawkins, *The Life of Samuel Johnson, LL.D.* (1787), ed. Bertram H. Davis (New York: Macmillan, 1961), 275.
35. "Mr. Tyers' Additional Sketches Relative to Dr. Johnson," *Gentleman's Magazine* 55 (February 1785): 85. Boswell states that he could not make much use of Tyers's "biographical sketch of Johnson," compiled and reprinted from several magazine entries, because "he abounded in anecdote, but was not sufficiently attentive to accuracy." Boswell, *Life,* 3:308.
36. *The History of Xenophon,* trans. Henry Graham Dakyns (New York: Tandy-Thomas Co., 1909), vol. 5, *The Cyropaedia,* book 8, 305.
37. Ford Madox Ford sees Boswell as a direct descendant of Xenophon in a shared tradition of biographical self-effacement: "This writer with a simple and gentle style is almost the first of authors to impress his personality through his writings upon the world. . . . But Xenophon you can always see in khaki with gilt tabs—a man of quiet movements and of great wisdom as become one who had been the favorite pupil of Socrates and whose *Memorabilia* of the life and habits of that philosopher will stand for you beside Boswell's *Life of Johnson,* though all the writings of Plato may leave you cold." Ford Madox Ford, *The March of Literature: From Confucius' Day to Our Own* (New York: Dial Press, 1938), 97.
38. For the corpse as scientific curiosity and collectible fetish by means of which a national identity is formed, see Joel Reed, "Robert Hooke's Academic Patriarch, or, Fetishizing the Restoration State," *Genre* 28, no. 3 (Fall 1995): 225–54; and Joel Reed, "Monstrous Corpses and Corporeal Epistemology: Representing the National Body," in *"Defects": Engendering the Modern Body,* ed. Helen Deutsch and Felicity Nussbaum (Ann Arbor: University of Michigan Press, 2000), 145–76.
39. William Harvey, "The Anatomical Examination of the Body of Thomas Parr," in *Works of William Harvey, M.D.,* trans. Robert Willis, MD (London: Sydenham Society, 1847; repr., New York: Johnson Reprint Corporation, 1965), 589.

40. Ibid., 589–90.
41. Ibid., 590, 591–92.
42. On sacred anatomy, particularly in relation to the female body, see Katherine Park, "The Criminal and the Saintly Body: Autopsy and Dissection in Renaissance Italy," *Renaissance Quarterly* 47, no. 1 (Spring 1994): 1–33.
43. Although Boswell was not above publishing anonymous pornography to defame his rival biographer, as we will see in chapter 4.
44. Bainbridge, in her fictional account, notes that a slice of Johnson's scrotum was preserved (he suffered from a sarcocele and his final self-wounding involved the cutting of leg and scrotum). I have not been able to find evidence of this in any of my medical sources and suspect that this might be artistic license on her part, meant to foreground the autopsy as an act of desire. Bainbridge, *According to Queeney*, 3.
45. "Obituary of considerable Persons; with Biographical Anecdotes," *Gentleman's Magazine* 54 (December 1784): 957. These words are part of a quotation from "the ingenious editor of 'Biographica Dramatica,'" which immediately and abruptly follows the summary of the autopsy report.
46. Edward Ravenscroft, *The Anatomist; or, The Sham-Doctor: As it is now Acted at the Theatre Royal in Drury-Lane* (London, 1697), 18. Jonathan Sawday discusses this desire to be dissected alive in the context of seventeenth-century culture more generally and links it to sadomasochism; he notes that Garrick chose to perform this play on a double bill with *Romeo and Juliet* in 1756, a tragedy that similarly depends on a fantasy of living death. Jonathan Sawday, *The Body Emblazoned: Dissection and the Human Body in Renaissance Culture* (London: Routledge, 1995), 39–53. My emphasis is less on the violence of sadomasochism than on the sublimation of that violence into the visual register of fetishism that governs the autopsy's print dissemination and reception.
47. "Extracts from Some Anatomical Reports by the late James Wilson Esq. Surgeon and read at the College by his son James Arthur Wilson, Professor of Anatomy to the Royal College of Surgeons," February 1, between 1821 and 1869. MS 1045/38, Historical Collections of the Royal College of Physicians of London. Johnson's address to Falstaff, which anticipates the address of centuries of readers to Johnson himself, can be found in his notes to *2 Henry IV* in *Johnson on Shakespeare*, ed. Arthur Sherbo, *The Yale Edition of the Works of Samuel Johnson*, 16 vols. (New Haven: Yale University Press, 1968), 7:523.
48. On the English admiration of authors and actors as manifested particularly in Westminster Abbey, see Voltaire, *Letters Concerning the English Nation* (1733; repr., London: Westminster Press, 1926), 166–67; and A. P. Stanley, *Historical Memorials of Westminster Abbey* (1867), 6th ed., 3 vols. (Philadelphia: George W. Jacobs, 1882), 1:214–15. Henderson's interment in the Abbey raises the question of how authors, along with actors, come to equal if not to supersede monarchs as national representative figures at the end of the eighteenth century, the era when the tombs of sovereigns gave way to monuments for poets, men of letters, performers, and statesmen. Joseph Roach has described the spectacle of the great Shakespearean actor Thomas Betterton (c. 1635–1710) in performance as a "dichotomy [which] provokes a constant alternation of attention from actor to role, from vulnerable body to enduring memory, in which at any moment one or the other ought to be forgotten but cannot be. This makes the effigy a monstrosity. As a monstrous double, it reconnoiters the

boundaries of cultural identity, and its journey to the margins activates the fascination and the loathing that audiences feel for its liminality." Roach, *Cities of the Dead,* 73–118, quotation from 82. Like Johnson, Betterton, best known for his playing of tragic Shakespearean roles, was at once central and marginal to the culture he represented; like Johnson, his body was at once aberrant and transparent—to witness either performance (in Betterton's case of the role of Hamlet played by an old and deformed man, in Johnson's case the role of "Caliban of Literature" claiming authority while seesawing uncontrollably in the polite drawing room) demanded a complex act of disavowal. Such uncanny liminality also characterizes Betterton's funeral, suspended between the complementary poles of art and life, and life and death, just as it typifies the death and subsequent afterlife of Johnson, himself a surrogate monarch and performer of self-made authority. The bodies of both men, one at the beginning and one at the end of the century, serve as linchpins for fantasies of cultural authority and purity; both possessed the power of "summoning an imagined community into being." Roach, *Cities of the Dead,* 16.

49. Johnson himself in his 1755 *Dictionary of the English Language* defines "surgeon" as "one who cures by manual operation." Surgeons were lower on the ladder of the medical profession than physicians, a ladder whose rungs were in part articulated by virtue of the degree of the practitioner's actual contact with the body. William Hunter, for example, left the Company of Surgeons in order to join the Royal College of Physicians in 1756 and was fined by the Company of Surgeons in return. He encouraged his brother and successor in the school of anatomy, John, to cultivate the study of classical languages and literature in order to follow in his footsteps. John decided to confine himself to surgery, declaring, "They wanted to make an old woman of me; or that I should stuff Latin and Greek at the University; but . . . these schemes I cracked like so many vermin as they came before me." Here surgery's link to the body is positively gendered as a freedom from effeminate scholarly pursuits, the same pursuits that James Wilson Jr. affects. William Ottley, *The Life of John Hunter, F.R.S.,* in *The Works of John Hunter, F.R.S.,* ed. James Palmer, 4 vols. (London, 1835), 1:14. For a discussion of shifting class hierarchies within the medical profession and the class status of the medical profession more generally, see Geoffrey Holmes, *Augustan England: Professions, State and Society* (London: George Allen & Unwin, 1982), chaps. 6, 7. For a brilliant analysis of the efforts of nineteenth-century surgeons to raise (and avenge) their social status through the creation of a new form of middle-class male disease, see Ellen Rosenman, "The Spermatorrhea Panic: The Penis and the Phallus in Victorian Britain," unpublished paper delivered at the annual meeting of the International Society on Narrative, Evanston, IL, April 1998. For an influential argument about the influence of aristocratic patronage on medical knowledge more generally in the eighteenth century that pits the need for physicians' "conformity to the norms of upper class life" with "individual struggle for recognition by means of personal display and publicity," see N. D. Jewson, "Medical Knowledge and the Patronage System in 18th-Century England," *Sociology* 8, no. 3 (September 1974): 369–86. Also germane here are several meditations on the ambiguities of agency of the surgeon's hand, a hand that at once imitates the hand of God and displays—while undoing—God's creation. See Katherine Rowe, "'God's handy worke': Divine Complicity and the Anatomist's Touch," in *The Body in Parts: Fantasies of Corporeality in Early Modern Europe,* David Hillman and Carla Mazzio (New York: Routledge,

1997), 285–309, on the dissection of the hand as exemplary of the interconnections of human and divine agency in anatomy; Paul Valery, "Address to the Congress of Surgeons," *Occasions,* trans. Roger Shattuck and Frederick Brown (Princeton: Princeton University Press, 1970), 141–45, on the hand as inventor of language and creator of reality and, in the case of the surgeon, abstract and violent intervener in the processes of nature and therefore paradigmatic of modernity; and Simon Chaplin, "John Hunter's Museum of Human Miseries," paper delivered at the International Society for Eighteenth-Century Studies, Los Angeles, August 2003, on Hunter's virtuosic display of his own human art in self-conscious evocation of the hand of God.

50. Boswell, *Life,* 2:326–27 and note. On rhetoric as music, see also Jay Fliegelman, *Declaring Independence* (Stanford: Stanford University Press, 1993), 14–20.
51. Boswell, *Life,* 1:13.
52. George J. Squibb, "Last Illness and Post-mortem Examination of Samuel Johnson, the Lexicographer and Moralist, with Remarks," *London Medical Journal* 1 (1849): 615–23.
53. For an excellent survey of the copious medical literature on Johnson, see John Wiltshire, *Samuel Johnson in the Medical World* (Cambridge: Cambridge University Press, 1991). For a typical detailed overview of Johnson's medical history, see Peter Pineo Chase, "The Ailments and Physicians of Dr. Johnson," *Yale Journal of Biology and Medicine* 23, no. 5 (April 1951): 370–79. For an example of the "Xhumation" section of the *Lancet,* see Harold D. Attwood, "A Dissertation upon the Lung of the Late Dr. Samuel Johnson, the Great Lexicographer," *Lancet* (December 21/28, 1985): 1411–13. Finally, for a fascinating summary of William Heberden's own case notes on Johnson, which contain a description of the lung, see Lawrence C. McHenry Jr., "Medical Case Notes on Samuel Johnson in the Heberden Manuscripts," *JAMA* 195, no. 3 (January 17, 1966): 89–90.
54. Squibb, "Last Illness," 621n.
55. "The lungs are sometimes, although I believe very rarely, formed into pretty large cells, so as to resemble somewhat the lungs of an amphibious animal. . . ." Matthew Baillie, *The Morbid Anatomy of Some of the Most Important Parts of the Human Body* (London, 1793), 51. For the image of the lung, see figure 2.
56. I am indebted to Irene Fizer for this analogy. Similarly, Johnson's kidney was also taken away at the autopsy and preserved by the observing Mr. White, and new details about its "granular" status show up in Leslie Stephen's life of Johnson in the *Dictionary of National Biography.* For these details, see Rolleston, "Johnson's Medical Experiences," 544. For the debate on the identity of Johnson's lung, in addition to Squibb and Rolleston, see P. James Bishop, "Samuel Johnson's Lung," *Tubercle* 40 (December 1959): 478–81; Lawrence C. McHenry Jr., "Dr. Samuel Johnson's Emphysema," *Archives of Internal Medicine* 119 (January 1967): 98–105; and Attwood, "A Dissertation upon the Lung."
57. For Hawkins, see Rolleston, "Johnson's Medical Experiences," 544; Hawkins, *Life of Johnson,* 276. For similar uses of metaphor, see, for example, William Cheselden, *The Anatomy of the Human Body* (Boston: Manning and Loring, 1795): ". . . the coat of the liver about a quarter of an inch thick, which contained about five gallons of a gross yellowish fluid, in which were many hydatids about the size of gooseberries" (212). This text was popular throughout the eighteenth century; I am using the first American edition. Frank Nicholls, in his report of the autopsy of George II, refers to

"hyatides" found in the kidneys, none of which "exceeded the bulk of a common walnut." *Gentleman's Magazine* 32 (November 1762): 520.

58. Michel Foucault, *The Birth of the Clinic,* trans. A. M. Sheridan Smith (New York: Vintage, 1975), xiii. See also his characterization of the postmortem description of "the first cirrhotic liver in the history of medical perception" (169–70).
59. Wilson, "William Harvey's *Prelectiones,*" 86.
60. The examples of Johnson's distrust of detail (which, as noted in the introduction, exists in tension with his adherence to empirical observation) are legion. We might think of his pronouncement in the *Preface to Shakespeare* to which I allude here: "Nothing can please many, and please long, but just representations of general nature" (a declaration that uneasily contradicts his praise of Shakespeare's representation of "the real state of sublunary nature"); or the sage Imlac's definition of the poet's vocation in chapter 10 of *Rasselas:* "The business of a poet . . . is to examine not the individual but the species; to remark general properties and large appearances: he does not number the streaks of the tulip, or describe the different shades in the verdure of the forest." Samuel Johnson, *Selected Poetry and Prose,* ed. Frank Brady and W. K. Wimsatt (Berkeley: University of California Press, 1977), 301, 303, 90.
61. For an analysis of the Cartesian violation of cosmic interconnection, see David Le Breton, "Dualism and Renaissance: Sources for a Modern Representation of the Body," *Diogenes* 142 (Summer 1988): 47–69, esp. 66. For a Bahktinian take on dissection's violation and autopsy's reinstatement of the hierarchy of bodily discourse, see also Young, *Presence in the Flesh,* 109–19.
62. Young, *Presence in the Flesh,* 123.
63. Fineman, "History of the Anecdote." Fineman's essay speculates about "how it happened that historiography gave over to science the *experience* of history, when the force of the anecdote was rewritten as experiment" at a moment of "specifically Renaissance historiographic crisis" (63). At our late eighteenth-century historical moment when the relation between "history" and "fiction" was in a process of redefinition, John Bender's argument about an affinity in difference between Enlightenment science and the novel becomes germane. John Bender, "Enlightenment Fiction and the Scientific Hypothesis," *Representations* 61 (1998): 6–28.
64. Young, *Presence in the Flesh,* 125.
65. Fineman, "History of the Anecdote," 56.
66. In this regard, see Lorraine Daston and Peter Gallison, "The Image of Objectivity," *Representations* 40 (Fall 1992): 81–128, on the nineteenth-century separation of art and science in service of self-punishing "objectivity," opposed to the eighteenth-century utilization of the aesthetic in service of the ideal or the typical.
67. For a list of these references, see Lawrence McHenry Jr., "Samuel Johnson's Tics and Gesticulations," *Journal of the History of Medicine and Allied Sciences* 22 (April 1967): 152.
68. We might also add Korshin's characterization of the "clinical gaze" of twentieth-century critics provoked by the multiple narratives of Johnson's final days. Korshin, "Johnson's Last Days," 73.
69. Greenblatt's thinking on *Hamlet* in chapter 5 of Catherine Gallagher and Stephen Greenblatt, *Practicing New Historicism* (Chicago: University of Chicago Press, 2000), as a meditation on the "problem of the leftover"—a skeptical contemplation of a newly Protestant universe devoid of transubstantiative power—is crucial to my

thinking throughout the book. On Hamlet (and Shakespeare) as skeptical anatomists, see David Hillman, "Visceral Knowledge: Shakespeare, Skepticism, and the Interior of the Early Modern Body," in *Body in Parts,* especially note 33's meditation on a possible link between "Protestantism's move away from the Eucharistic incorporation of Christ's body and the approximately contemporaneous rise of what Jonathan Sawday calls the 'culture of dissection.'" Hillman sees a common demystification of "the mystery of human embodiedness" linking the Protestant need to "spirit away" the body (a need that saw anatomy as morally instructive) to the anatomist's reduction of "the self to an objectivized, mechanistic picture of the body" (101–2).

70. René Descartes, *Meditations on First Philosophy,* in *Descartes: Selections,* ed. Ralph M. Eaton (New York: Scribner's, 1927), 98.

71. For more on Johnson's inimitable contagion, see chapter 2.

72. Sawday, *Body Emblazoned,* 1–16. Sawday quotes Cecil Helman: "'The dissecting room shows us the difference between erotic art and pornography, between human experience and the worship of parts.' That difference rests on a parallel reduction of wholeness into fractured partition" (11). The "romance" I discuss recuperates the pornography of anatomy into a different kind of erotic reverie of personal wholeness.

73. John Bender, "Impersonal Violence: The Penetrating Gaze and the Field of Narration in *Caleb Williams,*" in *Vision and Textuality,* ed. Stephen Melville and Bill Readings (Durham, NC: Duke University Press, 1995), 256–81. For realism's defining nonreferentiality and the safety of identifying with fictional characters, see Catherine Gallagher, *Nobody's Story* (Berkeley: University of California Press, 1994), chap. 5; and Adela Pinch, *Strange Fits of Passion* (Stanford: Stanford University Press, 1996), chap. 1.

74. Peter Pindar [John Wolcot], *Bozzy and Piozzi, or, the British Biographers, A Town Eclogue. By Peter Pindar, Esq.* (London, 1786), 51–52.

75. Ruth Richardson notes "in 1752, an Act of Parliament for 'better Preventing the horrid Crime of Murder' gave judges discretion in death sentences for murder, to substitute dissection for gibbeting in chains. Hanging in chains was consciously designed as a grim fate. The corpse of the victim was treated with tar, enclosed in an iron framework, and suspended from a gibbet—either at the scene of the crime, or at some prominent site in the vicinity. . . . The gibbet with its creaking human-scarecrow corpse occupied an important place in popular imaginative apprehension of 'justice' and judicial retribution. As an exemplary punishment it was exceeded in power only by dissection. The intention of both punishments was to deny the wrongdoer a grave." Ruth Richardson, *Death, Dissection and the Destitute* (London: Routledge, 1987), 35–36.

76. Critics of Johnsonian immortalizers thus undo Luke Wilson's characterization of anatomy as a reparative reversal of both dissection and the punishment that preceded it. Wilson, "William Harvey's *Prelectiones,*" 89. For more on anatomy as class war and its relation to criminality, see Peter Linebaugh, "The Tyburn Riot against the Surgeons," in *Albion's Fatal Tree: Crime and Society in Eighteenth-Century England,* ed. Douglas Hay et al. (New York: Pantheon, 1975), 65–117; and for the history of the Anatomy Act of 1832, see Richardson, *Death, Dissection.* For a history of the autopsy in seventeenth-century England that makes a crucial distinction between normal anatomy (for which no medical history is needed and that is dependent on bodies of the condemned) and pathological anatomy (usually resulting from permission

granted by middle- or upper-class patients to doctors and originating in publicized postmortems on monarchs, the model with which Johnson's case seems to coincide), which historically limits Foucault's characterization of the clinical gaze, see David Harley, "Political Post-Mortems and Morbid Anatomy in Seventeenth-Century England," *Social History of Medicine* 7, no. 1 (April 1994): 1–28. For the fashionability of anatomy in the seventeenth and eighteenth centuries, originating with the aristocracy and often with the need to transport a corpse and becoming at once increasingly "scientific" as well as "existential" in its purpose, and for the resistance that arose in relation to this interest, see Phillipe Aries, *The Hour of Our Death* (New York: Oxford University Press, 1981), chap. 8. For an interesting counterexample involving another member of Johnson's Club, see Richard Wendorf, "Burying Sir Joshua," in *After Sir Joshua* (New Haven: Paul Mellon Center for Studies in British Art, 2005), chap. 1, for detailed analysis of Sir Joshua Reynolds's dissection and funeral.

77. Anon., *The Life of Samuel Johnson, LL.D.* (1786), reprinted in *The Early Biographies of Samuel Johnson,* ed. O. M. Brack Jr. and Robert E. Kelley (Iowa City: University of Iowa Press, 1974), 223.
78. On the history of the gibbet as a metaphor for the soul's plight within the body and a particular reading of Andrew Marvell's "Dialogue Between the Soul and the Body" as representative of Protestant responses to the new science, see Sawday, *Body Emblazoned,* 21–22.
79. Alan Liu, "Local Transcendence: Cultural Criticism, Postmodernism, and the Romanticism of Detail," *Representations* 32 (Fall 1990): 96.
80. Edward Hitschmann, "Samuel Johnson's Character, a Psychoanalytic Interpretation," *Psychoanalytic Review* 32 (1945): 207–8, 209–10.
81. I am grateful to Michael Meranze for this observation.
82. Rolleston is quoting Walter Raleigh, *Six Essays on Johnson* (Oxford: Clarendon Press, 1910), 32, in "Medical Aspects of Samuel Johnson," *Glasgow Medical Journal* 4, no. 101 (April 1924): 173. Unlike Hitschmann, Rolleston cites his source but would not have done so in oral performance.
83. Squibb, "Last Illness," 622.
84. Hitschmann is quoting without citing; he is also leaving the question of whether it is possible to analyze an English national character at all unanswered—though he does cite two analysts who have done so. He does, however, with an analyst's zeal not unlike that of the anatomist, want to penetrate Johnson's "facade" in order to explain the kind of identification I am describing. Hitschmann, "Samuel Johnson's Character," 216.
85. David Simpson, *The Academic Postmodern and the Rule of Literature* (Chicago: University of Chicago Press, 1995), 55–58. Simpson also cites Daston on the social construction of objectivity as a device with which to battle theory's divisiveness. In culture and literature, "the same desiderata become the mechanisms for social and aesthetic consensus" (58).
86. See Russell G. Maulitz, *Morbid Appearances: The Anatomy of Pathology in the Early Nineteenth Century* (Cambridge: Cambridge University Press, 1987), for a comparative study of the British and French cases, especially chap. 5. Nietzsche's remark, characteristic of nineteenth-century medicine, that "it is the value of all morbid states that they show us under a magnifying glass certain states that are normal—but not easily visible when normal," erases the distinction between health and disease

through a death's-eye perspective that simultaneously affords a new understanding of disease's individuality. This is more in accord with Bender's view of the realist gaze, the larger story of which the anecdote denies. Quoted from *The Will to Power* as illustrative of Claude Bernard's influence on French pathology in general by Georges Canghuilem, *The Normal and the Pathological,* trans. Carolyn R. Fawcett in collaboration with Robert S. Cohen (New York: Zone Books, 1991), 45. I am indebted to Canghuilem for his analysis of a pathology based on the rules of false resemblance. For the death's-eye view of pathology that sees individuality, see Foucault, *Birth of the Clinic,* chaps. 8 and 9, particularly 170.

87. Hawkins, *Life of Johnson,* 274, 275.
88. Another important philosopher of moral sentiment, Francis Hutcheson, refers to the Roman gladiator shows as examples of spectacles of human suffering made aesthetically pleasing by moral edification. At these spectacles, "the People had frequent Instances of great *Courage,* and *Contempt* of Death, two great *moral Abilitys,* if not *Virtues,*" and gained "Opportunitys of following their *natural Instinct* to *Compassion.*" Francis Hutcheson, *An Enquiry into the Original of our Ideas of Beauty and Virtue; In Two Treatises. I. Concerning BEAUTY, ORDER, HARMONY, DESIGN. II. Concerning MORAL GOOD and EVIL,* 3rd ed., corrected (London, 1729), 241–42. Considering Johnson as a gladiator torn between contempt of death and fear of it, and aware of his audience's moral gain at his expense, illuminates one side of this many-faceted anecdote.
89. Rowe, "God's handy worke," 305.

## Chapter Two

1. Samuel Johnson, "On the Death of Dr. Robert Levet," line 16, in *The Complete English Poems,* ed. J. D. Fleeman (New Haven: Yale University Press, 1971), 140.
2. While Johnsonian critics have rightfully and somewhat defensively observed that Johnson's style varied significantly from work to work, the necessity of such a claim itself indicates that few eighteenth-century voices were more distinctive, more consistent, or more frequently parodied than Johnson's. See, for example, the still-definitive W. K. Wimsatt, *The Prose Style of Samuel Johnson* (1941; repr., New Haven: Yale University Press 1972), especially chap. 5, "Consistency of Johnson's Style"; and Wimsatt, *Philosophic Words: A Study of Style and Meaning in the "Rambler" and "Dictionary" of Samuel Johnson* (New Haven: Yale University Press, 1948); as well as William Vesterman, *The Stylistic Life of Samuel Johnson* (New Brunswick, NJ: Rutgers University Press, 1977); and Boswell's "exhibiting specimens of various sorts of imitations of Johnson's style," both emulative and parodic, at the close of *The Life of Samuel Johnson, LL.D.,* ed. George Birkbeck Hill, rev. L. F. Powell, 6 vols. (Oxford: Clarendon Press, 1887), 4:385–92. For a review of the literature on Johnson's style and a refutation of the standard construction of it as abstraction, see Howard D. Weinbrot, "Samuel Johnson and the Domestic Metaphor," *The Age of Johnson* 10 (1999): 127–63; for "colonial master," who in Irishman Robert Burrowes's phrase "disdain[s] to associate with the natives," see 127. For "fetish object of the literary world" and the contemporary politics of audience definition surrounding that construction of John-

son, see Jonathan Brody Kramnick, *Making the English Canon: Print Capitalism and the Cultural Past, 1700–1770* (Cambridge: Cambridge University Press, 1998), 217.

3. Lennard Davis, "Dr. Johnson, Amelia, and the Discourse of Disability in the Eighteenth Century," in *"Defects": Engendering the Modern Body,* ed. Helen Deutsch and Felicity Nussbaum (Ann Arbor: University of Michigan Press, 2000), 54–74.
4. The phrase is Lucy Porter's, upon first meeting Johnson. Boswell, *Life,* 1:95.
5. A 1786 review of Thrale's *Anecdotes* in the *Monthly Review* criticizes both Thrale and Johnson himself in his *Life of Pope* for abuse of personal detail; reprinted in John Ker Spital, ed., *Contemporary Criticism of Dr. Samuel Johnson* (New York: E. P. Dutton, 1923), 11.
6. Samuel Johnson, *Life of Pope,* in *Selected Poetry and Prose,* ed. Frank Brady and W. K. Wimsatt (Berkeley: University of California Press, 1977), 532, 531.
7. The phrase "monstrous contingency" is coined by David Saunders and Ian Hunter, in "Lessons from the 'Literatory': How to Historicize Authorship," *Critical Inquiry* 17 (Spring 1991): 485. For a sustained reading of Johnson's description of Pope's body in the context of a larger cultural fascination, and for an elaboration of the couplet of deformity and form in Pope's career, see Helen Deutsch, *Resemblance and Disgrace: Alexander Pope and the Deformation of Culture* (Cambridge, MA: Harvard University Press, 1996), 1–39.
8. Samuel Johnson, *The Vanity of Human Wishes,* line 158, in *The Complete English Poems,* ed. Fleeman. Subsequent references to lines appear parenthetically in the text.
9. Johnson powerfully concludes the review of Jenyns with a different sort of "end": "the only end of writing is to enable the readers better to enjoy life, or better to endure it." Samuel Johnson, "Review of [Soame Jenyns], *A Free Enquiry into the Nature and Origin of Evil*" (1757), in *The Oxford Authors: Samuel Johnson,* ed. Donald Greene (Oxford: Oxford University Press, 1984), 536. I will return to Johnson's critique of Pope's *Essay on Man* as reflected in Vladimir Nabokov's *Pale Fire* in chapter 5.
10. Boswell's description of Garrick and More's comment in *Life,* 2:326–27 and note; "swung seconds," Letitia Hawkins, in George Birkbeck Hill, ed., *Johnsonian Miscellanies,* 2 vols. (Oxford: Clarendon Press, 1897), 2:142; Frances Reynolds, in Hill, *Miscellanies,* 2:254, observes Johnson "seesawing" while he reads in order to better memorize poetry; Cooke, in Norman Page, ed., *Dr. Johnson: Interviews and Recollections* (London: Macmillan, 1987), 27, or Hill, *Miscellanies,* 2:166.
11. Boswell, *Life,* 1:192, also see 2:15. Thomas Percy, in Page, *Interviews,* remarks that he "has often heard [Johnson] humming and forming periods, in low whispers to himself, when shallow observers thought he was muttering prayers, etc." (29). Here Johnson's tic of talking to himself in public is revealed as composition. For alternative interpretations of this mannerism, see also Hill, *Miscellanies,* 2:257 (struggling with "mental evil") and 2:273 (repeating past conversation).
12. See Jonathan Lamb, "Blocked Observation: Tautology and Paradox in *The Vanity of Human Wishes,*" in *Cutting Edges: Postmodern Critical Essays on Eighteenth-Century Satire,* ed. James E. Gill (Knoxville: University of Tennessee Press, 1995), 335–46. On Johnson's distrust of satire, see Walter Jackson Bate, *Samuel Johnson* (New York: Harcourt Brace Jovanovich, 1977), 380–81, 493–97.
13. See Thomas Reinert, *Regulating Confusion: Samuel Johnson and the Crowd* (Durham, NC: Duke University Press, 1996), 75–88, for the synecdochal character of objects

and agents in the poem. For an analysis of line 174 that unpacks Johnson's satiric sentimentality, see Lamb, "Blocked Observation," 339–40.

14. On the rhetorical contagion of personifications, see Paul de Man, *The Rhetoric of Romanticism* (New York: Columbia University Press, 1984), 78.
15. See, for example, Robert Burrowes, *Essay on the Stile of Doctor Samuel Johnson* (1787; repr., Los Angeles: Clark Library, 1984), 45: "Even in Johnson's hands this ornament has become too luxuriant, when affections, instead of being personified, are absolutely humanized."
16. Steven Knapp, *Personification and the Sublime: Milton to Coleridge* (Cambridge, MA: Harvard University Press, 1985).
17. Chester Chapin, *Personification in Eighteenth-Century English Poetry* (1954; repr., New York: Octagon, 1968), 112, 113.
18. Samuel Taylor Coleridge, *1811–1812 Lectures on Shakespeare and Milton,* lecture 6, ed. Kathleen Coburn, in *The Collected Works,* vol. 5, ed. R. A. Foakes (Princeton: Princeton University Press, 1987), 292. See also Fredric Bogel, "Johnson and the Role of Authority," in *The New Eighteenth Century: Theory, Politics, English Literature,* ed. Felicity Nussbaum and Laura Brown (New York: Routledge, 1987), 189–209: "The possibility of . . . self-mockery is in some sense written into an intensely sententious style" (203).
19. Oliver Sacks, *An Anthropologist on Mars* (New York: Alfred A. Knopf, 1995), 81.
20. See Christopher Ricks, "Samuel Johnson: Dead Metaphors and 'Impending Death,'" in *The Force of Poetry* (Oxford: Oxford University Press, 1984), 80–88.
21. Robert DeMaria Jr., *The Life of Samuel Johnson* (London: Blackwell, 1993), 139, argues that the Christian ending to the English imitation was added to appeal to a wider audience and to sell more copies of the poem.
22. There is a decidedly stoic tinge to these lines in the *Vanity,* which evoke the following passage (one which would have been familiar to Johnson) from Seneca's letter 24 to Lucilius "On Despising Death." Death from this perspective is not a single and final event but rather the end to a life defined by the repetition of loss. Seneca writes: "For every day a little of our life is taken from us; even when we are growing, our life is on the wane. We lose our childhood, then our boyhood, and then our youth. Counting even yesterday, all past time is lost time; the very day which we are now spending is shared between ourselves and death. It is not the last drop that empties the water-clock, but all that which previously has flowed out; similarly, the final hour when we cease to exist does not of itself bring death; it merely of itself completes the death-process." Seneca, *Ad Lucilium Epistulae Morales,* trans. Richard M. Gunmere, 3 vols. (Cambridge, MA: Harvard University Press, 1917), 1:177–78. Johnson, whose fear of damnation could not trust to death as the end, puts this movingly and personally toward the end of his life in his *Prayers and Meditations:* "I found Mrs. Aston sick, as I expected. Whether she will recover I cannot guess. She is old. At Birmingham I was told that Mrs. Roebuck, who was once Miss Camden, was dead, and at Lichfield I found Harry Jackson dead. I hoped sometime to have seen Miss Camden, and reckoned upon the company of Jackson. Miss Turton is dead too. De spe decidi. [I have given up hope.]" Samuel Johnson, *Diaries, Prayers, and Annals,* ed. E. L. McAdam Jr. with Donald and Mary Hyde, *The Yale Edition of the Works of Samuel Johnson,* 16 vols. (New Haven: Yale University Press, 1958), 1:268–69. This paradox of

death as either terrifying end (annihilation) or endlessness (extended into the possibility of damnation) will be considered further in chapter 3.

23. For Johnson's contemplation of the futility of exemplarity, see Reinert, *Regulating Confusion,* particularly 75–121.
24. Boswell, *Life,* 4:236, 3:19. For Johnson's conversation's relation to print norms, see Alvin Kernan, *Printing, Technology, Letters and Samuel Johnson* (Princeton: Princeton University Press, 1987), 205–6.
25. Boswell, *Life,* 1:194. Garrick contrasts the "lively and easy" *London,* written "when Johnson lived much with the Herveys," to the difficult *Vanity,* written in relative retirement.
26. Archibald Campbell, *Lexiphanes, a dialogue. Imitated from Lucian and suited to the present times. . . . [B]eing an attempt to restore the English tongue to its ancient purity, and to correct, as well as expose, the affected style, hard words, and absurd phraseology of our English Lexiphanes, the Rambler* (London, 1767). See Churchill's poem, *The Ghost* (1762), parodying Johnson's (he's called "Pomposo") rumored credulity in the story of the Cock-Lane ghost, discussed in Boswell, *Life,* 1:406.
27. Quoted in Linda R. Payne, "An Annotated *Life of Johnson:* Dr. William Cadogan on 'Bozzy' and His 'Bear,'" *Collections* 2 (1987): 12.
28. For a similar critique of Johnson's conversational style consisting of mere "surly tone of voice, and scholastic phrase," see William Temple, in Page, *Interviews,* 24.
29. Burrowes, *Essay on the Stile of Samuel Johnson,* 34.
30. Ibid., 30–31.
31. Ibid., 29. See also Archibald Campbell, repeating Swift's advice, on the necessity of imagining female and working-class readers in *Lexiphanes,* 143.
32. See, for example, the continuation of Burrowes's discussion of the *Rambler,* an ambiguous act of imitation that may or may not be conscious and that persists throughout his text, *Essay on the Stile of Samuel Johnson,* 31.
33. Ibid., 40.
34. Elizabeth Montagu to Elizabeth Carter, [October] 24, [1765], [Sandleford] [Berks], MO 3158. This and all subsequent references to the Montagu correspondence are taken from the Montagu papers held by the Henry E. Huntington Library and are published by kind permission of the Huntington Library, San Marino, California.
35. Neil Saccamano, "Wit's Breaks," in *Body and Text in the Eighteenth Century,* ed. Veronica Kelly and Dorothea Von Mucke (Stanford: Stanford University Press, 1995), 45–67.
36. James Thomson Callender, *Deformities of Dr. Samuel Johnson: Selected from His Works,* 2nd ed. (1782; repr., New York: Garland Press, 1974), title page.
37. On the allegorical nature of character in Johnson, even in his *Lives of the Poets,* and on the enlivening function of physical detail, see Reinert, *Regulating Confusion,* 105, 177n13.
38. John Courtenay, *A Poetical Review of the Literary and Moral Character of the Late Samuel Johnson* (1786; repr., Los Angeles: Clark Library, 1969), 13.
39. Ibid., 8–9. The text Courtenay is translating is quoted above in Latin, as Seneca, Epistle to Lucilius 101, who condemns Maecenas's fear: "Is it worth while to weigh down upon one's own wound, and hang impaled upon a gibbet, that one may but postpone something which is the balm of troubles, the end of punishment?" Seneca,

*Ad Lucilium,* 3:165. Like Johnson's self-wounding, an ambiguous gesture—"weigh down upon one's own wound"—signals not model stoic suicide but desperate clinging to life even to the point of enduring the spectacular torture of the gibbet.

40. Quoted in George Whale and John Sargeaunt, eds., *Johnson Club Papers by Various Hands* (London: T. Fisher Unwin, 1899), 3, 13.
41. Sigmund Freud, "The 'Uncanny,'" in *The Standard Edition of the Complete Psychological Works of Sigmund Freud,* translated under editorship of James Strachey, 24 vols. (London: Hogarth Press, 1968), 17:217–56; I'm also influenced here by Terry Castle's claim that Freud's essay is itself haunted by the "historical allegory" of the eighteenth century's "invention of the uncanny" at the moment of its "'disenchantment of the creative imagination,'" its turn toward reason and realism. Terry Castle, *The Female Thermometer: Eighteenth-Century Culture and the Invention of the Uncanny* (New York: Oxford, 1995), 3–20, quotations from 9, 11, 13.
42. On Boswell's innovations in the use of physical detail, particularly in the context of his rescuing of Johnson from the indiscriminate use of such particularity by Hester Thrale and John Hawkins, see Ralph Rader, "Literary Form in Factual Narrative: The Example of Boswell's *Johnson,*" in *Boswell's* Life of Johnson*: New Questions, New Answers,* ed. John A. Vance (Athens: University of Georgia Press, 1985), 25–53; Felicity A. Nussbaum, "Boswell's Treatment of Johnson's Temper: 'A Warm West-Indian Climate,'" *Studies in English Literature* 70 (1974): 421–33; and Nussbaum, *The Autobiographical Subject* (Baltimore, MD: Johns Hopkins University Press, 1989), 103–26.
43. Peter Pindar [John Wolcot], *Bozzy and Piozzi, or, the British Biographers, A Town Eclogue. By Peter Pindar, Esq.* (London, 1786), 5. For similar complaints, see O. M. Brack Jr. and Robert E. Kelley, eds., *The Early Biographies of Samuel Johnson* (Iowa City: Iowa University Press, 1974), 223.
44. Boswell, *Life,* 1:13.
45. Ibid., 1:222.
46. Sir James Mackintosh, private journal, December 1811, quoted in James T. Boulton, ed., *Johnson: The Critical Heritage* (London: Routledge & Kegan Paul, 1971), 350.
47. Lawrence C. McHenry Jr., "Samuel Johnson's Tics and Gesticulations," *Journal of the History of Medicine and Allied Sciences* 22 (April 1967): 168.
48. For the evolution of the definition of "character" from printed mark, to eccentric, to invisible depth (accompanied by a parallel shift in the meaning of the term "nondescript" from one so eccentric as to exceed description to one that evades description by typicality), see Deidre Shauna Lynch, *The Economy of Character: Novels, Market Culture, and the Business of Inner Meaning* (Chicago: University of Chicago Press, 1998), part 1.
49. Sir William Temple, "Of Poetry," in *Critical Essays of the Seventeenth Century,* ed. J. E. Spingarn, vol. 3, *1685–1700* (1908; repr., Oxford: Oxford University Press, 1957), 104.
50. Ibid., 105.
51. Laurence Sterne, *A Sentimental Journey through France and Italy. By Mr. Yorick,* ed. Gardner D. Scout Jr. (Berkeley: University of California Press, 1967), 166–67.
52. Coleridge deplored Johnson's style as yet another symptom of the debased appetites of an overstimulated public who formed a new general audience for literature in a commercial age: "Johnson's style has pleased many from the very fault of being per-

petually translateable; he creates an impression of cleverness by never saying any thing in a common way." "Lecture on Style" (1818), in Boulton, *Critical Heritage*, 355. Coleridge's Johnson writes commonplaces for a depraved crowd, dressed up in the clothing of individual excess, in an unnatural version of Pope's aphorism in *An Essay on Criticism*—"*True Wit* is *Nature* to Advantage drest" (297)—which Johnson himself would have deplored, but which highlights the shift to a Romantic apprehension of Johnson's style. Boswell, speaking of the *Rambler*, casts this sentiment quite differently: "He delighted to express familiar thoughts in philosophical language; being in this the reverse of Socrates, who, it was said, reduced philosophy to the simplicity of common life. But let us attend to what he himself says in his concluding paper: 'When common words were less pleasing to the ear, or less distinct in their signification, I have familiarised the terms of philosophy, by applying them to popular ideas.' . . . Johnson's comprehension of mind was the mould for his language. Had his conceptions been narrower, his expression would have been easier." Boswell, *Life*, 1:217–18, 222.

53. Anna Seward, *Letters* (1811), 4:155–60, in Boulton, *Critical Heritage*, 414. Critics and devotees alike characterize Johnson's stylistic singularity as a distinctly un-Shakespearean fixity that renders him unable to imitate another. Whether such individuality is deemed manly (Boswell) or effeminate (Seward, Montagu, Coleridge) depends on the class, gender, and historical moment of the critic.
54. The phrase is from Felicity A. Nussbaum's "Effeminacy and Femininity: Sarah Fielding, Elizabeth Montagu, and Johnson," in *The Limits of the Human: Fictions of Anomaly, Race, and Gender in the Long Eighteenth Century* (Cambridge: Cambridge University Press, 2003), 81. In a chapter to which I am much indebted, Nussbaum shows how Montagu, by deploring Johnson's ornate stylistic mannerisms as un-English and even inhuman, creates a public space for herself as female critic fighting an "effeminacy internal to England and integrally connected" (in a view not far from Coleridge's) to "fashionable and luxurious display" (77). The battle between men and women of letters over proper literary style in the newly booming British print marketplace, Nussbaum shows, was a battle for Britain itself (81). Shakespeare becomes the counter to Johnson for more practical reasons as well. Montagu argues: "Our rank in the Belles lettres depends a good deal on that degree of merit which is allow'd to Shakespeare, who is more than any other writer read by foreigners." Elizabeth Montagu to Elizabeth Carter, [October] 21, [1766], Sandleford [Berks], MO 3187.
55. "His Preface is so ingenious it terrifys me." Elizabeth Montagu to Sarah Scott [October 26,] 27, [1765]. Sandleford [Berks], MO 5830.
56. Elizabeth Montagu to Sir William Weller Pepys, November 3, 1781, Hill Street, London, MO 4070.
57. *Verses Address'd to the Imitator of the First Satire of the Second Book of Horace. By a Lady* (London: 1733), 2. This text was jointly and anonymously authored by Lord John Hervey, but the authors were quickly identified.
58. "His language is laboured, rather too fine for my taste, who love the negligences of genius better than the ornaments of study, but perhaps I judge so for want of refinement. I never love the velvet style, an equal pile as if cut by an engine, however in these effeminate days perhaps it will be admired for the very thing I dislike." Elizabeth Montagu to Elizabeth Carter, [October] 17, [1765], Sandleford [Berks], MO 3157. Nussbaum carefully distinguishes "effeminacy" in this context from its standard

and ahistorical associations with homosexuality in order to articulate an ambiguous range of meanings that cross a wide variety of differences. Nussbaum, "Effeminacy and Femininity," 78.

59. Horace Walpole, in Boulton, *Critical Heritage,* 324–25. For Montagu, Johnson's artifice and easy imitability is cast in political terms, marking him not just as unmanly and monstrous, but also as a threat to English sovereignty: "At the same time he always writes as if he was a parnassian born, his language is wonderfully dress'd & finical, & I own, I think if he was fashionable enough to be copied, he would ruin the english language for which I have a great respect, as it is noble, strong, & fit for a free people. Let ye trembling slaves of Despotes weigh & measure syllables, & cut their periods by a rule, but let the free born britten speak in a manly style!" Elizabeth Montagu to Elizabeth Carter, [November] 20, [1765], [London], MO 3162. Stylistic singularity in this account is figured paradoxically as potentially contagious enslavement to an abstract despotic rule. See also Nussbaum, "Effeminacy and Femininity," 76, 78.

60. Thomas Babington Macaulay, "Diary and Letters of Madame D'Arblay" (1843), in *Critical and Historical Essays,* 3 vols. (Boston: Houghton Mifflin, 1900), 3:331–95; quotation, describing Burney's English prose style after her residence in France, from 389; for Burney as caricaturist, see 380–87. For more on Johnson's intellectual influence on the women writers of his day, see Norma Clarke, *Dr. Johnson's Women* (London: Hambledon and London, 2000).

61. D. A. Miller, *Jane Austen, or The Secret of Style* (Princeton: Princeton University Press, 2003); on Austen's uniquely disembodied brand of narrative authority, see especially chapter 2. See also my comments on style and "nature" in Austen and Johnson in the introduction.

62. Horace Walpole, in Boulton, *Critical Heritage,* 324–25. The animal analogies used in describing (and delimiting) Johnson are common. See, for example, the comments of the nineteenth-century aesthete Edgar Allan Poe and twentieth-century psychoanalyst Edward Hitschmann in chapter 1. See also, for the title echoing a nickname in Boswell, C. E. Vuillamy, *Ursa Major: A Study of Dr. Johnson and His Friends* (London: M. Joseph, Ltd., 1946).

63. Cuthbert Shaw, *The Race,* 2nd ed. (London, 1766).

64. Walter Scott, extract from *Lives of the Novelists* (1825), in Boulton, *Critical Heritage,* 420.

65. "Monster" is used by Johnson's contemporary the physician William Cadogan in his marginalia to Boswell's *Life;* he also terms Boswell Johnson's "fond puppy" or "Shewman" and inscribes on his volume's pages Soame Jenyns's epitaph on Johnson, which terms him a "sleeping bear" and ends: "A Christian & a Scholar—but a Brute." Payne, "An Annotated *Life,*" 9, 17. For "savage," see Richard Cumberland, who turns the metaphor toward a praise of Johnson's conversational powers: "Some have called him a savage; they were only so far right in the resemblance, as that, like the savage, he never came into suspicious company without his spear in his hand and his bow and quiver at his back." Page, *Interviews,* 61. "Unlick'd cub" is used by Campbell in *Lexiphanes,* xxxix. John Hawkins, referring to Johnson's visible greed at table, states: "Such signs of effeminacy as these, suited but ill with the appearance of a man, who, for his bodily strength and stature, had been compared to Polyphemus." Sir John Hawkins, *The Life of Samuel Johnson, LL.D.* (1787), ed. Bertram H. Davis (New York:

Macmillan, 1961), 147. Here, as is often the case, Johnson's physical appearance is described as at once beyond (of exceptional strength and size) and beneath (compared to the monstrous Cyclops) the human (see n. 67). The portrait painter Ozias Humphrey remarked on first meeting: "I was very much struck with Mr Johnson's appearance, and could hardly help thinking him a madman for some time, as he sat waving over his breakfast like a lunatic." Page, *Interviews,* 30. For a very similar account of Hogarth's first impression of Johnson as an "ideot," see Boswell, *Life,* 1:146–47. Turning to terms of class, Arthur Murphy notes that "the exteriors of politeness did not belong to Johnson." Page, *Interviews,* 21; and William Temple, responsible for terming Johnson a chairman, porter, or giant, argues that his "contortions and grimaces may be the effect of solitude and low-breeding, to which he was long condemned in the early part of his life." Page, *Interviews,* 23. Although James Northcote counters this (giving evidence of the persistence of the charge) with a defense of Johnson's "delicacy" (Page, *Interviews,* 47), he also recounts the anecdote of the servant maid of the Miss Cotterells mistaking "this uncouth and dirty figure of a man" for a robber. Page, *Interviews,* 41–42. Here, as often, Johnson's poverty and unkempt appearance merge with his gesticulations to form an overall impression of social monstrosity. "Oddity" is the term used by a Chatsworth innkeeper to describe the "greatest writer in England" to a disingenuous Boswell. *Life,* 3:209.

66. On the contagious imitability of Johnson's style, see Boswell, *Life,* 4:385–92; Burrowes, *Essay on the Stile of Samuel Johnson,* 42. On his inability to take on the voices of others, see Boswell, *Life,* 2:231; see also Burrowes, *Essay on the Stile of Samuel Johnson,* 32; Campbell, *Lexiphanes;* and James Northcote, in Page, *Interviews,* 45.
67. Frances Burney and Frances Reynolds characterize Johnson by his bodily lack. Burney in her memoir of her musicologist father makes deafness central to Johnson's character and follows Johnson's description of himself: "All animated nature loves music except myself!" Page, *Interviews,* 50. Reynolds explains Johnson's unpolished "want of politeness" with reference to "the disqualifying influence of blindness and deafness" and "corporeal defects." Hill, *Miscellanies,* 2:276, 299. The debate such accounts provoke over the actual extent of these deficiencies continues to be considerable. Contemporaries similarly debated whether Johnson was physically deformed or exceptionally well-endowed. William Cooke encapsulates both extremes: "In respect to person, he was rather of the *heroic* stature, being above the middle size; but, though strong, broad, and muscular, his parts were slovenly put together." In Brack and Kelley, *Early Biographies,* 125.
68. While Naomi Schor argues in *Reading in Detail* (New York: Routledge, 1987) that attention to detail constitutes a feminine aesthetic, here the detail paradoxically undermines and asserts masculinity. The important distinction here is between the cultural construction of the feminine and the cultural power of male feminization. For Johnson's impossible project in the *Dictionary* of representing a national language riven with internal difference, see Janet Sorensen, *The Grammar of Empire in Eighteenth-Century British Writing* (Cambridge: Cambridge University Press, 2000), chap. 2.
69. Courtenay, *Poetical Review,* 11.
70. On the sublime's collapsing of author into text, see Neil Hertz, "A Reading of Longinus," in *The End of the Line* (New York: Columbia University Press, 1985),

1–20. For an essay that links the eighteenth-century sublime to the birth of "psychology" that accompanied the rise of empiricism (and with it, doubt as to the existence of objects—a doubt that the figure of Johnson in Boswell's *Life* is meant to combat), and that argues for the sublime critic's drive to be "a metonymy of his culture," see Frances Ferguson, *Solitude and the Sublime: Romanticism and the Aesthetics of Individuation* (New York: Routledge, 1992), chap. 2, quotation from 38.

71. Leslie Stephen, *Samuel Johnson* (London, 1878), 167, 167–68, 168, 173.
72. I owe this phrase to Jayne Lewis.
73. Henry Meige and E. Feindel, *Tics and Their Treatment*, trans. S. A. K. Wilson (New York: William Wood and Co., 1907), v–vi.
74. Ibid., xiv, 80, 74.
75. For the gamut of these phrases, see Boswell, *Life*, 1:144n.
76. For the extent of Johnson's contributions, see O. M. Brack Jr. and Thomas Kaminski, "Johnson, James, and the *Medicinal Dictionary*," *Modern Philology* (May 1984): 378–400.
77. Samuel Johnson, *A Dictionary of the English Language*, 3rd ed. (London, 1765).
78. For an argument that links the word "antick" to an ancient popular and comic conception of the grotesque body, see Paul Semonin, "Monsters in the Marketplace: The Exhibition of Human Oddities in Early Modern England," in *Freakery: Cultural Spectacles of the Extraordinary Body*, ed. Rosemarie Garland-Thomson (New York: New York University Press, 1996), 69–81.
79. Ibid.
80. Definition for "Chorea Sancti Viti," in Robert James, *A Medicinal Dictionary*, 3 vols. (London, 1743–45). When Boswell quotes Sydenham's definition of the disease, he omits the reference to children. See *Boswell's* Life of Johnson*: An Edition of the Original Manuscript*, 4 vols., general editor Claude Rawson, vol. 1 edited by Marshall Waingrow (New Haven: Yale University Press, 1994) 1:100n6.
81. In Robert Hooper's *Lexicon Medicum; or, Medical Dictionary*, ed. Klein Grant, 8th ed. (London, 1848), the cause of St. Vitus's dance is attributed to "an irritability of the nervous system, chiefly dependent on debility," and aggravated by "the general irritation that pervades the system" at puberty, making females "in weakly habits" particularly liable to the disease (379).
82. Bogel, "Johnson and the Role of Authority," 204.
83. Quoted in Thomas Clifford Allbutt, ed., *A System of Medicine by Many Writers*, 9 vols. (New York: Macmillan, 1905), 7:868. English translation my own; "spasm" could also be translated as "chorea."
84. Frances Reynolds, in Hill, *Miscellanies*, 2:274.
85. Frances Burney, *Memoirs of Dr. Burney*, 3 vols. (London, 1832), 2:91. For an earlier and less idealized version of this passage (1777), see Page, *Interviews*, 24–25.
86. Boswell, *Life*, 1:266.
87. Letter CCXII, quoted in Hill, *Miscellanies*, 1:384; also quoted in Callender, *Deformities*, 18–19, in note. While Boswell and many contemporaries identified Johnson as Chesterfield's subject, Hill argues in his note to this passage as well as in his notes to *Life*, 1:267–68, that Chesterfield indisputably was referring to Lord Lyttleton. Interestingly enough, Johnson's arguments with Lyttleton, a prominent Whig and guest in Mrs. Montagu's famous salon, were in part responsible for some of her frustration with his unmannerly disruption of her gatherings.

88. Boswell, *Life,* 4:322–23. See also ibid., 1:334, on *Idler* 90: "His unqualified ridicule of rhetorical gesture or action is not, surely, a test of truth; yet we cannot help admiring how well it is adapted to produce the effect which he wished. 'Neither the judges of our laws, nor the representatives of our people, would be much affected by laboured gesticulation, or believe any man the more because he rolled his eyes, or puffed his cheeks, or spread abroad his arms, or stamped the ground, or thumped his breast, or turned his eyes sometimes to the ceiling, and sometimes to the floor.'"
89. Ibid., 2:211. On Hume's distancing himself from the seductive power of ancient oratory, see Jerome Christensen, *Practicing Enlightenment: Hume and the Formation of a Literary Career* (Madison: University of Wisconsin Press, 1987), chap. 1. See also Adam Potkay, *The Fate of Eloquence in the Age of Hume* (Ithaca, NY: Cornell University Press, 1994).
90. Adam Smith, Lecture 30 (1763), *Lectures on Rhetoric and Belles Lettres,* ed. J. C. Bryce (Oxford: Clarendon Press, 1983), 196, 198.
91. Boswell, *Life,* 1:483–86; quotation from 1:485–86.
92. Boswell, *Life,* 4:183n2.
93. Meige and Feindel, *Tics and Their Treatment,* 82.
94. The term "enshrinement" is used by an unidentified woman with Tourette's to describe the unique psychic archival character of the syndrome's symptoms, noted by Sacks in *An Anthropologist on Mars,* 81. Meige and Feindel in *Tics and Their Treatment* repeatedly use the latter three descriptions.
95. The first chapter of Meige and Feindel's *Tics and Their Treatment* is "Confessions," a document praised by Sacks. See also Joseph Bliss's first-person narrative, "Sensory Experiences of Gilles de la Tourette Syndrome," eds. Donald J. Cohen, MD, and Daniel X. Freedman, MD, *Archives of General Psychiatry* 37 (December 1980): 1343–47.
96. See Meige and Feindel, *Tics and Their Treatment,* especially "Confessions," 1–24; Allbutt, *System of Medicine,* 875. For the persistence of the urge to imitate as a symptom of Tourette's and inherent in tic formation, see Ruth Dowling Bruun and Bertel Bruun, *A Mind of Its Own: Tourette's Syndrome: A Story and a Guide* (New York: Oxford University Press, 1994); and Sacks, *An Anthropologist on Mars,* 89, who links the Touretter's tics to caricature. For diagnoses of Johnson with Tourette's, see Bruun and Bruun, *A Mind of Its Own,* where it is noted that Tourette's is sometimes accompanied by excessive religiosity and a drive toward self-punishment (156–57); McHenry, "Johnson's Tics and Gesticulations"; Julia Epstein, *Altered Conditions: Disease, Medicine, and Storytelling* (New York: Routledge, 1995), 64–67; and the two essays by T. J. Murray cited below in note 118; for the diagnosis of tic, see Allbutt, *System of Medicine,* 885; and for a skeptical overview, see John Wiltshire, *Samuel Johnson in the Medical World* (Cambridge: Cambridge University Press, 1991), 24–34. For a historical account of Tourette's that sees an intrinsic connection between psychological and neurobiological accounts of the syndrome and that connects it to autoimmune disorders, see Howard I. Kushner, *A Cursing Brain? The Histories of Tourette Syndrome* (Cambridge, MA: Harvard University Press, 1999).
97. Boswell, *Life,* 1:483, 4:399–400n6. For a compelling analysis of Johnson's scruples that links them to his obsessive-compulsive turn of mind (itself linked by medical thinkers to Tourette's) and to his fear of satire, see Bate, *Samuel Johnson,* 380–81, 493–97.
98. Hill, *Miscellanies,* 2:254.

99. Ibid., 2:275.
100. Ibid.
101. Allbutt, *System of Medicine*, 7:877; Boswell, *Life*, 1:144–45, quotation from 145; Hester Thrale, in Hill, *Miscellanies*, 1:219, 231; Joshua Reynolds, in Hill, *Miscellanies*, 2:221.
102. Christopher Lawrence argues that Whytt's "principal contention was that the body's responses are purposeful and not the result of blind mechanism." "The Nervous System and Scottish Society," in *Natural Order: Historical Studies of Scientific Culture*, ed. Barry Barnes and Steven Shapin (Beverly Hills: Sage Publications, 1979), 25.
103. Michel Foucault, *Madness and Civilization*, trans. Richard Howard (New York: Vintage, 1988), 153–54.
104. Quoted in Boswell, *Life*, 1:144; see also Hill, *Miscellanies*, 2:222.
105. Hill, *Miscellanies*, 2:227–28.
106. Quoted in ibid., 2:228.
107. Johnson, *Rambler* 89, in *The Rambler*, ed. W. J. Bate and Albrecht B. Strauss, *The Yale Edition of the Works of Samuel Johnson*, 16 vols. (New Haven: Yale University Press, 1969), 4:108. On Johnson's skepticism about authority, see Bogel, "Johnson and the Role of Authority"; and Reinert, *Regulating Confusion*.
108. Quoted in Hill, *Miscellanies*, 2:222, 221.
109. Many accounts of Tourette's—including Allbutt, *System of Medicine;* Meige and Feindel, *Tics and Their Treatment;* and Bruun and Bruun, *A Mind of its Own*—note the ways in which people with Tourette's adopt "voluntary" tics in order to "mask the involuntary movements." Allbutt, *System of Medicine*, 877.
110. Boswell, *Life*, 4:320.
111. Ibid., 4:99.
112. What Bogel reads as a simultaneity, I read as a temporal sequence of *repetition*, an important distinction for my argument about the habitual ticlike nature of Johnson's conversation. See Bogel, "Johnson and the Role of Authority," on "the copresence of genuine authority and the histrionic affectation of authority . . . that makes this episode so revealing" (204).
113. Joshua Reynolds states: "The great business of his life (he said) was to escape from himself; this disposition he considered as the disease of his mind, which nothing cured but company." Boswell, *Life*, 1:144–45.
114. Hill, *Miscellanies*, 2:222.
115. For Joshua Reynolds's depiction of Johnson's gesticulations as a kind of oratory, see figure 11.
116. "Everything he says is as *correct* as a *second edition*: 'tis almost impossible to argue with him, he is so sententious and so knowing." Page, *Interviews*, 31.
117. Boswell, *Life*, 1:145–47.
118. T. J. Murray, "Dr. Samuel Johnson's Movement Disorder," *British Medical Journal* 1 (June 16, 1979): 1613. Murray cites John Hawkins's *Life* as the source for this anecdote, which he uses as one of many to confirm his diagnosis of Johnson with Tourette's, but in keeping with my theme of lost originals, I have not yet been able to trace its original source. See also T. J. Murray, "Doctor Samuel Johnson's Abnormal Movements," in *Gilles de la Tourette Syndrome*, ed. Arnold J. Friedhoff and Thomas N. Chase (New York: Raven Press, 1982), 25–30, in which the same anecdote is quoted and attributed to Hawkins (30) without page number.

119. Julian Barnes, *England, England* (New York: Vintage, 2000), 215; Beryl Bainbridge, *According to Queeney* (New York: Carroll & Graf, 2001), 130.

## Chapter Three

1. *Johnson on Shakespeare,* ed. Arthur Sherbo, *The Yale Edition of the Works of Samuel Johnson,* 16 vols. (New Haven: Yale University Press, 1968), 8:968–69.
2. Ibid., 8:970.
3. Ibid., 8:969, 970.
4. I draw throughout this paragraph on Michael MacDonald and Terence R. Murphy, *Sleepless Souls: Suicide in Early Modern England* (Oxford: Oxford University Press, 1990), which links the growing tolerance and secularization of suicide over the course of the period to the growth of print media (6) and the replacement of the art of dying with the art of living (1); while religious politics during the period increasingly focused on the "pernicious effects of predestinarian doctrines," effects that might be argued to be exhibited in Johnson's own excessive fear of damnation. The honorable suicide of that Whig republican idol Cato, while admired by some, was not something to be publicly endorsed without risk (64). For an interestingly apposite consideration of the demise of purgatory during the Reformation and the secularization and aestheticization of ghosts (whose primary province becomes the theater) at the end of the seventeenth century, see Stephen Greenblatt, *Hamlet in Purgatory* (Princeton: Princeton University Press, 2001). Greenblatt's narrative informs my own as the ghost of Hamlet's father, revenant of a Catholic past, is translated by Boswell into the ghost of the professional father of English letters.
5. Anon., *A Sad and Dreadful Account of the Self-Murther of Robert Long, alias Baker* (London, 1685), quoted in MacDonald and Murphy, *Sleepless Souls,* 43, 41.
6. Arnold Van Gennep, *The Rites of Passage,* trans. M. B. Vizedom and G. L. Caffee (Chicago: University of Chicago Press, 1960), 160–61, quoted in MacDonald and Murphy, *Sleepless Souls,* 46; MacDonald and Murphy, *Sleepless Souls,* 46.
7. James Boswell, *The Life of Samuel Johnson, LL.D.,* ed. George Birkbeck Hill, rev. L. F. Powell, 6 vols. (Oxford: Clarendon Press, 1934–50), 4:300.
8. Anna Seward to Mrs. Knowles, Lichfield, April 20, 1788, *Letters of Anna Seward: Written Between the Years 1784 and 1807,* 6 vols. (Edinburgh: George Ramsay & Co., 1811), 2:104.
9. Thomas Babington, Lord Macaulay, "Samuel Johnson," review of *The Life of Samuel Johnson, LL.D., by James Boswell, Esq.,* ed. John Wilson Croker (1831), in *Critical and Historical Essays,* ed. Hugh Trevor-Roper (New York: McGraw-Hill, 1965), 115.
10. James Northcote, *The Life of Sir Joshua Reynolds,* 2 vols. (London: Cornmarket, 1972), 1:234.
11. Quoted in ibid., 2:3; and George Birkbeck Hill, ed., *Johnsonian Miscellanies,* 2 vols. (Oxford: Clarendon Press, 1897), 1:313.
12. Boswell, *Life,* 1:144.
13. Quoted in Hill, *Miscellanies,* 2:274.
14. Lawrence C. McHenry Jr., "Samuel Johnson's Tics and Gesticulations," *Journal of the History of Medicine and Allied Sciences* 22 (April 1967): 152–68.

15. Duncan Robinson, "Giuseppe Baretti as 'A Man of Great Humanity,'" in *British Art 1740–1820: Essays in Honor of Robert R. Wark,* ed. Guilland Sutherland (San Marino, CA: Huntington Library, 1982), 91, 93. See also Kai Kin Yung, *Samuel Johnson, 1709–84: A Bicentenary Exhibition* (London: Arts Council of Great Britain and Herbert Press, 1984), 103.
16. For a fascinating discussion of the cultural semiotics of the squint during this period in relation to a radically different political figure, John Wilkes, see Shearer West, "Wilkes's Squint: Synecdochic Physiognomy and Political Identity in Eighteenth-Century Print Culture," *Eighteenth-Century Studies* 33, no. 1 (1999): 65–84.
17. I should add here that Macaulay and Stephen are both part of a tradition, beginning with Edmund Burke (who claimed that Johnson would be better known for his talk than for all of his writings), that distinguishes Johnson's conversation stylistically from his writings and prefers the former to the latter. I'm interested in how the figure of Johnson can merge the two and how a Johnsonian like R. W. Chapman can enter into this critical fray in the early twentieth century, claiming that in Johnson's case, voice and page are identical—"Johnson is one." R. W. Chapman, *Johnsonian and Other Essays and Reviews* (Oxford: Clarendon Press, 1953), 14.
18. Paul de Man, "Autobiography as De-Facement," in *The Rhetoric of Romanticism* (New York: Columbia University Press, 1984), 76.
19. Ibid., 70, 76, 78.
20. James Boswell, *The Journal of a Tour to the Hebrides,* in *Life,* 5:18.
21. Quoted in J. W. Croker, ed., *Johnsoniana; or, Supplement to Boswell* (London, 1836), 113.
22. Kevin Hart remarks of this passage, "Death separates the true Johnson, his soul, from his body, revealing him to be a classical figure, not unlike John Bacon's monument to him in St Paul's Cathedral. His end does not make [as Hannah More had put it] 'a kind of era in literature' so much as continue a classical tradition." Kevin Hart, *Samuel Johnson and the Culture of Property* (Cambridge: Cambridge University Press, 1999), 40.
23. Boswell, *Life,* 4:465 (appendix I).
24. John Bacon, *Gentleman's Magazine* 66, no. 1 (March 1796): 180, quoted in Boswell, *Life,* 4:469.
25. Hart, *Johnson and the Culture of Property,* 16.
26. Rev. John Campbell to James Boswell, April 19, 1793, in Marshall Waingrow, ed., *The Correspondence and Other Papers of James Boswell Relating to the Making of* The Life of Johnson (New York: McGraw-Hill, 1966), 519.
27. De Man, "Autobiography as De-Facement," 78, 81.
28. Ibid., 75
29. Marjorie Garber, *Shakespeare's Ghostwriters: Literature as Uncanny Causality* (New York: Methuen, 1987), 124–76.
30. Boswell, *Life,* 1:391–92. For apposite readings of this passage and the larger symbolic framework it suggests, see Lisa Berglund, "'Look, my Lord, it comes': The Approach of Death in the *Life of Johnson,*" in *1650–1850: Ideas, Aesthetics, and Inquiries in the Early Modern Era,* vol. 7 (Brooklyn: AMS Press, 2002): 239–55; Greg Clingham, "Double Writing: The Erotics of Narrative in Boswell's *Life of Johnson,*" in *James Boswell, Psychological Interpretations,* ed. Donald J. Newman (New York: St. Martin's Press, 1995), 203; and Hart, *Johnson and the Culture of Property,* chap. 1.

31. Boswell, *Life,* 1:392; see also Berglund, "'Look, my Lord, it comes,'" 244–45.
32. Robina Napier, ed., *Johnsoniana,* 2nd ed. (London: George Bell & Sons, 1884), v.
33. The pictorial equivalent to Napier's literary imaginings in the Victorian period would be W. P. Frith's "Subject Pictures," discussed in this book's introduction.
34. *Johnson's Works,* ed. John Hawkins (London, 1787), 11:204, quoted in Boswell, *Life,* 4:450. See also Augustine Birrell, who at the end of the nineteenth century is so confident that the portrait he alludes to is universally known that he does not identify it: "Of his friend Sir Joshua's two most famous pictures I need not speak. One of them is the best known portrait in our English world. It has more than a trace of the vile melancholy the sitter inherited from his father, a melancholy which I fear turned some hours of every one of his days into blank dismay and wretchedness." George Whale and John Sargeaunt, eds., *Johnson Club Papers by Various Hands* (London: T. Fisher Unwin, 1899), 14–15. I am particularly interested in how Johnson's melancholy, in its visible traces, becomes a vehicle for group identification with an otherwise indecipherably individual subject.
35. Boswell, *Life,* 1:250.
36. *Horace: Satires, Epistles and Ars Poetica,* trans. H. Rushton Fairclough (Cambridge, MA: Harvard University Press, 1966), 35.
37. The fate of Reynolds's 1778 portrait is exemplary in this regard. See Boswell, *Life,* 4:180–81, 451–52, for an account of the recurring appearance and disappearance of the Horatian inscription, an account that serves as an apt metaphor for the continual appearance and reappearance of considerations of Johnson's physical difference, and thus the continual need for supplementation, in the construction of his 200-year-old cultural imago.
38. Soame Jenyns, May 1786, quoted in Robert E. Kelley and O. M. Brack Jr., eds. *Samuel Johnson's Early Biographers* (Iowa City: University of Iowa Press, 1971), 10; can also be found in *European Magazine* (May 1786): 306.
39. De Man, "Autobiography as De-Facement," 81.
40. I am indebted here to Lorna Clymer's situation of de Man's essay in the context of classical rhetoric and the eighteenth-century didactic poetic tradition of which Wordsworth was a part. In this account, prosopopoeia is read as a form of reciprocal animation between the dead and the living; the epitaph thus becomes the didactic vehicle of community in the face of death. Lorna Clymer, "Graved in Tropes: The Figural Logic of Epitaphs and Elegies in Blair, Gray, Cowper, and Wordsworth," *ELH* 62 (1995): 347–86.
41. The phrase from Horace's Ode 1.31 is used by Richard Brocklesby to describe Johnson's works; quoted in Hart, *Johnson and the Culture of Property,* 15.
42. Arthur Murphy, *Essay on Johnson's Life and Genius* (1792), reprinted in Hill, *Miscellanies,* 1:439.
43. John Milton, *Paradise Lost,* ed. Merritt Y. Hughes (New York: Odyssey Press, 1935), 2:146–47.
44. Sir John Hawkins, *The Life of Samuel Johnson, LL.D.* (1787), ed. Bertram H. Davis (New York: Macmillan, 1961), 273, 275.
45. Joseph Addison, *Cato. A Tragedy* (1713), act 5, scene 1, reprinted in *The Beggar's Opera and Other Eighteenth-Century Plays,* selected by John Hampden (London: J. M. Dent & Sons, 1981), 45–46. See Coppelia Kahn's analogous account of Antony's suicide, modeled on Cato's, in Shakespeare's *Antony and Cleopatra*—both are botched—both

are nevertheless heroic. Coppelia Kahn, *Roman Shakespeare: Warriors, Wounds, and Women* (London: Routledge, 1997), 110–43.

46. Hawkins, *Life,* 274.
47. Samuel Johnson, *The Rambler* no. 78, ed. W. J. Bate and Albrecht B. Strauss, *The Yale Edition of the Works of Samuel Johnson,* 16 vols. (New Haven: Yale University Press, 1969), 4:47.
48. Both extremes are encapsulated in Belial's continuing words, "Those thoughts that wander through Eternity / To perish rather, swallowd up and lost / In the wide womb of uncreated night, / Devoid of sense and motion?" *Paradise Lost* 2.148–51. The ongoing debate in Johnsonian circles on the topos of Johnson's fear of death revolves around what he feared most—annihilation or damnation—and was revived in a question-and-answer session following the delivery of a earlier version of some of this material at the Clark Library. For its beginning, see Johnson's friend, the Anglican divine Charles Taylor, *A Letter to Samuel Johnson, LL.D., on the subject of a Future State* (London, 1787), which was written in response to Johnson's claim "that he would prefer a state of torment to that of annihilation," out of concern "that such a declaration, coming from a person of his weight and character, might be productive of evil consequences." Johnson is quoted in this text as clarifying his original statement by stating "that life was indeed a great thing; and that [he] meant nothing more by [his] preference of a state of torment to a state of annihilation, than to express at what an immense value [he] rated vital existence" (6). Demanding more precision in such a statement, Taylor argues for the immortality of the soul and the resurrection of the body. His greatest fear, it seems, is that his defense against Johnson's remarks might lead some to suicide in hope of abandoning the sufferings of this life for the happiness of the next, and he ultimately argues for abandoning any attempt to see into futurity in favor of faith. Oddly, once again, despite suspicions of suicide around his own death, it is Johnson's attachment to life that seems a greater transgression than that of the suicide who hopes for redemption in the afterlife. For the stoic Roman version of Taylor's refutation of Johnson's preference of painful life over potential nothingness, see Seneca, Letter 101, "On the Futility of Planning Ahead," *Ad Lucilium Epistulae Morales,* trans. Richard M. Gunmere, 3 vols. (Cambridge, MA: Harvard University Press, 1917), 3:159–67. John Courtenay, quoted in chapter 2 of this book, also refers to this letter when he translates Seneca's reference to Maecenas's preference for crucifixion and deformity over death into Johnson's own fear. Seneca calls this preference "womanish and indecent" (3:165), arguing: "We must get rid of this craving for life, and learn that it makes no difference when your suffering comes, because at some time you are bound to suffer." Seneca, *Ad Lucilium,* 3:167. For classic examples of Johnsonian criticism on this topic, see J. H. Hagstrum, "On Dr. Johnson's Fear of Death," *ELH* 14 (1947): 308–19; and Max Byrd, "Johnson's Spiritual Anxiety," *Modern Philology* 78, no. 4 (May 1981): 368–78. This may also be part of what Samuel Beckett praised as Johnson's vision of "positive annihilation."
49. Slavoj Žižek's reading of Kierkegaard's "notion of 'sickness unto death'" is suggestive of Johnson's dilemma here: "The 'sickness unto death' proper, its despair, is to be opposed to the standard despair of the individual who is split between the certainty that death is the end, that there is no Beyond of eternal life, and his unquenchable desire to believe that death is not the last thing—that there is another life, with its promise

of redemption and eternal bliss. The 'sickness unto death,' rather, involves the opposite paradox of the subject who knows that death is not the end, that he has an immortal soul, but cannot face the exorbitant demands of this fact . . . and therefore, desperately wants to believe that death *is* the end—that there is no divine unconditional demand exerting its pressure upon him." Slavoj Žižek, *The Plague of Fantasies* (New York: Verso, 1997), 90.

50. Samuel Johnson, *Selected Poetry and Prose,* ed. Frank Brady and W. K. Wimsatt (Berkeley: University of California Press, 1977), 642.
51. Boswell, *Life,* 2:328. In the paragraph preceding Murphy's anecdote, Gray's "Elegy" is quoted as Johnson "casts one longing lingering look behind" at Streatham.
52. Roger Lonsdale, ed., *The Poems of Gray, Collins and Goldsmith* (London: Longman, 1969), 132–34.
53. Neil Hertz, "Dr. Johnson's Forgetfulness, Descartes' Piece of Wax," *Eighteenth-Century Life* 16 (November 1992): 167.
54. Boswell, *Life,* 4:392n1.
55. Hertz, "Johnson's Forgetfulness," 179; René Descartes, *Meditations on First Philosophy,* in *Descartes: Selections,* ed. Ralph M. Eaton (New York: Scribner's, 1927), 97–98.
56. Hertz, "Johnson's Forgetfulness," 179.
57. Ibid.
58. Michael Joyce, *Samuel Johnson* (London: Longman's, 1955), vi.
59. John Bailey, *Dr. Johnson and His Circle,* 2nd ed. with assistance of L. F. Powell, vol. 64 of the Home University Library of Modern Knowledge (1913; repr., Oxford: Oxford University Press, 1945), 8–9, 109–10, 11, 9–10.
60. Bertrand H. Bronson, "The Double Tradition of Dr. Johnson," in *Johnson Agonistes and Other Essays* (Berkeley: University of California Press, 1965), 156, 157.
61. Ibid., 170, 173.
62. Ibid., 176.
63. James Boswell, *Boswell: The Ominous Years, 1774–1776,* ed. Frederick A. Pottle and Charles Ryskamp (New York: McGraw-Hill, 1963), 80.
64. Boswell would later attempt to distance Johnson from this anti-American pamphlet. See Boswell, *Life,* 3:312–13.
65. Boswell in Pottle and Ryskamp, *Boswell: The Ominous Years,* 80.
66. Interestingly, the flesh resurfaces almost immediately in the journal entry: Boswell's merging with Johnson's spirit is followed by a meditation on the desirability of polygamy and its permissibility for monarchs ancient and modern. Ibid., 81.
67. H. J. Jackson, *Marginalia: Readers Writing in Books* (New Haven: Yale University Press, 2001), 165. Frustrated by her inability to find an intellectual pattern of response to the *Life,* Jackson concludes: "Boswell's readers were looking for help with their own lives and were most struck by those places in which there was something at stake for them personally" (178).
68. Ibid., 169–70, 172.
69. Walter Raleigh, *Six Essays on Johnson* (Oxford: Clarendon Press, 1910), 31. Raleigh first delivered this as the Leslie Stephen Lecture in the Senate House, Cambridge, 1907.
70. *Transactions of the Johnson Society* (Lichfield, 1949–50), 12–13. Johnson's birthplace, a house and bookshop on the Lichfield Market Square, was purchased by the city of Lichfield in 1887 and turned into a museum in 1901.

71. Arthur B. Platt, "The Johnson Anthem" (London: Weekes and Co., 1909), 2000.87, Samuel Johnson Birthplace Museum. The full text of the prayer appears in Boswell, *Life,* 4:417; also in Hill, *Miscellanies,* 1:121.
72. This prayer has been the source of controversy among Johnson scholars, who have debated whether or not it documents a "late conversion" to evangelical Christianity. While the consensus seems to be that Johnson remained a High Church Anglican to the end, it is generally agreed that he did experience a marked "turn" (in the literal sense of conversion) toward repentance and grace during the last months of his life. What interests me about such narratives of Johnson's religious life is that, whatever their differences, all are structured as romances of quest and Christian resolution that the conflicting endings to his story refuse to grant. For one of many examples, see Charles E. Pierce Jr., *The Religious Life of Samuel Johnson* (Hamden, CT: Archon Books, 1983).
73. Samuel Johnson, *The Complete English Poems,* ed. J. D. Fleeman (New Haven: Yale University Press, 1982).
74. Johnson would have understood the Anglican sacrament of the Eucharist as meant to provide "spiritual nourishment" and "medicine that provided to the soul needed grace." Most important in Johnson's case, the Eucharist "gave assurance of the resurrection to eternal life, calming fears concerning one's eternal state." Robert D. Cornwall, *Visible and Apostolic: The Constitution of the Church in High Church Anglican and Non-Juror Thought* (Newark: University of Delaware Press, 1993), 139.
75. Edward Young, *Conjectures on Original Composition* (1759), in Edmund D. Jones, ed., *English Critical Essays (Sixteenth, Seventeenth, and Eighteenth Centuries)* (London: Oxford University Press, 1922), 359.
76. Ibid., 360, 363, 360. The context of the line from the *Aeneid* is intriguing—Aeneas, about to kill Helen of Troy in retribution for the damages of the war, recognizes his mother, Venus, revealed to him for the first time as immortal.
77. Addison's persona of disembodied *Spectator* comes to mind here, as if Young's project were to provide a supplement to such elusiveness by celebrating the author at the moment of death.
78. Young, *Conjectures,* 362–63.
79. Ibid., 315.
80. For the controversy provoked by the publication of the *Prayers and Meditations,* a text that revealed, among other things, the possibility of Johnson's unorthodox and superstitious belief in purgatory, see Maurice J. Quinlan, "The Reaction to Dr. Johnson's *Prayers and Meditations,*" *Journal of English and Germanic Philology* 52, no. 2 (April 1953): 125–39.
81. *Transactions of the Johnson Society* of Lichfield for 1988 contains an essay by the bishop of Oxford proposing Johnson as an Anglican saint, complete with a drafted commemoration service. See Hart, *Johnson and the Culture of Property,* 66–67. Such canonization (underwriting the literary canon as a religious one) was ironically proposed in *The St. James Chronicle* soon after Johnson's death under the heading DEIFICATION. Undated clipping, Samuel Lysons's *Book of Cuttings.* All quotations from this source are published by permission of the Rare Book and Manuscript Library, Columbia University.
82. In this regard, Johnson's embattled Christian death was often compared to the peaceful death of the skeptic David Hume (Boswell went to witness the latter). See, for

example, Rev. William Agutter, A.M., *On the Difference between the Deaths of the Righteous and the Wicked, Illustrated in the Instance of DR. SAMUEL JOHNSON, and DAVID HUME, Esq. A Sermon, Preached before the University of Oxford at St. Mary's Church on Sunday, July 23, 1786* (London, 1800). In his defense of Johnson's fearful death, Agutter addresses the failure of exemplarity by exposing it as inherently theatrical. See also Michael Ignatieff, *The Needs of Strangers* (New York: Viking, 1985); and Stephen Miller, *Three Deaths and Enlightenment Thought: Hume, Johnson, Marat* (Lewisburg: Bucknell University Press, 2001). For Johnson's lifelong obsession with the parable of the talents (Matthew 25, 14–30) and the punishment of the "unprofitable servant," as well as the text from Luke 12, 48, "For unto whomsoever much is given, of him shall be much required," see Boswell, *Life,* 4:427 and note; Johnson, *Rambler* no. 77, 4:38–44. and "Verses on the Death of Dr. Robert Levet," whose "single talent" was "well employed." Samuel Johnson, "Verses on the Death of Dr. Robert Levet," in *Complete English Poems,* line 28.

83. Lionel Johnson, "At the Cheshire Cheese," in *Johnson Club Papers by Various Hands,* 276. For an earlier Johnsonian Last Supper poem, also in doggerel verse, by a member of the original Literary Club, see Oliver Goldsmith, "Retaliation," written in response to his friends' earlier witty compiling of a series of epitaphs on him. In Goldsmith's poem, epitaphs merge with menu entries as each fellow author joins the table as at once object of eulogy and dish to be consumed: "Let each guest bring himself, and he brings the best dish" (115). To give one of the better-known examples, "Our Garrick's a salad, for in him we see / Oil, vinegar, sugar, and saltness agree" (116). Johnson is not included. *Poems, Plays, and Essays by Oliver Goldsmith, M.B. with a Critical Dissertation on His Poetry by John Aiken, M.D. and an Introductory Essay by Henry T. Tuckerman, Esq.* (New York: Thomas Y. Crowell, n.d.), 115–23.
84. Boswell, *Life,* 2:261–62.
85. Greg Clingham, *James Boswell: The Life of Johnson* (Cambridge: Cambridge University Press, 1992), 59–60. Clingham notes in addition that Boswell's partly uncomprehending novelistic "hyperbole" in this passage "moves away from burlesque towards nightmare" (54).
86. Peter Brown, *The Cult of the Saints: Its Rise and Function in Latin Christianity* (Chicago: University of Chicago Press, 1981), 1. Brown's book reminds us of how radically alien, even monstrous, the worship of the bodies of saints seemed to pagan minds in late antiquity. Rendering bodily resurrection (previously unimaginable to the sophisticated pagan imagination and conceivable only in the distant future to Jews and early Christians) literally commonplace, the saint crossed the seemingly immutable boundaries of the late antique universe, "between those beings who had been touched by the taint of human death and those who had not," between mortal heroes and immortal gods, between dead matter and living soul. The saint, in short, permanently altered the pagan "familiar map of the relations between the human and the divine, the dead and the living" (5).
87. On the new physical proximity of the dead in the Christian cult of the saints, see ibid., 4–5. In *Cities of the Dead: Circum-Atlantic Performance* (New York: Columbia University Press, 1995), Joseph Roach discusses the emergent eighteenth-century practice of segregation of the dead from the living as symptomatic of complex problems of imperial memory, denial, surrogation, and fetishization. The print frenzy that immortalized Johnson before and (especially) after his death demonstrates be-

yond doubt that the eighteenth-century world of print is also an uncanny world that brings the dead, at once marginal and representative, back into contact with the living.

88. Boswell's foremost critic is the Johnsonian Donald Greene, who played the Puritan text-based counterpart to Boswell's Anglican mode of author worship. See in particular his "The Logia of Samuel Johnson," in *The Age of Johnson: A Scholarly Annual* 3, ed. Paul J. Korshin (New York: AMS Press, 1990), 1–33, which models its dissection of the "truth" of Boswell's *Life* upon the philological study of scripture. The *Life*'s status as Johnsonian bible is paradoxically bolstered by this attack. See also Greene's vituperative "The World's Worst Biography," *American Scholar* 62, no. 3 (Summer 1993): 365–82.
89. Northrop Frye, *The Secular Scripture: A Study of the Structure of Romance* (Cambridge, MA: Harvard University Press, 1976). On questions of truth and fiction in Boswell's *Life* as mediated by Johnson's character and animated presence, see Fredric V. Bogel, "'Did You Once See Johnson Plain?': Reflections on Boswell's *Life* and the State of Eighteenth-Century Studies," in *Boswell's* Life of Johnson*: New Questions, New Answers,* ed. John A. Vance (Athens: University of Georgia Press, 1985), 73–93; Ralph Rader, "Literary Form in Factual Narrative: The Example of Boswell's *Johnson,*" in ibid., 25–53; Clingham, *Boswell: Life of Johnson;* and William C. Dowling, *Language and Logos in Boswell's* Life of Johnson (Princeton: Princeton University Press, 1981), particularly on "perspective as moral choice" (158–59).
90. See, for example, George Hoffmann, "Anatomy of the Mass: Montaigne's 'Cannibals,'" *PMLA* 117, no. 2 (March 2002): 207–21.
91. This repetition, in the form of Thomas Tyers's allusion to the story of the Ephesian matron, is the subject of chapter 4.
92. Stephen Greenblatt, "Remnants of the Sacred in Early Modern England," in *Subject and Object in Renaissance Culture,* ed. Margreta de Grazia, Maureen Quilligan, and Peter Stalleybrass (Cambridge: Cambridge University Press, 1996), 342, 344. See also Greenblatt's "The Mousetrap," in Catherine Gallagher and Stephen Greenblatt, *Practicing New Historicism* (Chicago: University of Chicago Press, 2000), 136–62.
93. "It may be said, the death of Dr. Johnson kept the public mind in agitation beyond all former example. No literary character ever excited so much attention." Murphy, *Essay on Johnson's Life and Genius,* in Hill, *Miscellanies,* 1:356.
94. *English Review* VI (April 1786): 259. The article begins: "The love of anecdote is one of the most prevailing passions, or rather appetites, of the present age" (254).
95. Francis, Barber [pseudonym], *More Last Words of Dr. Johnson: consisting of Important and Valuable Anecdotes, and A Curious Letter from a Medical Gentleman: Now published, for the first time, from the Doctor's Manuscripts, with some original and interesting Stories of A Private Nature, relative to that* great man. *To which are added, Several singular and unaccountable Facts relative to his Biographical Executor, formerly Chairman of the Quarter-Sessions* (London, 1787), 29. The pseudonym alludes to Johnson's ward and servant, the Jamaican-born African Francis Barber, whom Johnson adopted, educated, and, scandalously, made his executor. Marian Pottle, in her assemblage of Boswellian ephemera for the Yale collection, speculates over the identity of the author of this pamphlet to no avail. In his edition of John Hawkins's *Life,* Bertram H. Davis reads *More Last Words* as an attack on Hawkins, who bore a special animus toward Barber

and was even accused of stealing from him. Hawkins, *Life,* xvii. For a reading of Swift's materialism in *A Tale of a Tub* that resonates with my thinking on the double nature of Johnson's corpse, see James Noggle, *The Skeptical Sublime* (Oxford: Oxford University Press, 2001), 71–96. The puns in this particular quotation, we might recall, echo Johnson's own inadvertent puns in the *Life,* discussed in chapter 2.

96. Francis, *More Last Words,* 30, 33–34.
97. The medical gentleman's recording of scientific experiments involving the application of authorial feces to his skull are also worth noting—apparently, this made him smarter. This seems to me to be a secular literary/medical version of the magical power of saints' relics.
98. Francis, *More Last Words,* 32–33.
99. Ibid., 31.
100. Ibid., 51, 52, 53–54.
101. I owe this phrase to Robert Griffin, personal communication.
102. J. Hillis Miller, "The Critic as Host," in Harold Bloom et al., *Deconstruction and Criticism* (New York: Seabury Press, 1979), 220–21.
103. Donna Heiland, "Remembering the Hero in Boswell's *Life of Johnson,*" in *New Light on Boswell: Critical and Historical Essays on the Occasion of the Bicentenary of* The Life of Johnson, ed. Greg Clingham (Cambridge: Cambridge University Press, 1991), 199–200.
104. Gillian Feeley-Harnik, *The Lord's Table: Eucharist and Passover in Early Christianity* (Philadelphia: University of Pennsylvania Press, 1981), chap. 5. In this regard, it is worth noting that Anglican theology emphasized the ecclesiastical nature of the Eucharist, its standing for and creation of the collective body of the English church.
105. Lysons's *Book of Cuttings,* unidentified newspaper clipping.
106. *Morning Post,* March 12, 1788, quoted in Kelley and Brack, *Early Biographers,* 11.
107. Croker, *Johnsoniana,* 476.
108. See also the following epigram: "Dr. Johnson's fate resembles that of the *Great Ox.*—They were both gigantic . . . both shown by their *friends* for a time, and both, after being knocked on the head, have *cut up well!*" Kelley and Brack, *Early Biographers,* 3. During Johnson's own lifetime, such cannibalistic discourse articulated a struggle for a representative national body and language. Consider a Glasgow paper's response to the author's visit to the Isle of Skye, quoted toward the end of Boswell's *Tour to the Hebrides:* "We are well assured that Dr. Johnson is confined by tempestuous weather to the isle of Sky. . . . Such a philosopher, detained on an almost barren island, resembles a whale left upon the strand. The latter will be welcome to every body, on account of his oil, his bone, &c. and the other will charm his companions, and the rude inhabitants, with his superior knowledge and wisdom, calm resignation, and unbounded benevolence." Boswell, *Journal,* in *Life,* 5:344. Likening the "consumption of Johnson's conversation to an image of corporeal decimation," this passage in Orrin N. C. Wang's reading also evokes "an anxiousness that it be Scotland's citizens who first and foremost make off with the benefits of [union] with England." Johnson's conversation and intellect provide a feast of cultural capital figured here as the rewards of national community. Orrin N. C. Wang, "The Politics of Aphasia in Boswell's *Journal of a Tour to the Hebrides,*" *Criticism* 36, no. 1 (Winter 1994): 80.
109. James Boswell to Anna Seward, April 30, 1785. Waingrow, *Correspondence and Other Papers of Boswell,* 96.

110. Elizabeth Moody, *Poetic Trifles* (London, 1798), 59–62; this is also included in Croker, *Johnsoniana,* 474–75, original source cited as *Gentleman's Magazine* 56 (1786): 427.
111. It's important to remember, however, that the eighteenth century was the first era that saw fit to embody Shakespeare in a full-length anthropomorphic monument, while founding a Shakespeare industry that rivaled Johnson's in its author worship. Shakespeare's disembodiedness served him better in later periods since it allowed him to translate empire through print in a way that Johnson's resolutely corporeal ghost could not. I am indebted to Coppelia Kahn's work in progress for this point, as well as a reminder from Jennifer Davidson. For Shakespeare's monumentalization, see Michael Dobson, *The Making of the National Poet: Shakespeare, Adaptation, and Authorship, 1660–1769* (Oxford: Clarendon Press, 1992).
112. Boswell, *Life,* 2:330–31.
113. James Boswell to Joseph Cooper Walker, July 1, 1785, in Waingrow, *Correspondence and Other Papers of James Boswell,* 111–12. "When people today choose Boswell as a guide to Johnson, they pick up his biography little realizing it is a sepulchre." Hart, *Johnson and the Culture of Property,* 32.
114. Beckett's "Human Wishes," as well as the notebooks in which he researched the play, are discussed by Ruby Cohn in *Just Play: Beckett's Theater* (Princeton: Princeton University Press, 1980), 143–62. Cohn also prints the dramatic fragment itself in an appendix (295–305). The orange peel reference, my own discovery, appears in RUL MS 3461/1, page 3 (Beckett marks it with an "X" for special emphasis), the Archive of the Beckett International Foundation at the University of Reading. This and all further quotation and reproduction from the works of Samuel Beckett by permission of the Samuel Beckett Estate. See also Beryl Bainbridge, *According to Queeney* (New York: Carroll & Graf, 2001).
115. James Merrill, *First Poems* (New York: Alfred A. Knopf, 1951), 8. Merrill also alludes here to Hart Crane's watery landscape in "At Melville's Tomb," which similarly ponders, through Melville's eyes, human remains, concluding: ". . . High in the azure steeps / Monody shall not wake the mariner. / This fabulous shadow only the sea keeps." *The Complete Poems and Selected Letters and Prose of Hart Crane,* ed. Brom Weber (New York: Anchor Books, 1966), 34.
116. Augustine Birrell, *Self-Selected Essays: A Second Series* (London: Thomas Nelson and Sons, 1916), 86–87.
117. "The contradiction between life and writing where the modern critic ends is the point at which Johnson begins, the problem he sets out to solve. And if Johnson cannot be said to have solved that problem once and for all, neither can it be said that modern theories of intertextuality have proved very satisfactory at accounting for Johnson. We need some opening to life that texts do not close off." Lawrence Lipking, "Johnson and the Meaning of Life," in *Johnson and His Age,* ed. James Engell, Harvard English Studies, vol. 12 (Cambridge, MA: Harvard University Press, 1984), 19.
118. Lipking suggestively continues, "Hence Johnson becomes the container and thing contained: simultaneously a mode of thinking and the object of that thought, an example of something and the something it exemplifies, an instance of life and the life that gives it meaning." Lipking, "Johnson and the Meaning of Life," 23. In his serious contemplation of Johnson's moral project, Lipking's evocation of the author rehearses (however unconsciously) a structural symbolism that is profoundly Christian.

119. From a Protestant perspective, to repeat the sacrifice of the Crucifixion would be sacrilege; from the pagan perspective, to sacrifice a god rather than to a god was the ultimate barbarity. See William R. Crockett, *Eucharist: Symbol of Transformation* (New York: Pueblo Publishing, 1989), 128–63, for a good summation of Reformation thinking on the Eucharist. In a parallel vein, Anglican doctrine held saints to be exemplary but not divine—models for imitation rather than worship. The translation of Johnson into the category of secular saint works well when we consider the ways in which the Protestant transformation of the saint also disavows a history of hagiographic embodiment (commemorated in Johnson's case by the preservation of his corpse by medical writers). See Paul Elmer More and Frank Leslie Cross, eds., *Anglicanism: The Thought and Practice of the Church of England, Illustrated from the Religious Literature of the Seventeenth Century* (London: Society for Promoting Christian Knowledge, 1935), 524–40.
120. Samuel Johnson Birthplace Museum, Lichfield, Flier, panel 1, "The Kitchen."

## Chapter Four

1. Nero, under whose reign Petronius wrote and who ordered Petronius's suicide, was a famous matricide. The editor of the Revels edition notes two references in Shakespeare: *King John* 5.2.152–53: "You bloody Neroes, ripping up the womb / Of your dear mother England"; and *King Lear*, 3.6.7–8: "Nero is an angler in the lake of darkness." Chapman's spin on Nero (who in Suetonius caresses his mother's corpse and assesses her charms) seems to conflate him with Caligula, who ripped the child he conceived from his sister's womb. George Chapman, *The Widow's Tears*, ed. Akihiro Yamada (London: Methuen, 1975).
2. Thomas Tyers, "Additional Sketches Relative to Dr. Johnson," *Gentleman's Magazine* 55 (February 1785): 85. Tyers published his biography in two installments, the first of which appeared almost immediately after Johnson's death in *Gentleman's Magazine* 54 (December 1784): 899–911, and "Supplement," 982. The entire text was reissued with revisions as a pamphlet in 1785, the text of which appears in O. M. Brack Jr. and Robert E. Kelley, eds., *The Early Biographies of Samuel Johnson* (Iowa City: Iowa University Press, 1974), 61–90, revised version of quotation on 62.
3. See Barbara J. Todd, "The Remarrying Widow: A Stereotype Reconsidered," in *Women in English Society, 1500–1800*, ed. Mary Prior (London: Methuen, 1985): 54–83, for the historical argument that over the course of the early modern period, widows desperate to remarry became increasingly rare in reality—an argument that makes the persistence of the stereotype all the more interesting.
4. Thomas Tyers, "Biographical Sketch," in Brack and Kelley, *Early Biographies*, 87; Sir John Hawkins, *The Life of Samuel Johnson, LL.D.* (1787), ed. Bertram H. Davis (New York: Macmillan, 1961), 273.
5. Historical research on the frequency of autopsies of celebrities who were not aristocrats or royalty—there is a separate tradition of the dissection of royalty outlined in Phillipe Aries, *The Hour of Our Death* (New York: Oxford University Press, 1981)—is sadly lacking. It was not completely uncommon for the results of the autopsies of famous actors and, occasionally, authors to appear in print, but it was also true that

many Christians of all classes entertained the prospect with horror. For more on the history of popular attitudes toward anatomy, see chapter 1. What distinguishes Johnson's autopsy is not so much that it occurred but that it occurred in light of the patient's known fear of death and express desire not to have his body violated.

6. See chapter 3, n. 81.
7. Steven Mullaney's "Mourning and Misogyny: *Hamlet, The Revenger's Tragedy,* and the Final Progress of Elizabeth I," in *Centuries' Ends: Narrative Means,* ed. Robert Newman (Stanford: Stanford University Press, 1996), 238–60, is useful here both in terms of its shrewd analysis of the psychic dynamics that link mourning and misogyny in *Hamlet* and also for its historical situation of such a dynamic in relation to cultural ambivalence toward the death of Elizabeth. Also germane is Valerie Traub on Hamlet's misogyny as representative of Shakespearean male inability to cope with female desire and its association with death: "Jewels, Statues, and Corpses: Containment of Female Erotic Power in Shakespeare's Plays," *Shakespeare Studies* 20 (1988): 215–38. It would be interesting to consider the relation of *Hamlet*'s working through of the death of the "Virgin Queen" to Boswell's working through in the *Life* of the death of the monarch of English letters. For a fascinating analysis of the dramatic corporeality of Elizabeth's death and subsequent exploding corpse, see Catherine Loomis, "Elizabeth Southwell's Manuscript Account of the Death of Queen Elizabeth [with text]," *English Literary Renaissance* 26, no. 3 (Autumn 1996): 482–509, which ends with the same question Hamlet asks of the ghost, the question that is the epigraph to chapter 3.
8. A. Aarne, *The Types of the Folk Tale,* trans. and enlarged by S. Thompson (Helsinki, 1928), quoted in Peter Ure, "The Widow of Ephesus: Some Reflections on an International Comic Theme," in *Elizabethan and Jacobean Drama: Critical Essays by Peter Ure,* ed. J. C. Maxwell (Liverpool: Liverpool University Press, 1974), 221. I am hugely indebted to Ure's essay, which is the most comprehensive and literarily sensitive treatment of the story's history I have yet come across.
9. The Indian version has been linked to the ancient Indian *Book of Sindibad* and entered the Western tradition as part of an early thirteenth-century translation into Hebrew, translated into Latin as the *Historia septem sapientum Romae* (The History of the Seven Wise Masters) by a thirteenth-century French monk. The earliest printed Latin version on record appears in 1490; the first English version appears in 1520. The frame of the text in which this version appears is itself significant, since the narrative is motivated by the desire to prove a treacherous woman false. "A young prince, falsely accused by the wife of the king, his father, of having attempted to offer her violence, is defended by seven sages, who relate a series of stories to show the deceits of women, the queen at the same time urging the death of the accused prince by the example of stories told by herself." *The History of the Seven Wise Masters of Rome,* printed from the edition of Wynkyn de Worde, 1520, and edited, with an introduction, by George Laurence Gomme, F.S.A. (London: Villon Society, 1885), ii. For a convincing account of Petronius's use of ancient Near Eastern and Arabic sources that also feed the Seven Sages tradition, see Graham Anderson, *Ancient Fiction: The Novel in the Graeco-Roman World* (London: Croon Helm, 1984), chap. 12.
10. The story appears both in John Ogilby, *The Fables of Aesop Paraphrased in Verse: Adorn'd with Sculpture and Illustrated with Annotations* (London, 1665), and Ogilby, *Aesopics: or, A second collection of fables,* 2nd ed. (London: By the author, 1673).

11. Charleton's tale is larded throughout by another author (whose story of "The Cimmerian Matron" is appended to Charleton's 1688 text) with misogynist quotations from Chaucer, additionally foregrounding the gendered specificity of what purports to be a general allegory of human nature. In Walter Charleton, *The Ephesian and Cimmerian Matrons: Two Notable Examples of the Power of Love and Wit* (London, 1688).
12. For a version of the Chinese story most popular during the period told by a fictional Chinese visitor to England, first published in serial form 1760–61, see letter XXVIII of Oliver Goldsmith, *The Citizen of the World; or, Letters from a Chinese Philosopher Residing in London to His Friends in the East* (1762), 2 vols., ed. Austin Dobson (London: J. M. Dent, 1891), 1:65–70. A more scholarly and exact translation under the title of "The Impatient Widow" appeared in *The Asiatic Journal and Monthly Miscellany*, vol. 1, 3rd series (May–October) (London: Wm. H. Allen & Co, 1843), 607–18.
13. Joseph Addison and Richard Steele, *The Spectator* (no. 11, March 23, 1711), ed. Donald Bond (Oxford: Clarendon Press, 1965), 47–51, 48. In this version, hugely influential in its own right, the ruthless merchant Inkle, for whom "the natural Impulses of his Passions" are stifled "by Prepossession toward his Interests" (50), is saved from an Indian attack by the virtuous Yarico but ultimately abandons her, pregnant with his child, and sells her into slavery. For examples of the multiple versions of this story over the course of the century, see Frank Felsenstein, ed., *English Trader, Indian Maid: Representing Gender, Race, and Slavery in the New World: An Inkle and Yarico Reader* (Baltimore, MD: Johns Hopkins University Press, 1999).
14. We might note in passing that the display of the criminal's corpse varies over time from the cross of antiquity to the gallows to the early modern gibbet, and that in a 1707 illustration to the Latin text of Petronius, the artist inverts the cross into the conventional Christian position rather than the Roman position in which the head would have been upside down.
15. For a fascinating eighteenth-century compilation by a famous antiquarian of multiple national versions of the Ephesian matron story, including a French version translated by George Etherege, the Chinese version, and a "medieval" version in middle English, see Thomas Percy, Bishop of Dromore, ed., *The Matrons: Six Short Histories* (London, 1762), which opens with an epigraph from Prior urging women to abandon dead flames for living ones.
16. In her essay on Holbein's painting of the dead Christ (an analysis with interesting resonances for our consideration of things in themselves), Julia Kristeva, relying on Pascal's *Pensées,* meditates suggestively on the space of the tomb as the abyss in the narrative of Christ's life that at once solicits and denies identification. Julia Kristeva, *Black Sun: Depression and Melancholia,* trans. Leon S. Roudiez (New York: Columbia University Press, 1989), 105–38.
17. Interestingly, Chapman sets his version in Cyprus, Venus's territory; while Ogilby rewrites the tale as an epic battle between Diana and Aphrodite in which the mortals are mere pawns.
18. Michael MacDonald's analysis of the relation of narrative to emotion in the early modern period in "*The Fearefull Estate of Francis Spira:* Narrative, Identity, and Emotion in Early Modern England," *Journal of British Studies* 31 (1992): 32–61, is particularly germane here. What he argues about the famous apostasy narrative—hugely popular in the early modern period—of Francis Spira, an Italian Protestant

who recanted and died of starvation in what many construed as a suicide, is equally true of the matron story. Both narratives provide lenses with which to discern what he calls a history of emotion. If the Spira story—deployed in a variety of ways over the course of the seventeenth and eighteenth centuries from Reformist propaganda to spiritual autobiography—provides us with a history of religious despair, the matron story can be seen as the Enlightenment's inversion of such despair into libertine materialist hope. This works even at the basic narrative level—Protestant infidelity results in remorse, despair, and suicide through a man's spectacular self-starvation; while a Roman woman's equally spectacular self-starvation in the cause of female virtue ends in infidelity and the betrayal of the dead in the service of the living.

19. "Fiction became antiquity's most eloquent expression of the nexus between polytheism and scripture." G. W. Bowersock, *Fiction as History: Nero to Julian* (Berkeley: University of California Press, 1994), 141.

20. Simon Goldhill, *Foucault's Virginity: Ancient Erotic Fiction and the History of Sexuality* (Cambridge: Cambridge University Press, 1995), 142.

21. We should note here that some critics have read the *Satyricon*, whose heroes are a homosexual couple who are serially unfaithful, as a parody of the novel's heterosexual plot of chastity. This is a much-contested theory that also shows us how completely novel and epic are imbricated in the literature of antiquity—it's been argued with equal plausibility that Petronius is parodying the *Aeneid* in general in the *Satyricon*, and book IV of the *Aeneid* in particular in the matron story. (It's also been argued that Virgil was the first epic poet who had to respond to the novel.) The *Satyricon* is also distinguished by its all-pervasive misogyny and its scopophilia. See Goldhill, *Foucault's Virginity*, and Bowersock, *Fiction as History*, as well as P. G. Walsh, *The Roman Novel: The 'Satyricon' of Petronius and the 'Metamorphoses' of Apuleius* (Cambridge: Cambridge University Press, 1970); and J. P. Sullivan, *The Satyricon of Petronius: A Literary Study* (Bloomington: Indiana University Press, 1968).

22. Johnson devoured romances in secret and demanded that the novel—whose attempts at realism he had argued in *Rambler* 4 were potentially more dangerous than romance's explicit unreality—repose on the stability of virtuous example. His composition of the end of Charlotte Lennox's *Female Quixote* in which the heroine is cured of her addiction to romance is particularly germane here. William B. Warner distinguishes this particular phase of prose fiction from romance (specifying shorter length, contemporary setting, and emphasis on sex) in order to revise while maintaining the "rise of the novel" narrative in *Licensing Entertainment* (Berkeley: University of California Press, 1998), while Margaret Anne Doody does away with that narrative altogether in the service of what Helen Thompson considers a part of a feminist undermining of possessive individuality. Helen Thompson, "Plotting Materialism: W. Charleton's *The Ephesian Matron*, E. Haywood's *Fantomina*, and Feminine Consistency," *Eighteenth-Century Studies* 35, no. 2 (Winter 2002): 195–214; Margaret Anne Doody, *The True Story of the Novel* (New Brunswick, NJ: Rutgers University Press, 1996).

23. See Thompson "Plotting Materialism"; and Doody, *True Story of the Novel*, 110–13. Charlton also uses the matron story to refute the chaste claims of the then-fashionable Platonic Lady, thus foregrounding romance's insistence on the claims of physical desire; at this early modern materialist moment in its history, romance undoes the lofty claims of its alternative form, courtly love. On this point, see George

Williamson, "The Ephesian Matron and the Platonic Lady," *Review of English Studies* 12 (1936): 445–49. My thinking about romance, underlying this book's consideration of "anecdotal errancy," owes a great deal to Maurice Blanchot's reading of the corpse in "Two Versions of the Imaginary," translated by Lydia Davis in *The Station Hill Blanchot Reader: Fiction and Literary Essays* (Barrytown, NY: Station Hill, 1999), 417–28; and to Patricia Parker's reading of error in *Inescapable Romance* (New Haven: Yale University Press, 1976).

24. We might also note that in his essay on "Family Romances," the male child's knowledge of sexual difference, and with it the uncertainty of paternity, results in fictions, "which remind one of historical intrigues" of maternal infidelity. While the father in such scenarios is idealized and monumentalized, the mother, whose identity cannot be denied, becomes the object of desire, curiosity, and narrative: "The child, having learnt about sexual processes, tends to picture to himself erotic situations and relations, the motive force behind this being his desire to bring his mother (who is the subject of the most intense sexual curiosity) into situations of secret infidelity and into secret love-affairs. In this way the child's phantasies, which started by being, as it were, asexual, are brought up to the level of his later knowledge . . . and he often has no hesitation in attributing to his mother as many fictitious love-affairs as he himself has competitors. An interesting variant of the family romance may then appear, in which the hero and author returns to legitimacy himself while his brother and sisters are eliminated by being bastardized." The author of romance is always his own hero, propelled by desire to know feminine secrets of his own making, to claim as his own the place in which he lay. Sigmund Freud, *The Standard Edition of the Complete Psychological Works of Sigmund Freud,* trans. James Strachey in collaboration with Anna Freud, assisted by Alix Strachey and Alan Tyson, vol. 9 (1906–1908) (London: Hogarth Press, 1953–74), 239.
25. Anonymous screed by James Boswell, published (Boswell's attribution) in the *Morning Post,* 1790; Beinecke Rare Book and Manuscript Library, Yale attribution: c. March 1, 1790. [*(10) P117.4.]
26. Greenblatt writes of Hamlet's disgust at Gertrude's remarriage: "At issue is not only, as G. R. Hibbard suggests, an aristocratic disdain for a bourgeois prudential virtue, but a conception of the sacred as incompatible with a restricted economy, an economy of calculation and equivalence. Such calculation has led Gertrude to marry Claudius, as if he were his brother's equal: 'My father's brother,' Hamlet protests, 'but no more like my father / Than I to Hercules' (1.2.152–53). Her remarriage, like the reuse of the funeral baked meats, is a double defilement: it has sullied Gertrude's flesh, which becomes a leftover to be gobbled up by the loathsome Claudius, and, since 'Father and mother is man and wife; man and wife is one flesh' (4.3.53–54), it has retroactively stained old Hamlet by identifying his noble spirit with the grossness of the 'bloat King' (3.4.171)." Stephen Greenblatt, "The Mousetrap," in Catherine Gallagher and Stephen Greenblatt, *Practicing New Historicism* (Chicago: University of Chicago Press, 2000), 155–56. Jane Gallop's reading of Blanchot on Sade in *Intersections: A Reading of Sade with Bataille, Blanchot, and Klossowski* (Lincoln: University of Nebraska Press, 1981), is germane here in its connection of excrement (the material of critiques of Johnsonian communion in chapter 3) to the corpse and to the prostitute (figured here as Johnson's remains and the enduring figure of the matron). Both the corpse and the turd are unassimilable leftovers; neither can be subsumed into any

sort of "universal process" (46), to what Greenblatt would term a sacred economy, but instead lead to endless recycling, "infinite repetition, coprophagia" (55). All three of these figures—corpse, turd, prostitute—evade and undo identity: they are "that which is not self-identical." The prostitute "is precisely [what Blanchot calls] the 'corpse's presence' which is totally at your disposal, and yet always exceeds you" (60).

27. William Cooke attributes these words to Johnson upon learning of Thrale's remarriage. William Cooke, *The Life of Samuel Johnson, LL.D.,* in Brack and Kelley, *Early Biographies,* 126. This famous phrase is from book IV of Virgil's *Aeneid,* to which, as we'll see below, Petronius directly alludes in his telling of the matron story. At this moment in Virgil's text, Mercury defames Dido—a widow compelled by the gods to betray her dead husband by falling in love with Aeneas—in order to persuade Aeneas to leave Carthage and follow his destiny, Rome. Chapman's version emphasizes how a widow's remarriage figuratively murders a husband after his death. Virtuous widowhood in his play is a male fantasy of an afterlife experienced through the "glory" of feminine devotion. Chapman, *Widow's Tears,* 2.3.50–58. Mullaney, in "Mourning and Misogyny," develops this point at length, stressing that the prospect of a widow's remarriage forces husbands to imagine their own death and imminent replacement.
28. Ure ends his essay with a characterization of Petronius's tale as a joke that many misogynist responses to the story have missed. I am also indebted to Nicole Horejsi's unpublished essay "Steele's *Spectator* 11: Misogyny, Slavery, and the Limits of the Classical Tradition," which discusses how Arietta's refusal of the joke leads to a feminine critique of slavery.
29. William Blake, "An Island in the Moon" (1784), dialogue between "Suction the Epicurean" and "Quid the Cynic" (Blake's brother Robert and Blake himself) quoted in James T. Boulton, ed., *Johnson: The Critical Heritage* (London: Routledge & Kegan Paul, 1971), 363–64.
30. James Boswell, *The Life of Samuel Johnson, LL.D.,* ed. George Birkbeck Hill, rev. L. F. Powell, 6 vols. (Oxford: Clarendon Press, 1934–50), 4:418.
31. It is worth remembering that the first entry for "interesting" as an adjective in the *OED* is from Laurence Sterne's *Sentimental Journey* (1768). Thus, Langton's response to the "interesting solemnity" of the spectacle of Johnson's body is charged with sensibility, an ambiguous solicitation and investment of feeling that links self to other through an ironic awareness of their distance, and of the profit to be gained from their relation.
32. Boswell, *Life,* 4:418.
33. Chester F. Chapin argues that Johnson's self-wounding was an attempt to repeat a "miraculous experience" of several months earlier, in which he was relieved of several quarts of fluid by a similar surgical operation and wonderfully revived. This experience has been thought to have given rise to Johnson's much-disputed "late conversion." From this perspective, Johnson's act is one of faith in God's grace and consonant with the religious romance of conversion that for many scholars makes narrative sense of his death. Chapin's essay closes with this suggestive observation from William Windham's diary: "Although one of the self-inflicted incisions 'was a deep and ugly wound,' the others 'were not unskillfully made.'" Chester F. Chapin, "Samuel Johnson's 'Wonderful Experience,'" in *Johnsonian Studies,* ed. Magdi Wahba (Cairo: Oxford University Press, 1962), 51–60. Science and religion, surgery and

martyrdom merge here in the marks on Johnson's body. We might also link the "deep and ugly wound" to that which marks Dido at the end of this section.

34. Tyers, "Additional Sketches," 86.
35. In Samuel Johnson, *The Complete English Poems,* ed. J. D. Fleeman (New Haven: Yale University Press, 1982), 141.
36. "How will it help you," she asked the lady, "if you faint from hunger? Why should you bury yourself alive, and go down to death before the Fates have called you? What does Vergil say?—

    *Do you suppose the shades and ashes of the dead*
    *are by such sorrow touched?*

    No, begin your life afresh. Shake off these woman's scruples; enjoy the light while you can." Petronius, *The Satyricon,* trans. William Arrowsmith (New York: New American Library, 1983), frag. 111, p.119. The line from Vergil is *Aeneid* 4.34.
37. The same language of proving/testing a coin is used earlier at the beginning of act 3 in response to the epigraph from Lycus that begins this chapter.
38. James Boswell to Joseph Walker, in Marshall Waingrow, ed., *The Correspondence and Other Papers of James Boswell Relating to the Making of* The Life of Johnson (New York: McGraw-Hill, 1966), 96.
39. Might this perhaps echo the story of the novel in which nobody's story could be anybody's?
40. Jeremy Taylor, *The Rule and Exercises of Holy Dying* (V.viii, "Of Visitation of the Sick"), vol. 2, *Holy Living and Holy Dying,* ed. P. G. Stanwood (Oxford: Clarendon Press, 1989), 229.
41. See also Elizabeth Moody's 1786 poem "Dr. Johnson's Ghost": having reprimanded Boswell for capitalizing on his death, Johnson "ceas'd, and stalk'd from Boswell's sight / With fierce indignant mien, / Scornful as Ajax' sullen sprite, / By sage Ulysses seen." In J. W. Croker, ed., *Johnsoniana; or, Supplement to Boswell* (London, 1786), 475.
42. "But she had turned / With gaze fixed on the ground as he spoke on, / Her face no more affected than if she were / Immobile granite or Marpesian stone. / At length she flung away from him and fled, / His enemy still, into the shadowy grove / Where he whose bride she once had been, Sychaeus, / Joined in her sorrows and returned her love." Virgil, *Aeneid,* 6.630–37, trans. Robert Fitzgerald (New York: Vintage, 1985). Quotations in text, 6.615, 626–28.
43. I mean "conservative" here in its literal sense of saving: many Johnsonians—whether humanists or postmodernists—see themselves as conserving fundamental literary values (however those values might differ from one generation to the next) through their devotion to the figure of Johnson.
44. Boswell, *Life,* 1:30.
45. Walter Scott, *Lives of the Novelists* (1821–24), quoted in Boulton, *Johnson: The Critical Heritage,* 420.
46. I am indebted to Sean Silver for his equally virtuosic meditation on the anecdotal form of Fineman's essay, in his own oral presentation turned unpublished paper, "Hodge in New Wye: The History of an Anecdote" (March 2003). Silver's thinking also inspired my own consideration in chapter 5 of the anecdote from Boswell that serves as the epigraph to Vladimir Nabokov's *Pale Fire.*

47. Joel Fineman, "History of the Anecdote: Fiction and Fiction," in *The New Historicism,* ed. H. Aram Veeser (New York: Routledge, 1989), 61.
48. Kate Marshall, "Voiding History: The Anecdotal Break and the Literature of Modernity," unpublished essay (March 2003): "The part that the anecdote represents functions as a hole within the whole; a puncture in a totality" (6). I am particularly indebted to Marshall for her brilliant linkage of Martin Heidegger's essay "The Thing" to Wallace Stevens's poem "Anecdote of the Jar" in a consideration of anecdotal form, function, and history.
49. Kenneth Burke, *A Grammar of Motives* (New York: Prentice-Hall, 1952), 60, 61.
50. Ibid., 325, 326.
51. For more on the differences between Johnson's and Boswell's biographical methods, see Jennifer Snead, "*Disjecta Membra Poetae:* The Aesthetics of the Fragment and Johnson's Biographical Practice in the *Lives of the English Poets,*" *The Age of Johnson* 15 (2004): 37–56.
52. Boswell, *Life,* 4:427–28.
53. For another of many examples, once again see *Rambler* 4, which contrasts the remote fantasies of prose romance with the dangerously compelling examples of the more realistic brand of fiction—the novel—just coming into its own.
54. Bacon's *Essays,* extremely influential in the eighteenth century and on Johnson in particular, are a good case in point of the deliberate and didactic ambiguity of the aphorism. See, for example, Stanley E. Fish, "Georgics of the Mind: The Experience of Bacon's Essays," in *Seventeenth-Century Prose: Modern Essays in Criticism,* ed. Stanley E. Fish (New York: Oxford University Press, 1971), 251–80.
55. Fineman, "History of the Anecdote," 56–57.
56. Jane Gallop, *Anecdotal Theory* (Durham, NC: Duke University Press, 2002), 3, 164.
57. Boswell, *Life,* 1:13. In a recent address to the MLA, Stephen Greenblatt characterized the experience of literature itself as "speaking with the dead"; the literary figure he summons to body forth this definition is, interestingly enough, the ghost of old Hamlet. "Presidential Address 2002: 'Stay, Illusion'—On Receiving Messages from the Dead," *PMLA* 118, no. 3 (May 2003): 417–26.
58. Greg Clingham, *James Boswell: The Life of Johnson* (Cambridge: Cambridge University Press, 1992), 79–97.
59. For a meditation on Fineman's exemplification of a truly literary history, one that sees literary theory, history, and literature commonly rooted in the same narrative and rhetorical structures and that allows for historical change and individual agency, see Marshall Grossman, *The Story of All Things* (Durham, NC: Duke University Press, 1998).
60. Fineman is following Husserl, "for whom the Renaissance marked the inauguration of what he called 'The Crisis of European Science' because this is when the language of science, exemplified by Galileo's mathematical physics, grows so formal that the sedimented meanings embedded within it grow too faint to recall . . . their origination in the natural life-world. . . . For Husserl, this forgetfulness nestled within formal scientific language marks the beginnings of crisis—a crisis no doubt conceived in medical terms—for the European spirit, for this is what initiates a specifically European tradition of empiricist, technicist scientism that works to divorce knowledge from experience. It is possible, however—indeed, given Husserl—it is *almost* necessary to imagine a corresponding crisis in historical consciousness inaugurated by the

emergence in the Renaissance of a correspondingly technicist historicism that carries as its cost its unspoken sense of estranged distance from the anecdotal real. This . . . suggests why the New Historicism is at once a symptom of and a response to a specifically Renaissance historiographic crisis. It is my suspicion or hypothesis that we can identify in the passage from the early to the late Renaissance, from the first humanist enthusiasm for the revival or rebirth of the past to the subsequent repudiation of the ancients in the name of science—e.g., Francis Bacon—how it happened that historiography gave over to science the *experience* of history, when the force of the anecdote was rewritten as experiment." Fineman, "History of the Anecdote," 62–63.

61. Fineman explicitly enjoins historicism—new or old—to follow the example of the history of science in undoing its certainties: "What *is* clear, however, is that the aporias embedded in [Bacon's] idea of the crucial experiment—which I take to unpack the aporetic, chiastic structure of the anecdote . . .—have obliged science to replace the idea of an inexorable *démarche* of scientific history with, instead, an ethics of the real. This was clearly not the case for Bacon, whose simple faith in inductive reasoning led him to see in the crossroads of the crucial experiment a way of establishing the inevitable narration of scientific history. . . . Contemporary philosophy of science, which possesses the courage of the convictions that give it its intellectual coherence, no longer believes in such 'memory of the future'—[unlike Bacon] it certainly distinguishes between prediction and prophecy—and this is a loss of faith that properly responds to the paradoxes embedded in the idea of the crucial experiment. It remains for historicism, which still for the most part believes in the crux of the crucial experiment, to do the same with its equally pious fidelity to the memory of the past." Ibid., 71–72.

62. In his essay "The Body's Moment," Jean Starobinski sees Montaigne's self-admittedly subjective and infinitely variable essays as characteristic of the Renaissance chapter of an endlessly repeating historical struggle for possession of the body between the speaking subject and scientific discourse. Jean Starobinski, "The Body's Moment," *Yale French Studies* 64 (1983): 276.

63. In *Bodies of Evidence: Medicine and the Politics of the English Inquest, 1830–1926* (Baltimore, MD: Johns Hopkins University Press, 2000), Ian A. Burney discusses the scientific professionalization and legitimation of pathology, along with the growing privatization of the inquest in the nineteenth century: "This apotheosizing of things-in-themselves was the pathologist's stock in trade, a rigorous standard that, in restraining impulses to abstract theorizing, guaranteed the direct transmissibility of his findings. Half a century later, the society's newly installed president professed continuing support for this founding principle. Limiting discussion to things actually present before the society, Joseph Payne observed in his 1898 inaugural lecture, functioned 'like the rules of admission and evidence in courts of law, which, though sometimes excluding materials useful for arriving at a verdict, yet ensure that the evidence which is admitted shall be unchallengeable. There is a vast difference between the statements of any observer, however competent and veracious, about an object, and the view, handling, or examination of the object itself. The subject matter of science,' Payne gravely observed in closing, 'is not statements about things, but the things themselves'" (114–15). For the first chapter of this ascendance of the "real" of the scientific object of anatomy, a process that culminated in Vesalius's establishment of dissection as "didactic and investigative" method and "the only possible guide to

the trustworthy description of parts of the human body," superseding the long-established authority of ancient anatomy texts, see Andrea Carlino, *Books of the Body: Anatomic Ritual and Renaissance Learning,* trans. John Tedeschi and Anne Tedeschi (Chicago: University of Chicago Press, 1999), 1. My thinking on phenomenology's relationship to "things in themselves" is based both on Fineman's reading of Husserl (specifically his "The Origin of Geometry" and *The Crisis of European Science and Transcendental Phenomenology*) and on Heidegger's essay "The Thing," in *Poetry, Language, Thought,* trans. Albert Hofstadter (New York: Perennial Classics, 2001), 163–84. As Albert Hofstadter puts it, Heidegger's attempt to unthink the thing as object and instead to conceive of "the thing in its own thingness" "represents a movement away from the thin abstractions of representational thinking and the stratospheric constructions of scientific theorizing, and toward the full concreteness, the onefoldness of the manifold, of actual life-experience. This is the sort of response that Heidegger has made to the old cry of Husserl, 'Back to the things themselves!'" Hofstadter, introduction to Heidegger, *Poetry, Language, Thought,* xvii. I should add that I am less interested in phenomenological exegesis than I am in the resonance of the words "things in themselves" across the disciplines and in their relationship to the embodied "real."

When the phrase "things in themselves" reemerges in Samuel Beckett's 1948 essay on the painters of the Parisian school that included the Van de Velde brothers, it provides a context for his own translation of the painterly mandate he identifies here into his version of writing without object: "Rouault's Christ, the stylized chinoiseries of the still-life paintings by Matisse, a group by Kandinsky painted in 1943 or 1944, these too derive from the same effort, which is to express the manner in which a clown, an apple, and a red square are the same, drawn out of the same confusion, in the face of the sheer resistance of things to represent them. For these things are all alike in this alone: that they are things, *the* thing, 'thingness.' It seems absurd to speak, as Kandinsky did, of painting freed from objects. What painting has freed itself from is [precisely] the illusion that there exists more than one object of representation, or perhaps even the illusion that an imagined unique object can be represented." Samuel Beckett, "Peintres de l'Empêchement" [Endgame Painters], in *Disjecta: Miscellaneous Writings and a Dramatic Fragment,* ed. Ruby Cohn (New York: Grove Press, 1984), 135–36. Unpublished English translation by Lowell Gallagher. Seen from the perspective of art as "an unveiling that approaches the undisclosable, the nothing, the thing itself," Boswell's orange peels, reappearing in his notebooks for an unfinished play on Johnson, "Human Wishes," take on a very different set of resonances.

64. Isaac D'Israeli, *A Dissertation on Anecdotes* (1793) (New York: Garland, 1972), 29–30, 30.
65. Ibid., 27–28.
66. Ibid., 26.
67. Ibid., 50.
68. Ibid., 81, 82.
69. We might trace here, via John Bender's recent work, a crisis of knowledge in the late eighteenth century that parallels Fineman's consideration of a historiographic crisis in the late Renaissance. If in Fineman's account, history gives over the anecdotal real of experience to scientific experiment, in Bender's, science gives over the hypothetical nature of its plots to the "real" of novelistic fiction. What is being signaled here is

the emergence of literature at this moment as both distinct discipline and separate ontology. John Bender, "Enlightenment Fiction and the Scientific Hypothesis," *Representations* 61 (1998): 6–28. Also germane is Bender's "Impersonal Violence: The Penetrating Gaze and the Field of Narration in *Caleb Williams,*" in *Vision and Textuality,* ed. Stephen Melville and Bill Readings (Durham, NC: Duke University Press, 1995), 256–81, on the relationship between third-person indirect discourse (which emerged in the novel during the late eighteenth century) and the anatomical gaze. In both of these essays, literature and science are linked through mutual curiosity and an anamorphic blurring of the distinction between "fiction" and "reality."

70. Chick's love of trivial anecdotal detail at once makes him the perfect biographer for Ravelstein (his sketches have the "color" that others lack), while making him vulnerable to the charm of less-than-exemplary acquaintances, in particular a Romanian Nazi sympathizer turned American exile.

> "Ravelstein wanted to know just what Grielescu's line was like and I told him that at dinner he lectured about archaic history, he stuffed his pipe, and lit lots of matches. You grip your pipe to keep it from shaking, and then the fingers with the match tremble twice as hard. He kept stuffing the pipe with the rebellious tobacco. When it didn't stay stuffed, he didn't have enough thumb-power to pack it down. How could such a person be politically dangerous? His jacket cuffs come down over his knuckles."
>
> Rosamund said, "My guess is that being seen in public with you was worth a lot to Grielescu. But this is how you do things, Chick: the observations you make crowd out the main point."
>
> "That's exactly what Ravelstein eventually told me. And how curious it was that I let myself be used like that." (165)

The detail of the pipe, trivializing yet signaling Grielescu's futile rage for power, haunts Chick throughout the novel. Saul Bellow, *Ravelstein* (New York: Penguin, 2001), 193. All further references to the novel will be given by page number in the text. One might also speculate about the choice of Bloom, a controversial and conservative figure in academic circles, as the Johnson equivalent.

71. Seneca to Lucilius, Letter 24, "On Despising Death," *Ad Lucilium Epistulae Morales,* trans. Richard M. Gunmere, 3 vols. (Cambridge, MA: Harvard University Press, 1917), 1:171, 173, 177.

72. "Science makes the jug-thing into a nonentity in not permitting things to be the standard for what is real." Heidegger, "The Thing," 168. The alternative to a scientific discourse that forces things to adhere to the abstract standard of facts is, for Heidegger, poetry.

73. Fineman, "History of the Anecdote," 61.

74. For a Johnsonian comparison of Boswell and Thrale that finds Thrale disconcertingly blunt (particularly in matters of the body) and Boswell the greater artist of the truth of Johnson's character, see Ralph Rader, "Literary Form in Factual Narrative: The Example of Boswell's *Johnson,*" in *Boswell's* Life of Johnson*: New Questions, New Answers,* ed. John A. Vance (Athens: University of Georgia Press, 1985), 25–53. For a convincing analysis of the patriarchal nature of Boswell's biography and the feminine alternative authority of Thrale's, see Felicity A. Nussbaum, *The Autobiographical*

*Subject* (Baltimore, MD: Johns Hopkins University Press, 1989), 103–26. For Boswell and Thrale's rivalry, see Mary Hyde, *The Impossible Friendship: Boswell and Mrs. Thrale* (Cambridge, MA: Harvard University Press, 1972); and Irma S. Lustig, "Boswell at Work: 'The Animadversions' on Mrs. Piozzi," *Modern Language Review* 67, no. 1 (January 1972): 11–30, for Boswell's working through of his attack on Thrale in the *Life;* and Lucyle Werkmeister, "Jemmie Boswell and the London Daily Press, 1785–1795," *Bulletin of the New York Public Library* 67, no. 2 (February 1963): 82–114, and 67, no. 3 (March 1963): 169–85, for a detailed analysis of Boswell's many attempts, anonymous, pornographic, and otherwise, to attack Thrale in print, as well as the attacks on him in print for his unprecedented anecdotal exposure of Johnson.

75. Hester Lynch Piozzi, *Anecdotes of the Late Samuel Johnson, LL.D., during the Last Twenty Years of His Life,* ed. Arthur Sherbo (London: Oxford University Press, 1974), 139–40.
76. Susan Stewart, *On Longing: Narratives of the Miniature, the Gigantic, the Souvenir, the Collection* (Baltimore, MD: Johns Hopkins University Press, 1984), 136. We might even think of Thrale, in her offering of an aesthetic alternative to Boswell's monumental ethos of completeness, as an inheritor of a feminine aesthetics of the fragment that originated with Sappho, suggestively delineated by Page duBois in *Sappho Is Burning* (Chicago: University of Chicago Press, 1995).
77. James Boswell, ventriloquizing Johnson, anonymously published a bawdy poem on the subject, "Ode by Dr. Samuel Johnson to Mrs. Thrale, upon their supposed approaching nuptials" (London, 1784), which reduced the pair's falling out to farce, thus venting his aggression toward both.
78. *The Odes of Horace,* bilingual edition, trans. David Ferry (New York: Farrar, Straus and Giroux, 1997), 188–91.
79. See Lisa Berglund, "'Look, my Lord, it comes': The Approach of Death in the *Life of Johnson,*" in *1650–1850: Ideas, Aesthetics, and Inquiries in the Early Modern Era,* vol. 7 (Brooklyn: AMS Press, 2002), 245–46; Thomas Macaulay writes that upon Thrale's marriage to Piozzi, "the newspapers and magazines were filled with allusions to the Ephesian Matron, and the two pictures in *Hamlet.*" Thomas Babington, Lord Macaulay, "Samuel Johnson," in *Miscellaneous Essays and the Lays of Ancient Rome* (New York: E. P. Dutton, n.d.), 253.
80. Hester Thrale, September 18, 1777, *Thraliana: The Diary of Mrs. Hester Lynch Thrale (Later Mrs. Piozzi), 1776–1809,* vol. 1, *1776–1784,* ed. Katherine C. Balderston (Oxford: Clarendon Press, 1942), 158.
81. Piozzi, *Anecdotes,* 156.
82. Part of my project in this book has been to return self-consciously to the single-author study in order to analyze the genre and its seeming inevitability, and to reflect upon the gendered embodiment of exemplary individuality itself. Johnson's complex, supportive, and (in the case of Anna Seward and Elizabeth Montagu) occasionally embattled friendships with a number of women writers of his day, including Frances Burney, Elizabeth Carter, Charlotte Lennox, and Hester Thrale stand in powerful contrast to the singularity with which he has come to dominate the literary canon. See Felicity A. Nussbaum, *The Limits of the Human: Fictions of Anomaly, Race, and Gender in the Long Eighteenth Century* (Cambridge: Cambridge University Press, 2003), chaps. 2, 3, for more on the bluestocking circle and their relation to masculine authorship.

83. For more on this critical speculation, see the introduction.
84. James Boswell, "A Thralian Epigram," Beinecke Rare Book and Manuscript Library *32, *Public Advertiser,* May 13, 1788.
85. One might say that Thrale's special claim to anecdotal intimacy threatens to unmoor the Johnsonian tradition therefore intent upon excluding her. Even Beryl Bainbridge, discussed in the introduction, and Samuel Beckett, whose "Human Wishes" starts as a love story about Johnson and Thrale and ends with Johnson alone with death, are more interested in Johnson as desiring subject than in Thrale, the supposed object and anecdotal source of that desire. Bainbridge filters her view of Thrale through the disapproving perspective of her daughter Queeney, leaving them forever estranged, and Thrale's own desire a partial mystery. For another attempt to supplement this void, see Kathleen Danziger, *Dr. Johnson's Mrs. Thrale: An Imaginary Monologue to Be Read or Acted Based on Mrs. Thrale's Own Diaries and Reminiscences of Dr. Johnson* (London: Century, 1984).
86. Tyers, in Brack and Kelley, *Early Biographies,* 89.
87. A survey of *OED* examples from Burton through De Quincey demonstrates how, over the course of the eighteenth century, the faculty of understanding is reduced to a barren "reason" that demands a supplement from the newly defined province of the soul (with which understanding had previously been joined), namely, the aesthetic.
88. Tyers, in Brack and Kelley, *Early Biographies,* 86. The editors are unable to identify this quotation.
89. For the founding example of this use of the word, see Catullus's programmatic 1.1.
90. The phrase also echoes and inverts (in the mode of Milton's Satan) the lament of Achilles' shade in *Odyssey* book 11, who longed to serve on earth rather than rule in the underworld. Both Petronius and Satan choose to rule in the wrong realm.
91. Samuel Johnson, *The Rambler* no. 60, ed. W. J. Bate and Albrecht B. Strauss, *The Yale Edition of the Works of Samuel Johnson,* 16 vols. (New Haven: Yale University Press, 1969), 3:321, 322.
92. Walter Benjamin, *The Origin of German Tragic Drama,* trans. John Osborne (London: Verso, 1985), 232–33.
93. Ibid., 233.
94. Ibid., 235. No material fragment exemplifies this beauty better for Benjamin than the corpse: "And if it is in death that the spirit becomes free, in the manner of spirits, it is not until then that the body too comes properly into its own. For this much is self-evident: the allegorization of the physis can only be carried through in all its vigour in respect of the corpse. And the characters of the *Trauerspiel* die, because it is only thus, as corpses, that they can enter into the homeland of allegory. It is not for the sake of immortality that they meet their end, but for the sake of the corpse. . . . There is in the physis, in the memory itself, a *memento mori;* the obsession of the men of the middle ages and the baroque with death would be quite unthinkable if it were only a question of reflection about the end of their lives. . . . In the *Trauerspiel* of the seventeenth century the corpse becomes quite simply the pre-eminent emblematic property. The apotheoses are barely conceivable without it." Ibid., 217–19.
95. Ibid., 218.
96. Ibid., 158.
97. For a powerful delineation of literature's transformational relationship to loss, see Susan Stewart, *Poetry and the Fate of the Senses* (Chicago: University of Chicago

Press, 2002). "The poet's tragedy lies in the fading of the referent in time, in the impermanence of whatever is grasped. The poet's recompense is the production of a form that enters into the transforming life of language" (2).

98. See Gilles Deleuze, "The Simulacrum and Ancient Philosophy," in *The Logic of Sense,* trans. Mark Lester with Charles Stivale, ed. Constantin V. Boundas (New York: Columbia University Press, 1990), 253–66, especially the following: "This simulacrum includes the differential point of view; and the observer becomes a part of the simulacrum itself, which is transformed and deformed by his point of view. In short, there is in the simulacrum a becoming-mad, or a becoming unlimited, . . . a becoming always other, a becoming subversive of the depths, able to evade the equal, the limit, the Same, or the Similar: always more and less at once, but never equal. To impose a limit on this becoming, to order it according to the same, to render it similar—and, for that part which remains rebellious, to repress it as deeply as possible, to shut it up in a cavern at the bottom of the Ocean—such is the aim of Platonism in its will to bring about the triumph of icons over simulacra" (258–59). Christianity for Deleuze is the culmination of the classical urge to limit representation by "endow[ing] it with a valid claim to the unlimited" (259). This idea of likeness without end also leaves open the possibility of tragedy, of Johnsonian nightmare, of what Samuel Beckett would call "positive annihilation."

99. Blake's first line is an allusion to William Collins's "Ode to Evening." William Blake, "An Island in the Moon," in *Johnson: The Critical Heritage,* ed. James T. Boulton (London: Routledge Kegan Paul, 1971), 9–10.

100. Hawkins, *Life of Johnson,* 274–75.

101. Michel Foucault, "Nietzsche, Genealogy, History," in *Language, Counter-Memory, Practice,* ed. Donald F. Bouchard, trans. Donald F. Bouchard and Sherry Simon (Ithaca, NY: Cornell University Press, 1977), 154. The anecdote shares with Foucault's "effective" history an affinity for what is most proximate and seemingly ahistorical, namely, the body. Foucault's metaphorical link of history to the practice of medicine can also be applied to my connection of anecdote to anatomy: "Effective history studies what is closest, but in an abrupt dispossession, so as to seize it at a distance (an approach similar to that of a doctor who looks closely, who plunges to make a diagnosis and to state its difference). Historical sense has more in common with medicine than philosophy; . . . since among the philosopher's idiosyncracies is a complete denial of the body." Ultimately, for Foucault, the task of history is not to serve philosophy's master narrative, "to recount the necessary birth of truth and values," but rather "to become a curative science" (156).

The anecdote's ability to puncture wholes rewrites anatomy as a pleasurable cutting, as Fineman indicates via Lacan's "account of the *objet a* and the mark of the real": "The very delimitation of the 'erogenous zone' that the drive isolates from the metabolism of the function (the act of devouring concerns other organs than the mouth—ask one of Pavlov's dogs) is the result of a cut (*coupure*) expressed in the anatomical mark (*trait*) of a margin or border—lips, 'the enclosure of the teeth,' the rim of the anus, the tip of the penis, the vagina, the slit formed by the eyelids, even the horn-shaped aperture of the ear. . . . Observe that the mark of the cut is no less obviously present in the object described by analytic theory: the mamilla, faeces, the phallus (imaginary object), the urinary flow. (An unthinkable list, if one adds, as I do, the phoneme, the gaze, the voice—the nothing.) For is it not obvious that this feature, this partial feature, rightly emphasized in objects, is applicable not because these objects are part of a total object, the body, but because they represent only partially the function that produces them? These objects have one common feature in my elaboration of them—they have no specular image, or, in other words, alterity. It is what enables them to be the 'stuff,' or rather the lining, though not in any sense the reverse, of the very subject that one takes to be the subject of consciousness. For this subject, who thinks he can accede to himself by designating himself in the statement, is no more than such an object." Jacques Lacan, "The Subversion of the Subject and the Dialectic of Desire in the Freudian Unconscious," in *Ecrits: A Selection,* trans. A. Sheridan (New York: W. W. Norton, 1977), 314–15, quoted by Fineman, "History of the Anecdote," 67–68. The mark of the cut, written on the body itself, constitutes the desiring subject by—as Foucault remarks of the cutting action of effective history—setting him apart from himself.

102. Foucault, "Nietzsche, Genealogy, History," 158.

103. I am indebted here to Coppelia Kahn's chapter "Antony's Wound," in *Roman Shakespeare: Warriors, Wounds, and Women* (London: Routledge, 1997), especially 121–37, which historicizes Antony's "botched" self-wounding in a gendered history of the morally ambivalent semiotics of suicide in the early modern period. Following Montaigne on Cato among others, Kahn argues that "smoothness of execution was not—for the Romans or for the Renaissance—a criterion of suicide's dignity or moral value"; instead she explores suicide in relation to "the problematic structures of masculinity; especially, those of emulation" (127). In Kahn's reading, Roman suicide in Shakespeare is "an affair between men" (131), and its botched performance in the case of Antony exposes it as emulatory performance, a performance she argues that can be embodied equally well by a woman (as the case of Cleopatra demonstrates). The wounded Antony's "ringing assertions of triumph over Caesar" are "purely ideological constructions built up of words disjunct from the wounded body they reconfigure" (134). In a different way, Johnson's wounded body exposes the care of the doctors as curiosity; he triumphs both rhetorically and physically over the doctors and the body he renders a corpse.

104. Chapin, in "Samuel Johnson's 'Wonderful' Experience," argues that "Johnson's religion, always a source of anxiety, was also a source of hope" (58–59). Chapin elaborates an important paradox inherent in the narrative of Christian salvation—anxiety can lead to hope, is in fact the stuff of hope. The experience of despond, of the loathing of one's own corruption and the fear of God's wrath, which can lead to despair and even suicide, is also the stuff of true repentance. Thus the very fear that distressed many of Johnson's contemporaries as potentially excessive doubt could also be read as proof of his faith. As Johnson writes in his diary in a paradoxical sentiment worthy of Beckett, "Faith in some proportion to Fear." Samuel Johnson, *Diaries, Prayers, and Annals,* ed. E. L. McAdam Jr., with Donald and Mary Hyde, *The Yale Edition of the Works of Samuel Johnson,* 16 vols. (New Haven: Yale University Press, 1958), 1:269. For Chapin, Johnson's final self-wounding is evidence not of his excessive terror of death but rather of the yearning in his final moments for another miracle. His final act is thus less an assertion of himself as agent and more of a prayer for, or imitation of, divine intervention. Chapin's final words are evocative: "Johnson's act, desperate as it appears, may have had its origin in what would have been for him a slim but not an irrational hope" (60). John Wiltshire observes that Windham avoids men-

tioning that the ugly wound was made to cure a sarcocele or swelling of the testicle, thus bringing it perilously close to self-castration. John Wiltshire, *Samuel Johnson in the Medical World* (Cambridge: Cambridge University Press, 1991), 61. Johnson's self-wounding thus becomes at once skillful self-dissection, self-castration, and (with the evocation of the wound juxtaposed with the punctures) displaced Crucifixion.

105. Kahn, *Roman Shakespeare,* 124.
106. Boswell is quoting Thomas David Brocklesby, the brother of Johnson's doctor. Boswell, *Life,* 4:418. This version of Johnson's death is illustrated by Ernest H. Shepard in *Everybody's Boswell; being The life of Samuel Johnson abridged from James Boswell's complete text, and from the "Tour of the Hebrides"* (London: G. Bell and Sons, 1930), and concludes the American A. Edward Newton's gift to the preservers of Dr. Johnson's House in London, *Doctor Johnson: A Play* (Boston: Atlantic Monthly Press, 1923). In Hannah More's account, Dr. Brocklesby remonstrates with Johnson that he is "not a Christian," and begs him to pray to become one "in my sense of the word." Johnson complies and orders Brocklesby to record his prayer, which was subsequently published in Strahan's *Prayers and Meditations.* More concludes, in another attempt at moral closure, "No action of his life became him like the leaving of it. His death makes a kind of era in literature; piety and goodness will not easily find a more able defender, and it is delightful to see him set, as it were, his dying seal to the professions of his life, and to the truth of Christianity." Hannah More, excerpts from *Memoirs of the Life and Correspondence of Mrs. Hannah More* (1834), by William Roberts, Esq., 4 vols., in George Birkbeck Hill, ed., *Johnsonian Miscellanies,* 2 vols. (Oxford: Clarendon Press, 1897), 2:206.

## Chapter Five

1. Luke Bresky locates *Our Old Home* in the context of the Anglo-American craze in the 1850s for tourism oriented around "homes and haunts" of "living authors," as well as Hawthorne's ambivalence about the artist's function as "representative man"—"Behind the question . . . to what extent does an English man of letters represent his people?—lies the question of whether the American man of letters feels (or ought to feel) closer to such a figure if the answer to the first question is affirmative or negative." In his discussion of Hawthorne's visit to Stratford, Bresky also notes that "the scanty annals of American literature explain the provincial's gift for a specific kind of bardolatry; for better *and* worse, as Hawthorne shows, authors seem like mythical beings in a nation that provides comparatively few opportunities to examine them up close." Marshall Luke Bresky, "Literature and the Nationalization of Heroism in Antebellum America" (PhD diss., University of California, Los Angeles, 1999), 255, 261.
2. Nathaniel Hawthorne, *Our Old Home: A Series of English Sketches,* ed. William Charvat, Fredson Bowers et al. (Columbus: Ohio State University Press, 1970), 131. All further references will appear in the body of the text.
3. Hawthorne himself had recounted this anecdote from Boswell earlier as a monitory tale in his *Biographical Stories for Children* (1842), in *True Stories from History and Biography* (Columbus: Ohio State University Press, 1972), 239–49. The story's moral—that Johnson refused his father's request to sell books in the marketplace

and was tormented by guilt for years after his infraction—brings together the two estranged brothers who listen to their father's rendition of the story, but (as Bresky also observes) is only implied in Hawthorne's telling in *Our Old Home;* the later text is more interested in the anecdotal, near-photographic moment of Johnson's penance itself.

4. James Boswell, *The Life of Samuel Johnson, LL.D.,* ed. George Birkbeck Hill, rev. L. F. Powell, 6 vols. (Oxford: Clarendon Press, 1934–50), 1:57.
5. Ibid.
6. Ibid., 1:49.
7. Hester Lynch Piozzi, *Anecdotes of the Late Samuel Johnson, LL.D. during the Last Twenty Years of His Life,* ed. Arthur Sherbo (London: Oxford University Press, 1974). Thrale adds that Johnson's "recollection of such reading as had delighted him in his infancy, made him always persist in fancying that it was the only reading which could please an infant. . . . 'Babies do not want (said he) to hear about babies; they like to be told of giants and castles, and of somewhat which can stretch and stimulate their little minds'" (65). For more on Johnson's reading habits and relation to romance, see Eithne Henson, *The Fictions of Romantick Chivalry* (Rutherford, NJ: Fairleigh Dickinson University Press, 1992); and Robert DeMaria Jr., *Samuel Johnson and the Life of Reading* (Baltimore: Johns Hopkins University Press, 1997).
8. Samuel Johnson, *Life of Savage,* in *Selected Poetry and Prose,* ed. Frank Brady and W. K. Wimsatt (Berkeley: University of California Press, 1977), 565.
9. Boswell, *Life,* 1:38.
10. Piozzi, *Anecdotes,* 66.
11. Boswell, *Life,* 4:373.
12. For an exposition of this temporality in Christian Renaissance painting, see "The Wound in the Wall," in Catherine Gallagher and Stephen Greenblatt, *Practicing New Historicism* (Chicago: University of Chicago Press, 2000), 75–109.
13. While I disagree with Bresky's claim that the Uttoxeter penance was a relatively unknown moment in Boswell's *Life* (a claim that helps to reinforce his larger argument about Hawthorne's contempt for a popular audience), I agree with him about that contempt: "In *Biographical Stories,* Johnson becomes a representative man by remembering and, at last, atoning for a distinctly English contempt toward the common people and the vulgar peddling of books. In 'Lichfield and Uttoxeter,' the narrator forgets about that contempt—and re-enacts it—while striving to commemorate the atonement." Bresky, "Literature and the Nationalization of Heroism," 266.
14. Ibid., 267, 251–52.
15. Piozzi, *Anecdotes,* 67. As mentioned in chapter 2, Johnson bowed to public opinion in his editorial notes to *King Lear,* noting, "If my sensations could add any thing to the general suffrage, I might relate, that I was many years ago so shocked by Cordelia's death, that I know not whether I ever endured to read again the last scenes of the play till I undertook to revise them as an editor." *Johnson on Shakespeare,* ed. Arthur Sherbo, *The Yale Edition of the Works of Samuel Johnson,* 16 vols. (New Haven: Yale University Press, 1968), 8:704.
16. Walter Benjamin, *The Origin of German Tragic Drama,* trans. John Osborne (London: Verso, 1985), 140.
17. I am indebted here to Jayne Lewis's unpublished work on Defoe, Glanville, ghosts, and the spectral nature of the print medium. The final chapter of Gallagher and

Greenblatt's *Practicing New Historicism* elucidates how Enlightenment skepticism and materialism paradoxically gave birth to a variety of suspensions of disbelief—not the stuff of faith but certainly the stuff of phantasm—in both science and the novel. Materialism reduces the world to matter but makes of all matter potential life, even or perhaps especially the creations of the imagination. We animate literature and it animates us; literature thus makes the border between what is dead and what is alive impossible to draw. Hamlet's repulsion at the unredeemed materiality, the leftover of the divine host, is no longer the pressing issue, rather modern literature's leftover is fetishism, the ever present specter of disavowal.

18. Mary McCarthy, "A Bolt from the Blue," in *The Writing on the Wall and Other Literary Essays* (New York: Harcourt, Brace & World, [1970]), 33–34.
19. Vladimir Nabokov, *Pale Fire* (New York: G. P. Putnam's Sons, 1962; repr., New York: Berkeley Publishing, 1975), 62. All further references will be given by page number in the text. Lines from Shade's poem *Pale Fire* will be referenced by line number.
20. Priscilla Meyer expounds on Nabokov's play with the Anglo-Saxon roots of "Bodkin" in *Find What the Sailor Has Hidden: Vladimir Nabokov's* Pale Fire (Middletown, CT: Wesleyan University Press, 1988), 95–96, quotation from 96. "From the very first I tried to behave with the utmost courtesy toward my friend's wife, and from the very first she disliked and distrusted me. I was to learn later that when alluding to me in public she used to call me 'an elephantine tick; a king-sized botfly; a macaco worm; the monstrous parasite of a genius.' I pardon her—her and everybody" (123).
21. Stephen Greenblatt parses the lines in their religious context: "Hamlet's words swerve [away from the immediate dramatic exchange about the proper treatment of actors] in the direction of larger questions that have a moral and ultimately a theological resonance. We are all sinners and breeders of sinners, and by our own deserts we should all be punished; only through unmerited grace can we hope to be treated well." Gallagher and Greenblatt, *Practicing New Historicism,* 228n59.
22. Adam Phillips, *Darwin's Worms: On Life Stories and Death Stories* (New York: Basic Books, 2000), 86–88. Phillips articulates the paradox at the heart of the Johnsonian dynamic beautifully when he comments that Freud's "heroic image of self-definition, of self-fashioning, is the notion that we want to die in our own way. The subject of a biography always dies in the biographer's own way" (108).
23. Ibid., 90.
24. This observation of Kinbote's appears in a particularly elegiac note to Shade's repetition (with a slight difference) of the poem's opening lines in the following couplet: "I was the shadow of the waxwing slain / By feigned remoteness in the windowpane" (131–32). Writing with the knowledge of Shade's death, he muses, "Today, where the 'feigned remoteness' has indeed performed its dreadful duty, and the poem we have is the only 'shadow' that remains, we cannot help reading into these lines something more than mirrorplay and mirage shimmer. We feel doom, in the image of Gradus, eating away the miles and miles of 'feigned remoteness' between him and poor Shade. . . . The force propelling him is the magic action of Shade's poem itself, the very mechanism and sweep of verse, the powerful iambic motor" (98).
25. In his memoir *Speak, Memory; An Autobiography Revisited* (New York: Putnam, 1966), chap. 9, Nabokov deflects the moment of his father's assassination onto an earlier childhood recollection of his father's near-death in an old-fashioned duel de-

flected at the last minute. The moment of the traumatic loss, whether in fiction or in "factual" narrative, is never confronted directly and is always articulated as a case of mistaken identity.

26. Michael Wood, *The Magician's Doubts: Nabokov and the Risks of Fiction* (Princeton: Princeton University Press, 1994), 28.
27. For the opposition of the "Popean" Shade and the "untamed, enormously fecund" Shakespearean Kinbote who is at once lunatic, lover, and poet, see Robert Alter, *Partial Magic: The Novel as Self-Conscious Genre* (Berkeley: University of California Press, 1975), 202.
28. Lawrence Lipking, "Johnson and the Meaning of Life," in *Johnson and His Age*, ed. James Engell, Harvard English Studies, vol. 12 (Cambridge, MA: Harvard University Press, 1984), 19.
29. Boswell, *Life*, 4:380.
30. Ibid., 3:415; *The Letters of Samuel Johnson*, vol. 4, *1782–1784*, ed. Bruce Redford (Princeton: Princeton University Press, 1994), 434.
31. For the narrator of *Ravelstein* turning to Macaulay's Johnson as a model for his own biography in a memory he links to his first encounter with *Hamlet* and the "weary, stale, flat and unprofitable" uses of the world, see Saul Bellow, *Ravelstein* (New York: Penguin, 2001), 6–7.
32. On Kinbote's "re-Englished version of Conmal's Zemblan translation of these lines," which he quotes "without, however, suspecting that they contain the phrase 'pale fire,'" see Alter, *Partial Magic*, 206–7. "The gauche mistranslation of this passage," Alter concludes, "serves as a counterpoint to the splendid mistranslation effected through the Commentary" (207).
33. On Nabokov's translation of *Hamlet*, see Meyer, *Find What the Sailor Has Hidden*, 113–14.
34. Often but not entirely. See Matthew Charles Evans Morris, "Parody in *Pale Fire:* A Re-reading of Boswell's *Life of Johnson*," *Dissertation Abstracts International* 57, no. 5 (November 1996): 2028A; and Michael Seidel, "*Pale Fire* and the Art of the Narrative Supplement," *ELH* 51 (1984): 837–55. I am grateful to Sean Silver for these references and for his unpublished essay "*Pale Fire* and Johnson's Cat: The Death of the Academic Conversation," in which he observes that with the exception of Seidel, structural accounts of the relationship between Boswell's biography and Nabokov's novel are few.
35. Roland Barthes, *Camera Lucida: Reflections on Photography*, trans. Richard Howard (New York: Hill and Wang, 1981), 96.
36. For one of many examples of Boswell's ability to put his penchant for "worshipping" Johnson into dialogue with less impassioned admirers, see *Life*, 3:331. He thus incorporates Johnson's monitory perspective into his own work while transcending it.
37. Samuel Johnson, *The Rambler*, eds. W. J. Bate and Abrecht B. Strauss, *The Yale Edition of the Works of Samuel Johnson*, 16 vols. (New Haven: Yale University Press, 1969), 3:79–80.
38. "We are not, [Nabokov] says, to 'search for "real life" in the dead ends of art,' where 'dead ends' is a compliment, a tribute, a perverse pointer to the finished object. 'In art, purpose and plan are nothing; only the result counts. We are concerned only with the structure of a published work'. . . . In one sense this argument confirms the

author's death as a creature apart from the work. But it also means that the text is where the author triumphantly revives." Wood, *Magician's Doubts,* 13–14.

39. Brian Boyd, *Nabokov's* Pale Fire*: The Magic of Artistic Discovery* (Princeton: Princeton University Press, 1999). Boyd closes his imaginative and passionate reading of the novel (in which he argues for a truth of art and immortality that builds on, rather than contradicts, scientific discovery) by arguing that Nabokov "does not 'prove' his case, but his very case insists on the unprovability, from within mortality, of modes of being beyond it" (261). Boyd also cites approvingly Michael Wood's characterization of Nabokov offering readers a "theology for skeptics" (261); Wood, *Magician's Doubts,* 190.
40. "Indeed, the very first occurrence of the phrase 'pale fire' in the text of the novel, where it is attached to the incinerator burning the rejected drafts of Shade's poem, which then ascend into the air as 'wind-borne black butterflies,' associates the term with the refining process by which art comes into being, consuming its own impurities—and that pale fire is no reflection at all." Alter, *Partial Magic,* 192.
41. Nabokov, Boyd writes, "allows for the possibility that anything we might imagine beyond, even all the worlds of discovery and benevolent design he invents in *Pale Fire,* might merely reflect the depth of our desire, might merely help to define the limits of our imprisonment within a chaotic life that prompt us to dream up a freedom and order beyond." Boyd, *Nabokov's* Pale Fire, 261.
42. Lipking, "Johnson and the Meaning of Life," 23; Boyd, *Nabokov's* Pale Fire, 5–6, argues that Nabokov is not an ironist at the expense of life; he is an ironist who returns us to life with a new sense of wonder.
43. In "Pope's *The Rape of the Lock* and Nabokov's *Pale Fire,*" in *Nabokov at the Limits: Redrawing Critical Boundaries,* ed. Lisa Zunshine (New York: Garland, 1999), 161–82, Lisa Zunshine amply proves the important point that Nabokov's novel is as much homage to Pope's work (e.g., the *Dunciad* is another important model for the book) as it is a refusal on the part of both John Shade and Nabokov himself to give Pope his due. Zunshine attributes Nabokov's stubborn denial of the power of a poetry he nevertheless found inspiring to the pro-Romantic, anti-eighteenth-century intellectual climate of Cornell (where Nabokov taught while writing *Pale Fire*) in the 1940s and '50s.
44. Alter, *Partial Magic,* 198.
45. Samuel Johnson, "Review of [Soame Jenyns,] *A Free Enquiry into the Nature and Origin of Evil*" (1757), in *The Oxford Authors: Samuel Johnson,* ed. Donald Greene (Oxford: Oxford University Press, 1984), 526.
46. Piozzi, *Anecdotes,* 111. The image appears in the injunction to piety from *Rambler* 69, in the warning against flattery of *Rambler* 104, and in Johnson's commentary on Edgar's description of the fall from Dover cliff to his blind father Gloucester in *King Lear.* In each case, critical activity is a kind of warding off of the vacuity of the fall by elaborating it in all its infinite variety. *Johnson on Shakespeare,* 8:695. See also Piozzi's account of Johnson's cure of melancholy—the calculation of the national debt, which "would, if converted into silver, serve to make a meridian of that metal . . . for the globe of the whole earth, the real *globe.*" Piozzi, *Anecdotes,* 87. Here the very substance of the chain that binds the earth—or of the art that punctuates the void—is not a national but a universal debt. The measures taken to guard against infinity terrify with the prospect of endless empty repetition.

47. Johnson, "Review of Soame Jenyns," 534–35. Emphasis Johnson's for the purpose of quotation.
48. Piozzi, *Anecdotes,* 66.
49. Johnson, "Review of Soame Jenyns," 535.
50. *Johnson on Shakespeare,* 8:704.
51. Wood, *Magician's Doubts,* 187, 187–88.
52. Walter Benjamin, *Gesammelte Schriften,* vol. 5, ed. Rolf Tiedemann (Frankfurt: Suhrkamp, 1982), 1094; translated and quoted in Richard Sieburth, "Benjamin the Scrivener," in *Benjamin: Philosophy, Aesthetics, History,* ed. Gary Smith (Chicago: University of Chicago Press, 1989), 23.
53. Samuel Beckett, RUL MS 3461/2, 17–20.
54. Ibid., 81; Beckett is paraphrasing Johnson's April 15, 1778, conversation with the Quaker Mary Knowles in Boswell, *Life,* 3:294–96.
55. Samuel Beckett, letter to Joseph Hone, quoted in Deirdre Bair, *Samuel Beckett: A Biography* (1978; repr., New York: Simon & Schuster, 1990), 254.
56. Christopher Ricks, *Beckett's Dying Words* (Oxford: Clarendon Press, 1993), 55.
57. Ibid., 15–16. The Eliot quotation comes, aptly enough, from "Shakespeare and the Stoicism of Seneca," *Selected Essays* (New York: Harcourt, Brace & Co., 1932), 134.
58. Terming Johnson's art that of "self-doubting equivocation. His sentences, paragraphs, and essays move in two directions at once," Dilks links both Beckett and Johnson to Augustine's statement: "Do not despair; one of the thieves was saved. Do not presume; one of the thieves was damned," a sentence that, precisely because of its loss through translation, "captures the essence of Beckett's connection to Johnson" and forms the ideas of his later work. Stephen Dilks, "How It Is How It Is Not: Samuel Beckett's Spirit of Positive Annihilation" (PhD diss., Rutgers University, December 1992), *Dissertation Abstracts International* 53, no. 6 (December 1992): 1906A, 225–26; first quotation in text, on sentences that have the "shape of doubt," from 225. For the next two quotations in text, see Stephen John Dilks, "Samuel Beckett's Samuel Johnson," *Modern Language Review* 98, no. 2 (April 2003): 290. For one of the most famous examples of that sentence's influence, see the discussion between Vladimir and Estragon on the fate of the two thieves on Golgotha (in which "hell" and "death" seem interchangeable) in Samuel Beckett, *Waiting for Godot. A Tragicomedy in Two Acts,* translated from the original French text by the author (New York: Grove Press, 1954), 6–7.
59. Mary Bryden, "Figures of Golgotha: Beckett's Pinioned People," in *The Ideal Core of the Onion,* ed. John Pilling and Mary Bryden (Reading: Beckett International Foundation, 1992), 53.
60. Ibid., 54. For Beckett on Rouault, see "Peintres de l'Empêchement" (1948), in *Disjecta: Miscellaneous Writings and a Dramatic Fragment,* ed. Ruby Cohn (New York: Grove Press, 1984), 133–37, relevant passage translated in chap. 4, n. 63. The "dramatic fragment" in this volume is "Human Wishes."
61. Ibid., 54.
62. Beckett was particularly fascinated with Johnson's relationship with Levet; he takes copious notes on the latter, who figures as almost entirely mute tragicomic relief in "Human Wishes."
63. Quoted in Bair, *Samuel Beckett,* 256.
64. Bryden, "Figures of Golgotha," 55, 62.

65. Ibid., 61.
66. Samuel Beckett, quoted in James Knowlson, *Damned to Fame: The Life of Samuel Beckett* (New York: Simon & Schuster, 1996), 249.
67. Bair, *Samuel Beckett,* 245.
68. Knowlson, *Damned to Fame,* 670, notes that Beckett had done a fair amount of research into Johnson's published writings, including *Rasselas* and the *Life of Ascham* and *Life of Dryden,* as well as Boswell's *Life,* during 1934–36. Stephen Dilks observes that by contrast, "The Notebooks from 1937 are dominated by quotations from biographies, diaries, letters, private papers, and miscellaneous anecdotes. . . . The idiosyncratic, highly personalized Johnson pieced together from this biographical and 'once private' material is, in retrospect, highly Beckettian." Dilks, "Samuel Beckett's Samuel Johnson," 291–92. Dilks also makes a comprehensive list of Beckett's sources in the notebooks. His argument, that the famous 1945 epiphany that redefined Beckett's aesthetic and allowed him to escape from Joyce's influence, was "preceded by an equally profound, though less dramatic, discovery of a Johnson-inspired aesthetic persona" (286), has been very important to my own. I am less interested in Dilks's reading of the notebooks as evidence of "the six-month period in 1937 when Beckett deliberately turned away from Joyce towards Johnson" (286), and more interested in Beckett's reading of Johnson not merely as "idiosyncratic" and "personalized," but rather as explicitly embracing the side of Johnson, death-obsessed and endlessly desiring, that has been disavowed by but nevertheless has haunted Johnsonians from Boswell to the present. I am grateful to Stephen Dilks for my first understanding of the riches to be found in Beckett's Johnson notebooks and for the conundrum of "positive annihilation."

For earlier important readings of Beckett's relationship to Johnson, see Lionel Kelly, "Beckett's Human Wishes," in *Ideal Core of the Onion,* ed. Pilling and Bryden, 21–44, upon which Dilks builds, which has a particularly evocative discussion of Beckett's debt in the heartrending monologue *Krapp's Last Tape* to the Johnson of the *Prayers and Meditations.* Ruby Cohn, *Just Play: Beckett's Theater* (Princeton: Princeton University Press, 1980), the first venue in which the fragment of "Human Wishes" was published, also includes a summary of the notebooks. Cohn seems baffled by Beckett's affinity with Johnson since her idea of Johnson, both stylistically and as a character, is closer to Macaulay's caricature than to the unpublished Johnson that is Beckett's subject. Frederik N. Smith argues for Beckett's stylistic debt to the Johnson of the *Dictionary* through an analysis of his diction in "'Pituitous Defluxion': Samuel Johnson and Beckett's Philosophic Vocabulary," *Romance Studies* 11 (Winter 1987): 86–95.

In *Beckett's Eighteenth Century* (Houndmills: Palgrave, 2002), Smith argues that Johnson shared with Beckett's characters what Boswell praised as "philosophick heroism" (*Life,* 4:190), and that Beckett "discovered the tragic in Johnson a few years before the critics" (128). Smith also suggestively compares Johnson's sarcocele to "Molloy's testicular difficulties" (124) and traces—among a host of other such migrations—Hawkins's anecdote of Johnson's death-inducing scarification of his legs and scrotum to the following passage from *Molloy:* "And, worse still, they [his testicles] got in my way when I tried to walk, when I tried to sit down. . . . So the best thing for me would have been for them to go, and I would have seen to it myself, with a knife or secateurs, but for my terror of physical pain and festered wounds, so that I shook." Samuel Beckett, *Molloy*, in *Three Novels by Samuel Beckett* (New York: Grove Press, 1979), 36; quoted by Smith, *Beckett's Eighteenth Century*, 125. Johnson's haunting of Beckett, it would seem, was very much an anecdotal and corporeal matter.

69. Samuel Beckett, *Proust*, quoted in Knowlson, *Damned to Fame*, 248.
70. Samuel Beckett to Mary Manning, July 11, 1937, quoted in Knowlson, *Damned to Fame*, 250.
71. Beckett to Manning, December 13, 1936, quoted in ibid., 249.
72. Samuel Beckett, German diary, quoted in ibid., 222.
73. Beckett, RUL MS 3461/1, 46, 47.
74. Beckett, RUL MS 3361/1, 47–48. "Queeney" is Hester Thrale's oldest daughter, Hester Maria Thrale, whose narrative perspective punctuates Bainbridge's eponymous novel.
75. Dilks puts Beckett's imaginative stake in Johnson's impotence in relation to his rejection of Joyce's omnipotence. Quoted in an interview with Israel Shenker in the *New York Times* for May 6, 1956, Beckett states: "The more Joyce knew the more he could. He's tending toward omniscience and omnipotence as an artist. I'm working with impotence, ignorance." Dilks, "Samuel Beckett's Samuel Johnson," 287.
76. Beckett to Manning, December 13, 1936, quoted in Knowlson, *Damned to Fame*, 249.
77. On Johnson's secret (which caused him great distress upon suspecting John Hawkins's theft of his private journals, which he subsequently destroyed), see Boswell, *Life*, 4:406.
78. "Harmless Dandy," Samuel Beckett to Arland Ussher, June 15, 1937, quoted in Bair, *Samuel Beckett*, 253; "en-Thraled," in Beckett, RUL MS 3461/1, 8.
79. Samuel Beckett to Thomas McGreevy, April 20, 1937, quoted in Bair, *Samuel Beckett*, 253–54.
80. Beckett is noting Hawkins. RUL MS 3461/2, 53. I have located the reference in John Hawkins's unabridged *Life of Samuel Johnson, LL.D.* (Dublin, 1787), 503–4.
81. Beckett, RUL MS 3461/1, 39 verso, 40.
82. Ibid., 12, emphasis Beckett's.
83. Ibid., 47.
84. Knowlson, *Damned to Fame*, 250; Beckett to McGreevy, August 4, 1937, quoted in ibid.
85. Beckett, RUL MS 3461/1, 32.
86. Dilks, "Samuel Beckett's Samuel Johnson," 297.
87. Smith, *Beckett's Eighteenth Century*, 130.
88. Ibid., 131.
89. Quoted in Bair, *Samuel Beckett*, 257.
90. Smith, *Beckett's Eighteenth Century*, 130.
91. I refer here to a passage in Beckett's German diary from earlier in 1937: "I am not interested in a 'unification' of the historical chaos any more than I am in the 'clarification' of the individual chaos, and still less in the anthropomorphisation of the inhuman necessities that provoke the chaos. What I want is the straws, flotsam, etc., names, dates, births and deaths, because that is all I can know. . . . Meier says the background is more important than the foreground, the causes than the effects, the causes than their representatives and opponents. I say the background and the causes are an inhuman and incomprehensible machinery and venture to wonder what kind of appetite it is that can be appeased by the modern animism that consists in rational-

ising them. Rationalism is the last form of animism. Whereas the pure incoherence of times and men and places is at least amusing." Quoted in Knowlson, *Damned to Fame,* 228.

92. "It was a question of putting it into the Irish accent as well as the proper language of the period. It would not do to have Johnson speaking proper language, after the manner of Boswell, while all the other characters speak only the impossible jargon I put into their mouths." Quoted in Bair, *Samuel Beckett,* 255.
93. Samuel Beckett, "Human Wishes," in *Disjecta,* 155. All further references will be cited by page number in the text.
94. Beckett, RUL MS 3461/2, 81. Quotations are from Beckett's Johnsonian source; words without quotations are Beckett's.
95. Ibid., 43.
96. See chapter 2, n. 22, for the catalog of death in Johnson's private writings and as stoic motif.
97. Jeremy Taylor, *The Rule and Exercises of Holy Dying,* vol. 2, *Holy Living and Holy Dying,* ed. P. G. Stanwood (Oxford: Clarendon Press, 1989), 24.
98. Ibid.
99. Lucretius, *De rerum natura,* 3.58. Editor's translation, in Taylor, *Holy Dying,* 2:243.
100. Samuel Johnson, *Preface to A Dictionary of the English Language,* in *Selected Poetry and Prose,* eds. Frank Brady and W. K. Wimsatt (Berkeley: University of California Press, 1977), 291, 280. The editors note that this aphorism, italicized by Johnson, might be of his own invention.
101. Beckett, RUL MS 3461/3, 84, verso; 3461/3, last page, verso, in crayon.
102. Dilks, "How It Is How It Is Not," 224.
103. Johnson to Bennet Langton, March 27, 1784, quoted in Cohn, *Just Play,* 158.
104. Repetition structures this "dramaticule": the sounds of breathing and the cry are first indicated as authentic, and then repeated as a "recording," while light and breath are "strictly synchronized" as they switch on and off. Samuel Beckett, *Collected Shorter Plays* (New York: Grove Press, 1984), 209–11.

INDEX